THE LIBRARY
ST. MARY'S COLLEGE OF MARYLAND
ST. MARY'S CITY, MARYLAND 20686

The Invented Universe

Einstein's draft of 1923 in which he withdrew his earlier objection to Friedmann's dynamic solutions to the field equations. The last bit of the last sentence was: "a physical significance can hardly be ascribed to them". He crossed this out before sending the note to print.

The Invented Universe

The Einstein–De Sitter Controversy (1916–17) and the Rise of Relativistic Cosmology

PIERRE KERSZBERG

CLARENDON PRESS · OXFORD
1989

Oxford University Press, Walton Street, Oxford OX2 6DP
Oxford New York Toronto
Delhi Bombay Calcutta Madras Karachi
Petaling Jaya Singapore Hong Kong Tokyo
Nairobi Dar es Salaam Cape Town
Melbourne Auckland
and associated companies in
Berlin Ibadan

Oxford is a trade mark of Oxford University Press

Published in the United States
by Oxford University Press, New York

© Pierre Kerszberg, 1989

All rights reserved. No part of this publication may be reproduced,
stored in a retrieval system, or transmitted, in any form or by any means,
electronic, mechanical, photocopying, recording, or otherwise, without
the prior permission of Oxford University Press

British Library Cataloguing in Publication Data
Kerszberg, Pierre
The invented universe: the Einstein-De Sitter
Controversy (1916–17) and the rise of relatavistic
cosmology.
1. Universe. Theories
I. Title
523.1'01
ISBN 0–19–851876–5

Library of Congress Cataloging in Publication Data
Kerszberg, Pierre.
The invented universe: the Einstein-De Sitter controversy
(1916–17) and the rise of relativistic cosmology/Pierre Kerszberg.
p.   cm.   Includes bibliographies and index.
1. Cosmology. 2. Relativity (Physics) 3. Einstein, Albert,
1879–1955. 4. Sitter, Willem de. I. Title.
QB981.K47 1989   523'.1—dc19   88-38563
ISBN 0–19–851876–5

Set by Colset Private Limited, Singapore
Printed in Great Britain by
Biddles Ltd,
Guildford & King's Lynn

Acknowledgements

> Mathematics and physics make the world appear more and more as an open one, as a world not closed but pointing beyond itself.
>
> Hermann Weyl, 1932 in *The Open World*

My sources for the Einstein–De Sitter correspondence are the Einstein postcards and letters which have been kept at the Leiden Observatory. Their counterparts are in the Einstein Archives at the Institute of Advanced Study in Princeton. I wish to thank Professor John Stachel (Boston University) for making available to me the Einstein–De Sitter postcards and letters, as well as Dr E. Dekker (Museum Boerhaave, Leiden) and Professor H. Van der Laan (Leiden Observatory) for permission to consult and quote from the De Sitter archives. Passages of Einstein's manuscripts and letters from (and to) Einstein are reprinted by permission of the Hebrew University of Jerusalem, Israel. Weyl's correspondence is quoted from the library of the Eidgenössische Technische Hochschule in Zürich, Switzerland. I am grateful to Dr E. Glaus for permission to use this material. All correspondence involving Lemaître has been kindly forwarded to me by Dr L. Moens from the University of Louvain, Belgium.

Two figures are reproduced with the kind permission of the following publishers: Hermann for Fig. 1.1 and the American Physical Society for Fig. 2.7. The De Sitter universe on Fig. 4.2 and its subsequent variations up to Fig. 5.12 have been adapted from figures first drawn by E. Schrödinger in his book *Expanding Universes* published by Cambridge University Press in 1956. I am grateful to Cambridge University Press for letting me use freely this material, as well as to W. Rindler (published by Springer in 1977) for Fig. 4.10 in particular.

A few sections of this book have also been adapted from parts of my articles which first appeared in the following journals: *Archive for History of Exact Science* (published by Springer), *Osiris* (published by the History of Science Society), *Studies in History and Philosophy of Science* (published by Pergamon), *The Review of Metaphysics* (published by Lancaster Press), and *The British Journal for the Philosophy of Science* (published by Aberdeen University Press). I wish to thank the editors and publishers for permission to reprint this material.

This book grew out of a Ph.D. in Philosophy which I submitted at the University of Brussels in 1982 while Research Assistant at the National Funds of Scientific Research (Belgium). I wish to reiterate here my profound gratefulness to Marc Richir. It was thanks to his generous and skilful supervision that the thesis could be brought to a successful issue, and I still feel that not even the plan for this book would have been conceived were it not for his enduring friendship and intellectual stimulation. I also owe a lot to the comfort and continuous support of those who followed my research ever after: Professor Joseph Kockelmans, who read an entire earlier draft with the sympathy which proved instrumental in achieving the final version; Professor Rod Home upon my first arrival in Australia at the University of Melbourne; Dr Keith Hutchison who has largely helped me to state some of my points with greater clarity than I could have mustered without him; and Professors John North and Gerald Whitrow who both strengthened my hopes that this book could stand beside their own histories of cosmology. In May 1986 I had the privilege of being invited by Professor John Stachel to the first international conference on the history of general relativity which was held in Boston; this was for me a unique opportunity to test the progress of my research and compare recent developments.

Since I was educated as a philosopher, it was inevitable that some formidable problems of mathematics would appear as my writing went ahead. I would like to thank Michel Ghins for permission to borrow inspiration from his presentation of De Sitter's argument on rotation. Michel Kerszberg has offered his assistance for the design of several figures and the solution of many mathematical puzzles.

A grant from the University of Melbourne in 1985 has greatly facilitated the progress in typing out the many stages of the manuscript.

Note: I use double quotes in citing from an author's own terminology or text, especially where this has a somewhat technical meaning. For all other purposes, I use single quotes.

University of Sydney   P.K.

Contents

Introduction   1

Four different senses of the universe   1
Cosmology and physics since the emergence of general relativity   10
Invention versus rediscovery   14
References   22

1 Cosmological enquiries into the nature of physical science   23

1 Views of the nature of physical science and their cosmological implications   23
2 Newtonian versus Einsteinian questions   25
3 Astronomical ideas on the universe as a totality   35
4 Theoretical conceptions of the universe   43
5 The concept of field: Physics and mathematics of boundary conditions   53
6 The cosmological significance of the path from the special to the general theory of relativity   57
7 Postulates, principles, and assumptions of general relativity   68
8 The problem of the planets   78
9 Variations on a non-Euclidean theme   84
References   92

2 From a universal physics to a physics of the universe   97

1 Introduction   97
2 Einstein's 1916 thought experiment and the problem of relative rotation   102
  (a) The philosophical background   102
  (b) A first version of Einstein's philosophical argument   106
  (c) The 1916 version   109
  (d) Antinomy of Mach's principle and the general relativity principle   117
3 De Sitter's early critique of the principle of relativity and the nature of boundary conditions   119
  (a) Introduction   119
  (b) The relativity of rotation and the principle of relativity   121
  (c) On the origin of inertia   127

4	The invention of cosmology	135
	(a) General relativity before cosmology	135
	(b) The implications for astronomy of the relativity of time	136
	(c) Einstein's very last doubts	141
	(d) Half-way to the solution: the true reasons for the impossibility of Newtonian cosmology	145
	(e) Boundary conditions: mathematical or physical?	153
	(f) "A method which does not itself claim to be taken seriously"	155
	References	167

3 The almost full and the almost empty 172

1	The changing picture of general relativity from September 1916 to February 1917. First reactions to the cosmological considerations	172
2	Cosmological consequences of the relativity of inertia	175
3	Paradoxes with time. First hints of the geometry of De Sitter's model	182
4	Mathematical and physical postulates of cosmology	191
5	Inertia and gravitation: ordinary and world matter	194
6	Properties of light. Motion and the size of the universe	201
7	Einstein's new reply	206
8	The scientific and philosophical outcomes of the Einstein–De Sitter controversies	208
	(a) From a physical point of view	208
	(b) The wider context of philosophical implications	224
	References	230

4 Matter without motion or motion without matter? 233

1	Questions of dynamic cosmology	233
2	New ideas of cosmic time	235
	(a) Difficulties with cosmic time in De Sitter's static universe	237
	(b) The philosophy of cosmic time	244
	(c) The search for a new expression for boundary conditions and the idea of cosmic time	247
3	Beyond the limits of both the Newtonian and the Riemannian world views	252
	(a) Weyl's statistical interpretation of Newtonian cosmology	252
	(b) Weyl and the project of unification of physics	257
4	Global implications of extended local physics	261
	(a) The impossibility of a completely empty world	261
	(b) On the physical nature of the mass-horizon	262

5	Weyl and his critics: the geometrical versus the physical approach	266
	(a) Felix Klein	266
	(b) Towards the most general cosmological form	275
	(c) Lanczos and the idea of a new type of combination of space and time	277
6	Eddington's solution of 1923: its foundations and limits	282
	(a) From physics to metaphysics and back	282
	(b) The four themes of Eddington's research	284
	(c) The challenge of a compromise: 1. Time and becoming	289
	(d) The challenge of a compromise: 2. Astronomical evidence versus theory	293
	(e) 'Reality' and 'appearance' of the redshift	297
	(f) The mass-horizon as a necessary illusion	301
7	Solutions of old problems, problems of new solutions	306
	References	308

5 The construction of a principle   311

1	Eddington's solution reconsidered	311
2	From metric to topology and causality	314
3	The redshift revisited	326
4	Weyl's principle physicalized: a) Weyl's own approach	330
5	Weyl's principle physicalized: b) The adventure of non-static cosmology	333
6	Towards the big bang	343
7	A paradigm and a paradox: the alleged equivalence between Newtonian and relativistic cosmology	354
8	Weyl's principle and the 'many-universes' problem	370
	References	376

Epilogue   381

Proliferating inventions	381
Local versus global	386
The philosopher's universe and the physicist's	390
References	393

Index   395

Introduction

Four different senses of the universe

In November 1931 Willem De Sitter, Director of the Observatory and Professor of Astronomy at the University of Leiden, delivered at the Lowell Institute in Boston a course of six lectures on the historical and conceptual development of astronomy and cosmology. These lectures formed part of his six-month tour of the United States and Canada. Not long after this the lectures appeared as a book—De Sitter's only book in fact—entitled *Kosmos*. Another highlight of De Sitter's trip was the course of Hitchcock Lectures he delivered at the University of California in January 1932 which took as their focus some of the technical aspects of a somewhat more sharply defined area: "The Astronomical Aspect of the Theory of Relativity". (See De Sitter 1932, 1933.)

His American tour was the culmination of a long career in astronomical research, a career in which De Sitter had distinguished himself and indeed acquired some notoriety on account of his quite epoch-making contributions in three distinct areas. In celestial mechanics, he was responsible for significant advances in our knowledge of the motions of Jupiter's satellites, and he also had an interest in systematically improving the determination of astronomical constants. And as early as 1911, he discussed those problems thrown up by the recently formulated special theory of relativity; in particular the small deviations in the motion of the Moon and other planets that the new theory had revealed. It was the latter which first sparked off De Sitter's restless passion to understand the astronomical implications of the theory of relativity. In the United States, De Sitter was very interested in visiting the world's largest telescope that had recently been built at Mount Wilson. It was there, only two years earlier (in 1929), that the astronomer Edwin Hubble had announced the definitive proof of what was to shock not only the entire scientific community but to confound the imagination of laymen as well: that the largest building blocks in the universe, the galaxies, were systematically receding from us with a velocity proportional to their remoteness.

De Sitter's *Kosmos* is the one work where this most meticulous and painstaking astronomer lets his mind range freely. In marked contrast to the style of his articles, he develops in this book at some length a number of philosophical, almost psychological, propositions about his conception of the highest pursuit of his science. The book's introduction is a sort of methodological apologia, designed to vindicate what De Sitter sees as the ultimate purpose of scientific research. He believes that the state of astronomical science in his day is very unusual, and that this in itself may lead to the kind of insight into the nature of science where philosophical and technical considerations become inextricably linked. As a scrupulous scientist, he holds the view that all science, whether it proceeds by generalization and induction or by a hypothetical and deductive method, begins and ends in observation. How, then, is it possible to account for those periods in the history of science when a variety of divergent directions are followed, as it were, haphazardly? De Sitter describes the occurrence of this state of affairs during any particular epoch in terms of a contrast between the *static* and *dynamic* views of the world—what he chooses to call an antithesis (1932, p. 8). When he is discussing ancient astronomy, for instance, De Sitter mentions conflicting systems like those of Heraclitus and the Eleatics or of Archimedes and Euclid. The unwearying succession of oppositional figures throughout the history of science works in De Sitter's opinion as a common denominator of the progress towards a more satisfactory conception of the real world beyond. However, casting his eye over the previous fifteen or twenty years, it seems to De Sitter that things are in a state of disarray, "as if. . .we could not speak any more of a chief direction, of one line in the progress of science" (1932, p. 4). Quite apart from the unexpected direction taken by the quest for a unified field theory on the basis of Einstein's concept of gravitation, and even bypassing quantum mechanics, astronomical science seems to have been impelled towards contradictory conclusions, not least because of the independent branches of the science which have sprung up. "Is then the unity of science lost?", asks De Sitter (p. 5). At this point, he appeals, very impressively, to the enduring belief in the realistic powers of any science. He speaks of the "solid faith in the existence of order and law", without which no science is even conceivable; he says that the ultimate impulse in scientific work derives from "the fundamental assumption that there *is* a reality behind the phenomena".

However, when De Sitter comes to the last of his six lectures and begins to touch upon what were then quite recent developments in cosmology, his own hope seems strikingly at variance with the concrete results. He analyses the particular theory of a dynamic universe which had just been formulated by the theoreticians in the wake of Hubble's discovery, that is, the idea of a *non-static* metric where space itself (rather than objects in space) becomes the subject of motion. He finds that the problem of the actual specification of the dynamic model is "indeterminate" (De Sitter 1932, p. 127), because we are left with too many variables when we have to examine the information yielded by the astronomical data. As De Sitter goes on to argue, it is because of this that the very idea of a radius of curvature for the whole world is merely a tool of mathematics, without any correlative in the real world.

In fact, as early as 1917, long before there was any hint of the nature of the universe being dynamical, De Sitter had already hit on a quite similar conception. Speaking of the whole universe, he said that its "curvature of space . . . only serves to satisfy a philosophical need felt by many . . . it has no real physical meaning" (De Sitter 1917, p. 1224). Such a statement would scarcely have been taken seriously by a single other well-informed specialist of the early 1930s. Today, the reality of the curvature is thought of as beyond question and the only problem is one of arriving at the exact form this curvature takes. However, it would be too easy to say that De Sitter has simply persisted in an initial error. Nowhere does De Sitter falter in his actual grasp of the bases of the relativity theories. It is, rather, that his 'scientific' position is at one with a philosophical commitment which is given a priority irrespective of the rapid evolution of relativity between 1917 and 1932. During that short interval, the science of the universe had of course undergone a very profound revolution from a static to a dynamic conception. It is apparent, then, that De Sitter's position is anchored in a very sharp recognition of the crucial importance of cosmology in any framework for an understanding of general relativity. If we can follow his argument without jumping on what looks like an error, we realize that what he really means is that the "antithesis" between a static and a dynamic picture of the universe has not been completely resolved, or transcended, by the discovery of non-statical metrics. The creative tension, which De Sitter sees as having dictated the most fruitful advances of astronomy, should be taken as a *continuing* feature of the cosmology, one that cannot be overlooked if we are to properly understand its nature, its new and contemporary face.

This lofty concern for a philosophical assessment of the revolutionary theory is of the utmost importance, especially because De Sitter deserves to be hailed as one of the great pioneers who tried to solve some of the technical problems in modern cosmology by highlighting the relevance of the dynamical picture. So, early in 1930, he wrote a crucial paper on the distances and radial velocities of the extra-galactic nebulae, where, for the first time, the emphasis was clearly placed on the fact that "the true solution represented in nature must be a dynamical solution" (De Sitter 1930, p. 482). The difficulty with a dynamical solution, though, when applied to the whole universe was that it could no longer represent the usual motion of objects in space but only the expansion of space itself.

In these years, 1931–32, which were marked by the unanimous acknowledgement on the part of the scientific community that the new cosmology *was* revolutionary, De Sitter was not the only leading European scientist to visit the United States. Arriving from the University of Göttingen, where he held the chair of mathematics, Hermann Weyl came to Yale University in April 1931 to deliver the Terry Lectures, and in doing so also helped to substantiate the new cosmology. He entitled the lectures "The Open World", and added a subtitle which served to underline their far-reaching significance: "Three Lectures on the Metaphysical Implications of Science". Weyl's orientation is markedly different from De Sitter's. The tension that Weyl sees as fundamental is, in his terms, "between subject and object" (Weyl 1932, p. 26). Given the subject/object dichotomy, Weyl wanted to take the physical science of his day off the pinnacle it enjoyed throughout the nineteenth century so that he could re-activate basic philosophical discussion meant to deal with the relation between our ideas of the world and the world itself. One-sided standpoints of idealism and realism are declared to be equally wrong: "in the transition from consciousness to reality, the ego, the thou, and the world rise into existence indissolubly connected and, as it were, at one stroke" (p. 27). The philosophy advocated here clearly cuts against any form of blind faith in realism, but, by the same token, realism is not dismissed as merely false. It is a partial truth, which can mean nothing unless it is bound to the human power of knowing. The consequence is clear: "The real world is not a thing founded in itself, that can in a significant manner be established as an independent existence" (p. 28). What underlies the real world is also, as an unassuageable part of its essential nature, *our* symbolical construction of it with the aid of mathematical apparatus. At this point, Weyl allows himself a form of speculation which contrasts markedly with the generally cool and scientific style of the lectures. He speaks of how scientists should be free to take part in what he calls "cosmic worship": "in our knowledge of physical nature we have penetrated so far that we can obtain a vision of the flawless harmony which is in conformity with sublime reason" (p. 29).

Certainly, Weyl's appeal to cosmic worship has very little to do with any proposed return to mythical forms of thinking. It comes from an extremely tough-minded appraisal of the contemporary developments in science. The congruence between our ideas and the world is, of course, related to and perhaps justified by another form of concordance whereby the sum total of the universe's discrete parts are united in an unexpected harmony of their own: "Not until now do we perceive the true perfection of the universe which springs from the relation of its parts to the whole" (Weyl 1932, p. 6). How does this come about? Seeing the appositeness of the non-static form of the metric in dealing with the universe, Weyl published in 1930 a paper in which he proposed something wholly distinct and novel, a *principle* capable of relating all the parts of the universe to one another. What is this principle? By totality we should understand something which also makes us comprehend the very occurrence of motion. Indeed, Weyl said that his "assumption" was "that in the undisturbed state the stars form . . . a system of common origin" (Weyl 1930, p. 939). The ideas of an undisturbed state and of mutual interaction inevitably interelate since they partake of the one "origin". The undisturbed state is thought of as preceding the upheaval, that is, motion. And Weyl identifies the source of this disturbance in his Terry Lectures with *matter*. Matter is dubbed the "spirit of unrest", but its fundamental character is to remain under the sway of what Weyl calls, following the German poet Hölderlin, "Father Ether" (1932, pp. 20–1). Part of the achievement of Einstein's general relativity theory was that it put paid to the old dualism between the inertial guiding field (i.e. the structural, geometric field) and the forces that are exerted upon a material body. This is not to mean that general relativity got rid of the inertial structure as such. Rather, by denying any validity to inertia as a separate and amorphous receptacle of gravity (by far the most powerful of all forces at the large-scale level), it served as a weapon to assault the rigidity that inertial structure assumed in classical physics. So, Weyl continues, if the structural field "were not disturbed by matter it would abide in the condition of rest". Rest, Weyl adds, is in the more accurate language of mathematics equivalent to homogeneity. Hence, motion becomes concomitant with the loss of homogeneity, to the extent that the disavowal of rigidity in the structural field is also a refutation of its Euclidean independence from matter-in-motion. Weyl goes on to describe as *the* deepest mystery of natural philosophy the apparently "powerful predominance of the ether [that is, the new, non-rigid structural field] in its interaction with matter". Of course, this predominance was, until Einstein, responsible for the belief that the rigid structural field really was a faithful and direct expression of the real.

The principle introduced by Weyl into cosmology is nothing less than an attempt to trace this predominance back to a supposed "origin". At the origin, the beginning of all things, there was no disturbance. But it was the inherent quality of matter, an essential expression of its "spirit", that is should be responsible for the first, irreversible occurrence of the motion/disturbance. At this stage, Weyl is not even concerned with an origin in time (or indeed of time). He is perfectly content to assume the course of the heavenly bodies "in the real world since eternity". Now, it is extremely difficult to give any kind of physical meaning to this conception of an origin. Like all mathematical physics since the advent of Newton's science of mechanics, the theory of relativity presupposes that the world-line of any body "is uniquely determined by the starting point and the initial direction of its motion in the world" (Weyl 1932, p. 15). But this starting point and this initial direction remain entirely arbitrary; there is nothing in our science that might constrain them to be this and not that, and it is even part of the normal methodology of physics to begin with an assumption of arbitrary initial conditions. Notwithstanding all this, the new vision offered by cosmology makes it inevitable now to see a question-mark hovering round the idea of "origin" and the integral role it might play in our understanding of both the foundations of physical science and the nature of the physical world itself. It seems that Weyl conceived of his principle less as an explanation than as an almost unmediated representation of the mystery he was grappling with.

In the eyes of the Plumian Professor of Astronomy at Cambridge, Arthur

S. Eddington, the establishment of the non-statical metric, even when supplemented by a principle intended to naturalize and domesticate it, could not be the whole story. In September 1932 Eddington, too, was invited to visit the United States. He delivered a public lecture at a meeting of the International Astronomical Union held at Cambridge, Massachussets, which was later expanded into his celebrated work, *The Expanding Universe*. Eddington confessed that he did not want to take up the challenge of the philosophical issues. He goes so far as to assert that, "to those whose interest in modern science is directed chiefly to the philosophical implications, the theory of the expanding universe does not, I think, bring any particularly new revelation" (Eddington 1933, p. 122). Coming from Eddington, this kind of judgement must be seen in relation to the grand design of his own research. It becomes a virtual rationale for engaging in philosophy at every step of the argument, for turning an essentially philosophical mode of thought into an essential part of scientific activity. As he progresses through his exposition of the new theory, Eddington raises a disconcerting series of "Why's" which lead to an unceasing series of speculations, often of the most provocative kind.

In the first instance, Eddington says that Einstein's law of gravitation "throws no light on why the nebulae are running away from us and from one another" (p. 21). Of course, because of universal attraction, the law should rather predict a tendency for every part of the system to incline towards every other part. Eddington poses as "a detective in search of a criminal" (p. 61) and in the preface of his book, he describes the universal dispersion of the galaxies as "the clue not the criminal". In Eddington's whodunnit, there is a "hidden hand" which he calls the *"cosmical constant"*. He sees physics proceeding inevitably to its "final capture and execution". However mysterious it may seem, the distinguishing quality of this cosmical constant is that it acts against the gravitational collapse predicted by general relativity as it had previously been understood. Allowing for this new element in the theory, how can we account for the high receding velocities of the nebulae? (p. 25). There are only two possible conclusions: either these large velocities have existed from the very "beginning", or they have been brought into being by the cosmical constant, which thus acts as a genuine force of repulsion. In the former case, the constant expresses the mystery; it remains a kind of mathematical function. In the latter case, the constant is a cause and therefore a physical reality. Eddington points out that the former possibility can scarcely be called an *explanation*, and therefore favours the latter. It is at this point that Eddington manages to jump the hurdle of Weyl's conclusions. Eddington is after more than a mere statement, or (at the very best) an evaluation of the problem; what *he* wants is "sufficient evidence to hang the criminal" (p. 62). To arrive at this, one should also be able to provide additional tests for the new theory. What we really need, in Eddington's words, is to be able "to predict the actual magnitude of the cosmical repulsion, and see if the observed motions of the nebulae confirm the predicted value" (p. 26). In Eddington's view, this necessitates a yet further extension of the theory, and in fact a recourse to quantum theory, which yields peculiar insights into the question. The stage is set for an enactment of "The Universe and the Atom", the title of Eddington's last chapter.

Here the realm of "pure theory" begins (p. 94). The guiding lines of Eddington's philosophical commitment appear in the persistent search for empirical verification of what is in the first instance pure thought. In contrast to De Sitter's pledge of realism and Weyl's call for a delicate balance between realist and idealist claims, Eddington's views incline quite frankly to the idealist side. So, there is his own emphatic rejection of the merely empirical: "*There are no purely observational facts about the heavenly bodies*" (Eddington's italics) (p. 17). He proceeds to explain what he means: "Astronomical measurements are, without exception, measurements of phenomena occurring in a terrestrial observatory or station." But a true theory, to merit the name, should say something more. It is, indeed, only by theory that these measurements can be "translated into knowledge of a universe outside". By the theory of relativity, we come to know something of the outside world, precisely because all observatories or stations are given an equal status; in this respect, the theory of the expanding space is relativistic, since it stems from the possibility of making similar observations (e.g. the recession of galaxies) from any station whatsoever. A wished-for cosmological synthesis lies, therefore, at the heart of any reconciliation between astronomy and the demands of physics. Unlike De Sitter, Eddington envisages convergence rather than divergence: "our theory comes from the welding together of different lines of physical research" (p. 20). Unlike Weyl, the transition to cosmology makes Eddington say, "I do not think that we should feel that we are stepping from solid ground to insecurity." The claim is supported by his unshakeable faith that, in calling in the cosmical constant, science has performed the most decisive step of all. His interpretation of the constant can be seen as an alternative to Weyl's rationale for the origin of motion in the universe, as it represents a substitute for the mysterious property that Weyl chose to locate within matter. Eddington is a good deal more radical than Weyl when he contends that the state of homogeneity (which might have prevailed if one were to look far enough back in time) is in reality *nothing at all*. As he puts it: "To my mind *undifferentiated sameness* and *nothingness* cannot be distinguished philosophically" (p. 57). Radicalizing Weyl's viewpoint to such a point of extremity makes cosmology seem a journey into the impossible. The new and all-encompassing physics takes up, with a scientific emphasis, the question which has haunted human minds ever since the rise of philosophy among the Greeks: Why is there something rather than nothing?

The clue to an understanding of the inhomogeneities, which are the sole physical realities, cannot be an initial state of homogeneity. Nor can it be nothing. Eddington will strive to identify the something which is their source by virtue of purely speculative constructs coined by the human mind, such as the cosmical constant. It follows from this that cosmology is that which reveals both the nature and the limits of physical science:

At first sight it seems a reasonable programme for science to tidy up the region of space and time of which we have some experience and not to theorise about what lies beyond; but the danger of such a limitation is that the tidying up may consist in taking the difficulties and inexplicabilities and dumping them over the border instead of really straightening them out (p. 28).

In an amazing reversal, Eddington suggests that the traditionally insuperable difficulties and unfathomables of the science of the whole universe should be taken as less frightening than those of ordinary physics.

From De Sitter to Eddington, via Weyl, there is a progression in the boldness of the arguments used to elaborate the nature of the universe. Or, at any rate, the reader certainly feels an increasing assurance in the actual perception of how the problem should be addressed. And there is a similar progression in the philosophical positions of the protagonists. De Sitter, the sceptic, is propped up by a blind faith in the realistic nature of scientific knowledge. At the other extreme, Eddington firmly believes that something as extraordinary as the whole universe remains a scientifically legitimate object of inquiry, while professing to a philosophy tempered with idealist overtones. Between the two, Weyl, the creator of a 'cosmological principle', is thoroughly critical in his account of the relations that hold subject and object together in every human act of knowing.

But wherever we look, the physics of the twentieth century bears the indelible imprint of one genius, Albert Einstein. And the area of cosmology is, in this respect, no exception. By 1917 he had built upon his recently completed theory of general relativity a great edifice that constituted nothing less than a science of the entire universe. Einstein is the real creator of modern cosmological theory. In the years that followed, De Sitter was to be the keenest critic of the new idea, Eddington was to try every possible way of accommodating the facts to a developing theory, and Weyl was to formulate the all-encompassing principle that laid down what is now the almost universally accepted form of the theory. At the time of their American peregrinations in the early 1930s, each of these men tried to strike some kind of balance between the revolutionary advance and an overall picture. They concentrated on what had been done in the previous fifteen years, paying special attention to the recent notion of expanding space that appeared to solve so many of the portentous mysteries. Significantly enough, the creator of the whole story seems to dismiss himself with a view on the relations between science and philosophy which disclaims any pretence of reliability. Thus, replying to some systematic comments on his "occasional utterances of epistemological content", as he called them, Einstein was quick to emphasize that "the facts of experience do not permit [him] to let [himself] be too much restricted...by the adherence to an epistemological system" (Einstein 1949, pp. 683-4). He went on to describe with grave sincerity how the scientist

> must appear to the systematic epistemologist as a type of unscrupulous opportunist: he appears as *realist* in so far as he seeks to describe a world independent of the acts of perception; as *idealist* in so far as he looks upon the concepts and theories as the free inventions of the human spirit (not logically derivable from what is empirically given); as *positivist* in so far as he considers his concepts and theories justified *only* to the extent to which they furnish a logical representation of relations among sensory experiences. He may even appear as *Platonist* or *Pythagorean* in so far as he considers the viewpoint of logical simplicity as an indispensable and effective tool of his research (p. 684).

This is perhaps the clearest statement of Einstein's devotion to his scientific ideal and I have quoted it at length because it contrasts so strikingly with the committed and clearcut views of the scientific ideal we have seen so far. It is almost a moral obligation for the scientist to appear unscrupulous to the epistemologist, for otherwise he would surrender his fundamental justification for his comprehensive overview of the facts. It might seem that the opportunist would stick, in a last ditch stand, to the safest form of pragmatism he could find. Not so Einstein. In his celebrated "Four Lectures" delivered at Princeton in May 1921, he described in terms of the human mind itself the equivalent of the quest for a total comprehension. He said that

> even if it should appear that the universe of ideas cannot be deduced from experience by logical means, but is, in a sense, a creation of the human mind, without which no science is possible, nevertheless the universe of ideas is just as little independent of the nature of our experiences as clothes are of the form of the human body (Einstein 1921, pp. 2-3).

In general, the amalgamation of different philosophic standpoints does not mean that each of them is entirely subsumed by the whole. On the contrary, the solidarity between the purely descriptive part of science and the creative part warrants the inexhaustible freedom of the scientist's capacity of inventing just as it forces him to enlarge the realm of experience.

There is little doubt that Einstein could set in motion a revolution in thinking and make the universe as a totality open to scientific investigation *because* he could use the concepts of physical science with a greater degree of freedom than anyone else. Einstein acknowledged in many places that his path to cosmology was determined by his critical attitude towards the great physicist and philosopher of science of the second half of the nineteenth century, Ernst Mach. Thus he once wrote to his beloved friend Michele Besso that Mach's influence on him had been very great in many areas, not just cosmology, but Mach "did not realise that [the] speculative character [of the theory of relativity] applies . . . to every conceivable theory." By speculative Einstein meant this: "[Mach] thought that theories are somehow the result of a *discovery* and not of an *invention*" (Einstein's letter of 6th January 1948, in A.Einstein-M.Besso 1972, p. 391). In turn Einstein's revolution ushered in a whole series of inventive speculations about cosmology on the part of other scientists. The aim of this book is to study the origin, nature, and impact of this revolutionary change, by attempting to reconstruct how it actually occurred in the minds of those who initiated it.

Cosmology and physics since the emergence of general relativity

When we look at the state of physical science today, the most striking thing is that the terms of how a theory of nature may be built up seem to be on the way of being articulated in an impressively unified fashion. Within that theory, all physical structures and processes may be described by, and subordinated to, one single principle which borrows as much from experience as it does from our mathematical constructs. And some prominent physicists tend to believe that such an achievement would represent the culmination of physical science. That is, the theory hovering on the horizon would not be just a new approximation to truth, but something bordering on truth itself. (See, for instance, an account of these ideas, free from technicalities, in P.C.W. Davies 1984.)

The basic idea is that all nature is ultimately under the control of a single force. Of course the quest for unification is not a new story: it is the world view that science expected to proffer almost from the time that it first took its modern form in the work of Newton. The advent of the theory of general relativity in the first two decades of this century has signally reanimated the aims and methods of the quest; the original form of this unification is Einstein's legacy to physics. What happened is that by 1916, when Einstein had laid down the essential tenets of his gravitation theory in which the large-scale mechanics of the universe was subsumed into a comprehensive space-time geometry, it was clear that the only fundamental forces in Nature are those which have their origin in gravitation and in the electromagnetic field. After Faraday and Maxwell had, in the nineteenth century, amalgamated electricity and magnetism and coordinated the effects of the electromagnetic field into laws of striking simplicity and clearness, it was felt desirable to attempt to explain gravitation also on the basis of electromagnetism, or at least to fit it into its proper place in the scheme of electromagnetic laws, in order to arrive at a unification of ideas. Several attempts were made during the first two decades of the twentieth century (H.A. Lorentz, G. Mie and others), but the success of their work was not wholly convincing. In virtue of general relativity, the terms of the problem changed quite dramatically: the nature of gravitation was now understood and the problem of unification was reversed. In the new picture, it was necessary to regard electromagnetic phenomena as well as gravitation as an outcome of the geometry of the universe. In order to make this possible, several ways were suggested by which the world-geometry on which Einstein based his original theory could be liberated from what made it suitable to gravitation only. Again this met with partial success only. Eddington summarized the situation in 1923 when he wrote, in a book which rapidly became a classic of general relativity, that "the possibility of the existence of an electron in space is a remarkable phenomenon which we do not yet understand" (Eddington 1923, p. 153).

For many years Einstein himself tried to find a way in which geometry could be equated more directly with the fundamental properties of matter, particularly the electromagnetic laws that govern the miniature domain of the atomic particle. But his quest for a unified theory never succeeded. In his search he found himself increasingly isolated during the last four decades of his life, out of sympathy with the philosophic views that he found impossible to square with his extraordinarily sophisticated pragmatism: those that derived from the formalism of quantum theory established in 1927, which had come to hold sway among his contemporaries in physics.

All in all, no fewer than four fundamental forces in nature have been singled out. Apart from gravitation and electromagnetism, so-called weak and strong interactions have been found to govern nuclear structure. More recently, eminent physicists have combined the electromagnetic and weak forces in order to produce an electroweak force. Grandly unified theories have also been proposed that unify the electroweak and strong interactions into a single hyperweak force. The crux of such theories is that the hyperweak force is conceived of as having ruled very early on in the universe's existence, when the energy concentration was extremely high, but in the present universe it operates in different ways that appear to us as three separate forces. No one seriously doubts that the fourth force, gravitation, will enter that scheme without too much delay.

Gravitation, however, again poses a particular problem because the way it could be integrated in the scheme of unified theories sends us back to the starting-point: will the theory unifying all forces still be geometrical in character? (See S. Weinberg 1982.) If it were not, what could possibly be substituted for it? And if it were, will this geometry still be that of Einstein? The alternative to understanding unification in terms of geometry lies in the idea of the quantum of gravity, which is a quantum with certain unusual properties. But so far all unifications which have proved successful (and which involve the other three known forces) have always been primarily geometrical. It is precisely in accordance with this that the need arises to test any possible theory of complete unification at scales of distance which exceed all direct experimental means. In this sense, the problem of gravitation changes little in the conviction that the early universe offers this laboratory which is crually missing in our here-and-now environment.

This is how most research in physics today came to include quite bold cosmological speculation almost as a matter of course. The key concept to the understanding of the nature of physical laws is nothing less than the *history of the universe*. The gradual fragmentation of a single primeval force is seen as providing both a concrete picture of the whole universe and a conceptual framework of, and meaningful foundation for, the whole of our physical science.

Ideas about taking a kind of a final step in science are by no means uncommon. For instance, the same kind of hope appears to have dominated physics at the end of the nineteenth century, when the conviction was widely felt that a series of ever-finer measurements of some constants was the one final task left for physicists. But what is entirely new about today's certitude is that this unified physics sees itself as concomitant with cosmology. Over the two centuries that marked the establishment of Newtonian mechanics, its continued success had rested upon its disregard of physical cosmology. Although it is true that Newtonian mechanics succeeded in integrating both the earth and the other planets of the solar system into one and the same conceptual schema, it simply relegated to the periphery of that system the ancient division between a world that is open to science and one that is not. Contemplation of a concrete representation of heavenly bodies beyond the solar system or even of the entire universe was for the most part the prerogative of purely speculative conjecture.

The turn to cosmology as a basis for the unified theory of today and tomorrow is fairly recent, originating in the early 1970s. And the search for the theory has taken on an aspect that even Einstein himself would hardly have predicted. Certainly his 'invention' of relativistic cosmology, as it may fittingly be called, had a tremendous impact upon his views of what general relativity was or should be. He immediately saw it, as he reported to Besso, as a "proof that general relativity can lead to a non-contradictory system" (A. Einstein's letter of 9 March 1917, in A. Einstein-M. Besso 1972, p. 104). But Einstein's initial model of the universe has little in common with present-day approaches to the problem of unification. The single feature which has contributed most to its being abandoned is the *static* metric used to describe geometrical properties at large. Of course, in the year 1917, Einstein had very good excuses for ignoring the dynamic solution which captures the idea of an expanding universe; until the investigations of Hubble by the turn of the 1930s there was simply no evidence of any astronomical velocities that were not very small compared with the velocity of light nor of any systematic cosmic motion. But even so, Einstein's behaviour in the period going from his first model to the establishment of expansion remains quite critical to the theoretical issue. For Einstein was so obsessed with the static form of the metric that he completely failed to estimate the extraordinary impact of the non-static metric, which was first proposed on purely theoretical grounds by the Russian, Aleksander Friedmann, in 1922. His early appraisal of Friedmann's work is quite astonishing: he first published a short note pointing out a mathematical error in Friedmann's calculations, then corrected his own note by confessing Friedmann's soundness. And that is all he did. Even more baffling is the complete silence of the whole scientific community with respect to the achievements of Friedmann, until they were rediscovered in the early 1930s by De Sitter, Eddington, and others, and came to form the basis of the theory of the expanding universe.

There are several reasons, both internal and external, à propos of the Friedmann papers, which might explain the delayed response to these early suggestions of an expanding universe, but none of them is really plausible. It has been suggested that Friedmann's mathematics was of too high a level and that, beyond that, his results lacked astronomical interest. (North 1965, p. 117.) Circumstances of publication might be added to the list: the editor of the most important journal *Astronomischer Jahresbericht* did not list a later and equally important contribution by Friedmann (1924), because he could see in it no astronomical import; and when the 1922 paper was mentioned, no one bothered to point out that it represented an alternative cosmological model. (Hetherington 1973, particularly pp. 24–5.) More recently, it has been argued that "Friedmann's work was carried out in Leningrad during the Russian civil war, presumably in complete ignorance" of some significant measurements that tended to prove its validity. On top of that, "some potential readers may have been deterred by Einstein's laconic remark, in the same journal, that Friedmann's non-stationary universes are incompatible with the field equations" (Torretti 1983, p. 204). But on that reasoning Einstein's emending note should have attracted even more attention, and Friedmann's possible ignorance of astronomical measurements serves only to highlight his genius rather than excusing the inattention of other scientists.

Beyond such vagaries, the delayed response has its own relevance to the foundations of general relativity. The historical development of relativistic cosmology in the years 1916–30 lays bare a kind of tension which dominates the establishment as well as the upshot of the theory. Inasmuch as the expanding universe has by now been hailed as the magical key to the understanding of the whole nature of physical laws, it is of the utmost importance to assess properly this historical development in terms of its relation to the ultimate questions raised by the formulation and entrenchment of general relativity. So my purpose in this book is twofold: first of all I want to examine in some detail the historical evidence about the developments of relativistic cosmology, from its earliest stages (1916–17) up to the time of the final recognition of a class of solutions according with different types of metrics (1932); secondly I will try to show how the delayed response to early suggestions of an expanding universe was by no means an accident but was closely connected with some essential aspects of the general relativity theory. I would like to think that this twofold approach will help clarify, in turn, the status of the ideas pertaining to a unified physics embedded in the notion of a history of the universe. This book is meant to clear the ground for a critical examination of these ideas.

Invention versus rediscovery

From the time when he was a young research assistant with Eddington, G.C. McVittie recalls "the day when Eddington, rather shamefacedly, showed me a letter from Lemaître which reminded Eddington of the solution of the problem which Lemaître had already given. Eddington confessed that, though he had seen Lemaître's paper in 1927, he had completely forgotten about it until that moment" (G.C. McVittie 1967, p. 295). This took place in 1930. The problem which had been puzzling astronomers, physicists, and mathematicians for more than a decade was how to give an adequate description of the whole universe which would tally with the relativity theory. Two models were proposed in 1917: apart from Einstein's model, De Sitter had put forward his own alternative, a very curious one since it was described as appearing to be 'empty'. And that seemed all the relativity theory could offer. In a 1927 paper, Lemaître perceived that a statical gravitational field, with spherical symmetry, admits of only these two solutions and noted the odd fact that there was simply no "fair compromise" (as he called it) between the two extremes. Lemaître's idea was that modifying the premise underpinning the whole line of reasoning might lead to the desired compromise: and that premise was the statical form of the metric. Interestingly enough, Lemaître did not know of Friedmann's earlier papers and his chief motivation for making the change to the new metric was, this time, observational—it would account for the observed redshifts of extra-galactic nebulae.

In *The Mathematical Theory of Relativity*, his fundamental synthesis first published in 1923 (and slightly revised in the second edition of 1924), Eddington pointed to what he called the "make-shift contrivance" of statical coordinates. There is little doubt that Lemaître had the Eddington book in mind when he delineated with the greatest clarity the two solutions that had been so far produced. That Eddington himself never took the step of exploring the non-statical form is therefore very strange indeed. When the time came for him to rediscover that form, he confessed to W. De Sitter, in a fascinating letter, that he himself was quite taken aback not to have realized it earlier.

Thus, if Friedmann was almost ignored, Lemaître, who originally formulated the idea of a 'primeval atom', was not given much attention either. The same is true of H.P. Robertson who (again quite independently, this was 1928) came to realize the oppositeness of a non-statical solution. What all this shows is that the dominating figures in the early stages of contemporary cosmology were not really the ones who perceived and anticipated the non-statical theory. In contrast to the heroic age of modern cosmology, the situation *today* is quite the reverse: Friedmann and Lemaître are credited with the invention of non-statical cosmology, while the pioneering debates between Einstein and De Sitter, as well as the profound reflections they provoked, have been far too easily forgotten. Those who persistently came to grips with the question of the limits associated with the statical picture have in fact opened the way to what is potentially a crucial reflection. I want to expound it as carefully and thoroughly as I can. By pointing our attention towards the authors who were actively engaged in this form of reflective speculation, we will first of all try to restore a true picture of the state of cosmology before the advent of the expanding universe.

But this is not all. Another, more philosophical intention, comes with the methodological difficulty of carrying out this project. Indeed, an historical excavation of this kind is in itself quite perilous, since it is continually jeopardized by a retrospective knowledge of the apparent 'naïvetés' which seem to pervade the pioneering discussions. Our sense of the *statical bias* of the whole period is part and parcel of this. Of course it is true that these discussions are in a way artificial because the authors work out what is ultimately an inadequate form of dynamism: they fail to perceive the possibility of an expanding space, and hold to the more traditional concept of a motion *in* space. However, the most important thing is the fact that the non-statical view has not only conquered by highlighting the now obvious inadequacies of the statical one, it has also tended to devalue the very type of reflection that was first promoted. And this devaluation springs from an approach to the early problem which is wholly inappropriate to the subject matter. In its own way, it echoes the uncritical stance adopted by most physicists today regarding the dominant myth of an evolutionary universe. What we need (and what I will attempt to give) is a step-by-step reconstruction of the ideas developed, giving prominence to the level of theoretical accuracy at each phase of developments. One obvious objection to this would be to point to the inevitable incompleteness of any reconstruction: surely we would be placing ourselves in such a position that no sharp line could be drawn between the *philosophical* analysis of the scientific thinkers of that time and the *technical* discussions of the models in their relevant aspects. My contention is precisely that any such attempt to distinguish too sharply between one aspect and the other will lead to an impoverishment of both. We have seen how radical and how diametrically opposed were the philosophical positions of the four major authors involved in our story. In retrospect, it appears that the radicalism of their positions was for a long time the major obstruction to a comprehensive solution. But the fact that the actual solution was a rediscovery rather than a sudden blinding breakthrough testifies to the powerful presence of some underlying philosophy. For the original breakthrough that Friedmann performed which so strikingly anticipates the systematic solution in what was virtually its definitive form, puts the mathematics in order and skips over the original purpose of matching the teachings of experience on a grand scale. To echo Einstein for a moment, Friedmann's breakthrough makes for an economy of philosophical scruples, but without really knowing that it does so. And the radicalism of our four above-mentioned philosophies is, in turn, the index of an unusual depth in the earlier approach.

Ignoring Friedmann, as his contemporaries seem to have done, shows the real difficulty connected with the advent of relativistic cosmology, *both* in its statical and the non-statical phases. The difficulty lies in each new development being a fresh attempt to spell out the implications of Einstein's original theory. This reflection demonstrates the increasingly important role played by essentially philosophical considerations *within* the realm of the technical discussions. Accordingly, the arduous formulation of an underlying philosophy has to be understood and expounded parallel to the focus on the technical aspects. It must be realized that each phase of the history serves to further underpin the foundations of scientific endeavour, in the sense that each contributor seems to displace his own philosophical conviction from the texture of technical analysis. One sees, then, how the scientific and the philosophical viewpoints at once supported and obfuscated each other; how, in consequence, the solution to technical difficulties at a given time does not nullify the totality of apparent 'mistakes' or 'fallacies' in the initial line of speculative enquiry. As a matter of fact, these errors are rooted in the philosophical bases of the *whole* endeavour of cosmology. What look like inconsistencies or omissions only testify to the extraordinary difficulty of carrying through the project of cosmology which was then springing up in the mind of Einstein, De Sitter, Eddington, and Weyl.

The non-statical form, as a theoretical solution to the conflict between the two models, looks quite natural, in the sense that we only had to *need* it in order to *get* it. But the apparent simplicity of the solution conceals its true difficulty, since the demand is itself the consequence of a very subtle intellectual journey. Taken in the form adopted in the Robertson–Walker metric (which was formalized in 1935), the solution entails a certain number of new difficulties, quite different in kind. These will not be completely intelligible so long as the primary and original problem has not been clarified by the right kind of historical and philosophical investigation. The difficulty that remains with the non-statical form is that it leads again to the emergence of a multiplicity of possible worlds. It is clear that the criteria which would enable us to single out the unique model corresponding to reality must lie *outside* mathematical formalism; they are the observations, but they are also what enables us in the last resort to *interpret* the observations, i.e. what brings out some natural connection between the world and our ideal constructs. Explicit hypotheses and *a priori* postulates, such as the well-known cosmological principle that states the equivalence of all viewpoints in the universe, play these roles of mediation. A variety of independence, consequent on this status of outsider for the cosmological principle, has gained increasing acceptance ever since the discovery of the non-statical solution. The so-called deductive cosmology, which originated with Edward A. Milne as early as 1932, and its extension into a steady-state cosmology in 1948, each represented first attempts to build an entirely new world view in which this degree of independence would be explicitly marked out. More recently, feeling the pressure of a putative unification of physics, orthodox relativistic cosmology has been nudged towards such all-embracing concepts as the Anthropic Principle. (On how this principle addresses probably the central problem facing both scientists and philosophers through the ages, namely the position and role of man in the universe, see J. Barrow and F. Tipler 1985.) Now, the paramount idea which serves to legitimize such a shift is what has become known as *Weyl's principle*. The way it came to be formulated can give us some understanding of how it might ultimately be vindicated.

Weyl's principle provides the natural justification for the idea of an expanding universe. It asserts that all the world lines of the universe, at any epoch and at any time, form a network of interacting geodesics which tend to orginate from a common origin in the past. In fact, the stated principle reflects Weyl's thought from as early as 1923 long before others saw its full significance as a foundation principle. Only the non-statical cosmology was capable of adopting Weyl's intuition so swiftly as a presiding principle; and cosmology *became* evolutionary when the shift was performed. Being formulated within the statical frame of reference, the idea represents the most radical and highly sophisticated response yet made to the problems raised by the early Einstein–De Sitter controversy. It also sowed the seeds for a type of cosmology the very possibility of which Weyl seems to have been oblivious to, and this certainly constitutes the problematical core of the development of relativistic cosmology. Indeed, the tension throws into high relief the intricate relation of cosmology to the deep-seated and covert motives which govern the structuring of the relativistic world view. The later and explicit independence of cosmic-scale postulates and principles from the relativistic theory of gravitation may even come to look like a straightforward consequence of resolving the strictly *technical* difficulty generated by Weyl's intuition.

To begin with, even a superficial glance makes it obvious that the idea of change of reference system from statical to non-statical coordinates seems to have been quite clearly grasped by Eddington and Weyl before it was explored analytically by other people. In those terms, there is an interesting distinction to be made right at the outset between the purpose of the change of reference system itself and its actual consequences at the analytic level. If the debates are then obfuscated in any way at all, it is not so much because of something like a statical bias as because the new concept of the universe was becoming known at the same time as the first attempts were being made at an encompassing and unified physics. The systematic study of the various *alternatives* to Einstein's quest for unification forms the appropriate complement to the rise of relativistic cosmology. The alternatives somehow seemed to fall short at the time, but they are now widely acknowledged as foreshadowing the type of unification which is envisaged today. I mean, in particular, Weyl's gauge theory and Eddington's amendments to it, as well as Eddington's original attempt at a synthesis between quantum theory and cosmology. In Chapter 4, I have tried to tell the story of how they were expounded during their crucial formative phases.

In fact, throughout the formative period of relativistic cosmology, not only after but also before the publication of Einstein's first model, De Sitter and Einstein himself engaged in a whole series of discussions which can be seen now as *the* decisive controversies on the subject. From 1916 to 1935 (the year of his death), De Sitter was a major contributor to the whole debate. As respondent to Einstein, he is both the star actor and the privileged spectator of almost all the significant ideas that contributed to the establishment of a relativistic cosmology. His personality and thought dominate each stage of its history, and towards the end of his life he was able to appreciate why and how the new kind of dynamic interpretation offered by cosmologists was one of the most astonishing episodes in the whole dialectical sequence of static and dynamic views.

Chapter 3 focuses on De Sitter's properly cosmological controversy with Einstein between 1917-20. It is written out of my own conviction that De Sitter had already seen, in his way, that the problems involved in the static picture could admit of not cut and dried demarcation with those of the dynamic one. I use the word 'properly' to describe De Sitter's cosmological controversy with Einstein, for the entire dispute sprang from an earlier controversy between the two scientists, which took place in the second half of 1916. The discussion then was mainly epistemological, and based on the different possible meanings that could be given to the phenomenon of rotation from the standpoint of general relativity—a phenomenon which had already raised much trouble in the Newtonian context. Chapter 2 deals with this discussion: in it I try to back up my hunch that Einstein was led to 'invent' physical cosmology in direct response to De Sitter's highly original and still very little known critique of rotation. From that point of view, it appears that Einstein's invention is just one among a number of responses to De Sitter's attacks. That the first comprehensive model of the universe was designed in order to deny as far as possible the substance of these attacks had the effect not only of constraining the subsequent developments of cosmology but also of shaping the very nature of how the question of the universe as a whole could be addressed. For the most striking feature of these attacks is their increasingly obdurate search for what it is that makes a theory like general relativity consistent with itself. It was Einstein's genius to realize that what gives his theory its 'cosmological' quality has great relevance to the question. In Chapter 1, the kind of consistency that only Einstein sought is related to the emergence of his new theory of gravitation, between 1905–16. I try to show how Einstein's resolution of the difficulties involved in Lorentz's conception of the ether and, somewhat later, his route from the special to the general theory of relativity both work to establish a new level of consistency for physical science. Because it is certainly no accident that general relativity (as opposed to Newtonian theory) could lend itself to cosmology, my interpretation of these early developments is coloured by what I contend in their 'cosmological' nature. For a more comprehensive (and less exclusive) exposition of the theory of general relativity, its antecedents and its subsequent impact, the reader may turn to a number of existing studies which provide ample detail. (A highly commendable one is S. Goldberg's (1984). I am following his recommendations with regard to the technical aspect of the difficulties: "Mathematical statements serve only as a benchmark for historically important mathematical foundations of key physical ideas" (p. 1). Also, "modernizing the mathematical notation. . .would obscure the motivations and intent of the subject" (p. 479).)

Higher consistency, of the kind required by De Sitter, was supposed to fill a gap that he had detected in the original form of general relativity. But it also pointed to a radically new form of physics, one that would incorporate straight away what was initially missing. From this point of view, the emergence of alternatives to orthodox relativistic cosmology, as well as the parallel development of unified physics, can be comprehended in certain essential aspects as deriving from the early Einstein–De Sitter controversies. All cosmological theories can now be said either to tally with some form of Weyl's principle (and it is very striking that the highly contentious steady-state cosmology has, in fact, pushed the principle to its extreme limits) or subvert it on radical grounds (as another very unorthodox theory, the Gödel cosmology, does by renewing a basic problem that, in Einstein's words, "disturbed me already at the time of the building up the general theory of relativity, without my having succeeded in clarifying it" (1949), p.687). This is the subject of Chapter 5 where more general issues, arising from Einstein's and De Sitter's positions, are discussed in the light of the lines they trace.

There is a key concept underlying the whole story, and that is the idea of cosmic time which, as anyone can see easily enough, was certainly quite alien to the theory of relativity as it was first constructed. Weyl's principle purportedly makes it seem a naturally sufficient condition in the context of the large-scale implications of general relativity. However, that it was also *necessary* is not quite so self-evident. That is, in fact, the chief reason why it is incumbent on the interpreter of this backdrop to come to some kind of reckoning with the purely philosophical evaluation at work in the technical discussions. Doing so will involve sticking to the belief that the primary creative impulse for a major and wholly new idea will somehow be concomitant with its ultimate validation. The conflict between the ideas of cosmic time and of general relativity was first perceived when De Sitter formulated his model of 1917 in which universal time partitions were rejected. As we shall see in Chapters 2 to 4, the whole development of cosmology throughout the subsequent fifteen years depended on the accuracy with which this model was interpreted, the upshot being that Weyl's principle represented the high water mark in the series of attempts made to incorporate cosmic time into the geometry of the model. The attempt required, in turn, that the very idea of cosmic time be transformed, in order for Lemaître to be able at last to write down the non-statical metric. But, when all is said and done, there is still the question of why cosmic time should be the ingredient without which any cosmological construct seems non-viable. The answer to this question will lead us to examine the relevance of De Sitter's *original* model to more recent issues. Certainly De Sitter's name crops up these days in quite a variety of contexts. It is to be found in connection with the familiar class of homogeneous and isotropic models, in which any point is geometrically like any other point and where 'rotation' about the origin leaves their properties unaltered. In mathematical terms this is expressed by saying that the group of isometries of this universe is a particular group of 'rotations' in which time remains essentially distinct from space; this group is often called the De Sitter group. Furthermore, the original form of the De Sitter metric is precisely where the exploration of that most dazzling form of cosmological speculation, steady-state cosmology, comes in. According to the steady-state cosmology, instead of having no matter at all, matter is being continuously created out of nothing. After a period of great enthusiasm in the 1950s, this is now widely regarded as mere heresy; it remains significant that the steady-state model emerged from what is almost certainly the most radical and definitive vision that has ever been communicated about the implications of cosmic time.

I delay my discussion of cosmic time and the variety of conceivable universes until Chapter 5, in order to show more precisely how they derive from the first, decisive stages of this story. I shall take most of the results proposed by the authors at face-value, commenting on them with hindsight only when this significantly enriches our understanding of the basic point at issue. If this may seem to stretch the reader's tolerance (for surely it would be hopeless to attempt to capture at any cost what Einstein or De Sitter *really thought*), I still do my best to hold on to the straight scientific activity of the period examined. So any difficulty in the style of this book is only the faint echo of a far more general problem. Relativity was originally conceived by Einstein as a refinement of Newtonian physics, where the measures (if not the concepts themselves) of the latter were supposed to remain valid in the larger view, in terms of greater scales of magnitude. Starting out with this original plan and confronting it with a view derived from classical science of certainty and causality, it is almost as though I would be applying one system to a problem where it has no relevance. Only if I am right in claiming the strict relevance of the static to the non-static world view will anyone realize that this course is inevitable; the interesting thing about this relevance is that it makes us realize not only where a departure from the totalized classical view was necessary but also why the departure could never be complete. For we will even come to see, in Chapter 5, that one of the strange things about contemporary cosmology is that its *formalism* has been viewed as by and large indistinguishable from the Newtonian frame of reference. This necessitates a detailed case study, which suggests that the essence of contemporary cosmology lies quite precisely in the still open field of potentialities thrown out by the various *interpretations* of that formalism.

There have been two detailed histories of contemporary cosmology so far, one by John North and one by Jacques Merleau-Ponty. By a curious coincidence both of them were published in the very year (1965) that a most decisive observation was made, the so-called background thermal radiation of the universe. This discovery was quickly taken as evidence of the relics of the big bang predicted by the proponents of general relativity in the wake of Lemaître's primeval atom theory. So, arguing the merits of the alternative cosmologies, as North and Merleau-Ponty did, is today somewhat out of fashion. For my purpose, though, these works remain essential. For instance, Merleau-Ponty says that Einstein's amazing confusion about the status of Friedmann's solutions may be explained by his attention being entirely focused on the development of quantum theories and the search for a unified field theory. Something like this is, in fact, true of all the major actors involved in the modern cosmological drama, apart perhaps from an astronomer like De Sitter, our privileged character. And our task of evaluating the two models then available should be seen in tandem with the various attempts to arrive at a comprehensive picture of the atomic world; each of them echoes in its own way those potentialities still in the store of formalism. (I may also mention the existence of a more recent investigation of the *observational* development of cosmology in the period I am most concerned with: this is the book by R.W. Smith (1982). A useful collection of original papers in twentieth-century cosmology is provided by J. Bernstein and G. Feinberg (1986).)

The kind of problem I want to address is exemplified by two recent comments on the impact of general relativity today. One comes from a biography of Eddington: speaking of the "supposed difficulty in understanding the general theory of relativity", the author goes on to argue that this difficulty "was greatly exaggerated: it contributed to the stagnation of the subject for several decades. Many of the developments of the sixties and the seventies could easily have taken place during the twenties or thirties" (S. Chandrasekhar 1984, p. 30). Such a smug judgement is somewhat blind to the actual nexus of problems involved in general relativity. Here, as in the more specific question of cosmology, the 'delay' only highlights the unusual importance of purely conceptual, and indeed philosophical, problems in the minds of the scientists themselves. But from the standpoint of science, philosophical troubles find their expression in *ignorance*. Wherever we look, we find that ignorance was constitutive of the emergence of relativistic cosmology: the Friedmann–Lemaître vision was ignored by the scientific community while Weyl's grand principle was established in ignorance of its very natural framework. And learning through ignorance seems to have made us willing to be oblivious of our roots. Thus, the other comment of importance in this context is the concluding remark of a masterful textbook on relativity: "How rich the actual discovery and how forgotten, by comparison, the original purpose" (W. Rindler 1977, p. 244). If the unusually rich resources of the actual discovery have always had the effect of keeping the philosophy of relativity alive, then it should be even less deniable that the historical excavation of the original purpose may keep pace with the whole scientific endeavour. As an eminent physicist of our own time has written, the significant advances in physical science over the past twenty-five years have been made possible by concentrating "almost entirely on Einstein's original 1915 theory. This research has revealed a number of desirable and remarkable properties, both physical and mathematical, that the original theory possesses" (R. Penrose 1979, p. 36). Most importantly, "many of these recent advances have been into the global rather than the local structure of the theory". It is primarily the link between the original form of Einstein's theory and what we may call global or indeed universal considerations that I wish to give substance to in the following pages.

References

Barrow, J. and Tipler, F (1985). *The anthropic cosmological principle*. Oxford University Press.
Bernstein, J. and Feinberg, G. (eds.) (1986). *Cosmological constants*. Columbia University Press, New York.

REFERENCES

Chandrasekhar, S. (1984). *Eddington*. Cambridge University Press.
Davies, P. C. W. (1984). *Superforce*. Heinemann, London.
De Sitter, W. (1917). On the relativity of inertia: remarks concerning Einstein's latest hypothesis. *Proc. Kon. Akad. Wet. Amst.*, **19**, 1217-25.
De Sitter, W. (1930). On the distance and radial velocities of extra-galactic nebulae, and the explanation of the latter by the relativity theory of inertia. *Proc. Nat. Acad. Sciences*, **16**, 474-82.
De Sitter, W. (1932). *Kosmos*. Harvard University Press, Cambridge, Mass.
De Sitter, W. (1933). The astronomical aspect of the theory of relativity. In *University of California publications in mathematics*, Vol. 2, No. 8, pp. 143-96, University of California Press, Berkeley.
Eddington, A. S. (1923). *The mathematical theory of relativity*. Cambridge University Press.
Eddington A. S. (1933). *The expanding universe*. Cambridge University Press.
Einstein, A. (1921). *The meaning of relativity*: four lectures delivered at Princeton University, transl. E.P. Adams, Methuen (1922), London.
Einstein, A. (1949). Reply to criticisms. In *Albert Einstein, philosopher-scientist* (ed. P.A. Schilpp), pp. 665-88, Open Court Publishing Company, Evanston.
Einstein, A. and Besso, M. (1972). *Correspondance 1903-1955* (ed. P. Speziali), Hermann, Paris.
Friedmann, A. (1922). Über die Krümmung des Raumes, *Zeitschr. Phys.*, **10**, 377-86.
Friedmann, A. (1924). Über die Möglichkeit einer Welt mit konstanter negativer Krümmung des Raumes, *Zeitschr. Phys.*, **21**, 326-32.
Goldberg, S. (1984). *Understanding relativity*. Clarendon Press, Oxford.
Hetherington, S. (1973). The delayed response to suggestions of an expanding universe, *J. Brit. Astr. Assoc.*, **84**, pp. 22-8.
McVittie, G. C. (1967). Georges Lemaître, *Qu J. Roy. Astr. Assoc.*, **8** pp. 294-7.
Merleau-Ponty, J. (1965). *Cosmologie du XXème siècle*. Gallimard, Paris.
North, J. (1965). *The measure of the universe*. Clarendon Press, Oxford.
Penrose, R. (1979). Recent advances in global general relativity: a brief survey. In *Einstein symposion Berlin*, Lecture notes in physics No.100 (ed. H. Nelkowski et al.) J. Springer, Berlin, pp. 36-45.
Rindler, W. (1977). *Essential relativity*. J. Springer, Berlin.
Smith, R. W. (1982). *The expanding universe*. Cambridge University Press.
Torretti, R. (1983). *Relativity and geometry*. Pergamon Press, Oxford.
Weinberg, S. (1982). Einstein and spacetime: then and now. In *Selected Studies*. (ed. Th. M. Rassias and G. M. Rassias), North-Holland Publishing Company, Amsterdam, pp. 383-392.
Weyl, H. (1930). Redshift and relativistic cosmology. *Philosophical Magazine*, **9** pp. 936-43.
Weyl, H. (1932). *The open world*. Yale University Press, New Haven.

1
Cosmological enquiries into the nature of physical science

1. Views on the nature of physical science and their cosmological implications

The seventeenth-century revolution in physical science, which led to the innovations of Galileo's experiments and the theoretical mechanics of Newton, began long before as a revolution in the heavens. It was primarily the Copernican view of the world, in which the status of the Earth shifted from being the centre of the universe to a mere runaway heavenly body, which prompted the first conception and subsequent realization of a universal mechanics like Newton's. In fact, Newton's mechanics was based on the rejection of almost all of the presuppositions embodied in a non-mathematical, largely Aristotelian view of the world (see A. Koyré 1957). Almost, but *not quite* all these old presuppositions. One aspect of the pre-Copernican world in particular was left quite untouched by Newtonian mechanics: the generally *static* conception of the universe, that is, the realm of the stars and, in general, of all those heavenly bodies that lie beyond the circuit of the planets revolving about our sun. The unification of all *motions* within the solar system was certainly the crowning achievement of Newton's science (and this unification was concomitant with the very advent of a consistent dynamics) but there are several indications in Newton's own writings that he believed motion would simply disappear, were it not for the miracle of God's handiwork in maintaining it: "Motion is much more apt to be lost than got, and is always upon the decay" (Newton 1730, p. 398). Newton believed in an active God, in a God who "being in all places, is more able by his will to move the bodies within his boundless uniform sensorium, and thereby to form and reform the parts of the universe, than we are . . . able to move the parts of our own bodies" (p. 402). This had many implications for an understanding of the nature of the motion described by the new dynamics, for when decay and change affront the body and eternity of God, God seems to be willing to maintain the status quo rather than oppose new forms of change; when He sets out to reform the parts of the universe, he does not 're-create' anything.

The subordination of motion to rest, particularly as far as heavenly bodies are concerned, is undoubtedly an inheritance from the Greek vision of cosmos. Thus, in Book VIII of his *Physics*, Aristotle made the point that since a circumference has no definite beginning, middle, or end, the movement of rotation of the stars "is stationary and motionless in one sense, and moves continuously in another" (265b, in 1934 edn, p. 401). It was two-and-a-half centuries after the first edition of Newton's celebrated *Principia* (1687) before this ancient conception of cosmic equivalence between motion and rest, over and above the achievements of modern celestial mechanics, could be overthrown. Here, as with any major change in scientific thinking, theory and observation actively helped each other. However, unlike most other drastic changes in science, the transition from a static to a genuinely dynamic view of the universe was effected in a quite abrupt way. In order to drop the assumption of statism, it sufficed that statism should be recognized as a powerful assumption in all previous attempts to come to terms with the cosmological problem; the assumption had been as powerful as it was hidden, and had simply not been questioned or ever made explicit. It is therefore important to realize at the outset that the static way of representing any such object as the whole universe may well be one of the deeply engraved structures of human thought which somehow impose themselves in the face of this representation. The example of Newton even suggests that the 'common sense' belief in statism crops up in reflecting on the nature of physical science, i.e. the essence of what is described by way of mathematization and experimentation.

Statism turns out to have resisted the drawn-out assaults of modern science to remove 'common sense' thinking from the actual understanding of nature in terms of mathematics and experiment. But what does it mean to retain the notion of common sense thinking when the questions addressed go far beyond any imaginable human scale? There is here a lesson which extends far beyond cosmology as a self-contained scientific discipline and which relates to the most general form of the human approach to an overall understanding of nature; the very word 'cosmology' suggests an intimate, yet overpowering kinship between the world (as order) and ourselves (as capable of reasoning). Not surprisingly, common sense thinking at the other end of the world scale, the world of microphysics, was heading for just this kind of pitfall at the same time that the expanding universe theory was being established: it was also in the 1920s that the quantum revolution effected its dramatic revision of classical determinism. It would always remain extremely hazardous to square statism and determinism with some definite view of the universe which would hold for its own sake, for only the very view which emerged in the second decade of the twentieth century put us in a position to look back at pre-relativistic and pre-quantum conceptions in those terms.

With regard to cosmology as a science, the situation is slightly peculiar, but perhaps typical enough of the problems involved in the revision of supposedly 'natural' ways of looking at the world. In the first place, the cosmological revolution which took place in the early twentieth century is already far more than a modification, however dramatic, of any pre-existent conceptions. It is certainly not an overstatement to say that cosmology as we understand it today then came into being all of a piece. But the disconcerting thing is that what Einstein invented in 1917 was indeed a static model of the whole universe, one in which the old presupposition of statism seems never to have been questioned at all. And even more surprising is Einstein's declaration in the opening section of his cosmological memoir that the very problem of cosmology arose from the very real parallels between the difficulties involved in the construction of a consistent Newtonian universe and those involved in the large-scale implications of general relativity. These parallels were seen as a last residue of 'classicism' in general relativity, and the development of relativistic cosmology derived from Einstein's wish to annihilate any analogy. The new model, therefore, owed much of its own inventiveness to a re-consideration of the *original* form of problems involved in Newtonian cosmology, even though what was then referred to by Newtonian cosmology was by no means a well-defined doctrine.

Throughout Chapter 1, I shall be arguing that the Newtonian conception of the universe in its original form has a deep relevance to any understanding of the early twentieth-century debates on cosmology. It follows that I shall not be concerned with any exhaustive chronicling of cosmology from Newton to Einstein; this history may be pieced together elsewhere, though to this day there is no single, comprehensive work (see S. Jaki 1972, M.A. Hoskin 1982, and J. Merleau-Ponty 1984). My real aim is to give an account of the various interactions between Newtonian and Einsteinian views on the nature of physical science, so that the realm of what had been classified in orthodox theory as not quite within the purview of physical science will gradually emerge as fundamentally important.

2. Newtonian versus Einsteinian questions

In December 1916, Einstein wrote to his beloved friend Michele Besso about a question of "great scientific significance" which, as he put it, "is not a product of my imagination" (Einstein and Besso 1972, p. 96). The problem at stake was how to interpret correctly the fact that the universe is in equilibrium. Einstein wanted to reply to a suggestion made by Besso that, for reasons of symmetry, it would be perfectly possible to conceive of matter uniformly filling all space to infinity; the claim was that such a distribution would produce no field at all. Einstein's argument against this was as follows. Take a sphere of matter K centered in P (see Einstein's own drawing,

Fig. 1.1). According to Gauss's principle, "a gravitational flux through the surface K must exist, being created by the matter contained in K". As a result, and quite independently of whether the space outside K was or was not filled with matter, "the matter must fall towards P with an acceleration increasing as the distance from P increases". In these conditions, is it possible for the universe to endure at all? Einstein considered for a moment the possibility that the universe might be capable of reproducing the structure of the solar system, that is, the possibility of centrifugal forces hindering the fall of matter, but he suggested that such a motion cannot exist for any length of time: "infinitely great differences of potential would give rise to very high stellar velocities, and these must have long since disappeared" (p. 100) precisely because they are so high. Einstein intended his argument to be a proof that the symmetry of a homogeneous, infinite distribution of matter is an insufficient condition for doing away with the cumbersome assumption that the field at infinity is infinite. Imagine that the problem is due to the *static* distribution: if you allow high velocities to exist, these cannot continue for a long time—the shortness of their existence is, as Einstein surmised, in proportion to their magnitude. We are thrown back on the static pattern, and the only way of avoiding infinite potentials is to set at zero the mean density of matter at infinity, in which case we would be left with the equally "distasteful conception" (1917a, p. 107) of a natural centre to the universe.

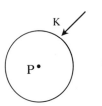

FIG. 1.1. Einstein's drawing in his letter of December 1916 to Besso where he explains that a symmetrical distribution of matter throughout all space up till infinity would not be sufficient to produce a stable universe.

In his argument, Einstein deploys the concept of field which marked the nineteenth century's great advance over Newton's original formulation of the law of gravitational force. In accordance with the differential expression of the law, Einstein follows a step-by-step procedure, starting with a finite volume of matter and ending up at infinity. However, strikingly enough, problems and even outright contradictions are *already* apparent in a whole variety of ways within the framework of Newton's original formulation. Newton had encountered exactly the same kind of hypothesis for explaining the equilibrium of the universe, and found it equally implausible. The suggestion came from the Reverend Richard Bentley who argued that, if the universe were a uniform, infinite distribution of matter in Euclidean space, infinite attraction exerted by all masses on one side of some central particle would be compensated for by an infinite attraction countering it on the other side. Newton replied, in his letters to Bentley, "that two equal infinites by the addition of a force to either of them, become unequal in our ways of reckoning" (in Cohen 1978, pp. 295-296). These ways of reckoning are none other than the basis of mathematical reasoning—the infinitesimal calculus does not make all infinites equal; on the other hand, the addition of a force is due to the occurrence of the slightest change of motion in the universe. If the infinites are equal, in accordance with Bentley's *physical* conception, then no dynamic motion at all is possible in the universe, even though equilibrium would be realized everywhere by virtue of Nature's own forces. But if the infinites are unequal, in accordance with *mathematics*, then there is dynamic motion but no equilibrium, i.e. gravitational collapse. In either event, we face the very same impossibility: the occurrence of any particular kind of finite force in the universe may be understood irrespective of its inclusion in the whole universe. As he expressed it to Bentley, Newton's argument leant heavily on the solar system taken as a model. What we find in this system is a variety of planets, comets, and stars; a common pattern of orbits for the planets around the sun, as opposed to the erratic motions of the comets; one luminous centre for a multiplicity of opaque bodies. All this, according to Newton, was bearing witness to God's handiwork. This kind of non-uniformity has been accounted for successfully enough by the application of the law of gravitation to the solar system; by contrast, Newton could not find a satisfactory response to the mathematical difficulties of the infinite, for if the whole starry universe were finite, the result should be, by rights, a central mass, blurred and undifferentiated. Newton went on to refuse a solution after the planetary model in which centrifugal forces would deflect the fall of stars from their straight paths, on the grounds that the particular kind of diversity exhibited in the solar system simply could not occur in this sphere (Cohen 1978, p. 306). It is certainly with this particular problem in mind that Newton reverted to a famous solution adumbrated in the General Scholium that ends the *Principia*: "if the fixed stars are the centres of other like systems, these must be all subject to the dominion of One. . .and lest the systems of fixed stars should, by their gravity, fall on each other, [God] hath placed those systems at immense distances from one another" (Newton 1729, p. 550). Of course, the minimization of gravitational effects does not rule out completely the occurrence of an infinite force at infinity. A typical perception of the contradiction that this alleged solution implies, is offered in one of the first popular expositions of Newton's theory on the Continent, that of Voltaire, which draws attention to the fact that Newton's theory "seems to contradict itself with regard to the fixed Stars". Indeed, according to this theory, the stars "attract one another, and yet remain immoveable; we must begin with explaining [Newton's] Sentiment, and Shewing, that it does not imply any Contradiction at all" (1738, p. 348). In fact, Voltaire made much bolder speculations than Newton, since he conjectured that "the reciprocal Gravitation of two fixed stars does not diminish precisely in the inverse or reciprocal Proportion of their Distances", that is, "the law of Gravitation may vary" (pp. 345–6).

These parallels (and contrasts) between Newton's and Einstein's discussions of the equilibrium of the whole universe testify to what are, by implication, the fundamental problems of formulating a consistent cosmology within the framework of Newtonian mechanics. In essence, both Newton and Einstein came to acknowledge the impossibility of combining any number of *centres* of forces and *motions* for all these centres. But while Newton abandoned cosmology in the highly precarious position of appearing to explain absolutely nothing at all, Einstein derived some positive stimulus from it. The reasons for this dramatic shift are both historical and conceptual. Before examining them in turn (as I shall do in the two next sections), the actual meaning of the persistent thematization of the universe in terms of a static conception must be clarified. For while the revolution achieved by Einstein touched explicitly upon the question of centres, the reflection on the static structure was guided by more or less implicit considerations.

Newton's letters to Bentley constitute a fundamental starting point for a proper clarification and understanding of the status of cosmology in Newton's establishment of a mathematical physics. Bentley, who later became Bishop of Worcester, consulted Newton before writing a series of lectures to be delivered in 1692 at St Martin-in-the-Fields in London. Through these lectures, entitled "A Confutation of Atheism from the Origin and Frame of the World", Bentley endeavoured to prove the existence of God on the basis of physico-theological arguments. To this end he set out to avail himself of an understanding as accurate as possible of the physical image of the world, the image that Newton defended.

Writing these letters was not Newton's sole reason for entertaining the cosmological question, for in them he concedes that he had carefully investigated the problem to a certain extent before Bentley invited him to do so. This remark is probably more than a polite gesture, although the exchange of letters did lead Newton to probe the problem in a decidedly more systematic manner during the early 1690s, when he was preparing a second edition of the *Principia*, as several other contemporaneous manuscripts of Newton's indicate.

These letters find Newton at a crucial stage in the evolution of his ideas on the foundations of mechanical philosophy (see D. Kubrin 1967). Starting from the belief generally held by English theologians of the time, that the cosmos, once left to itself, must decline in motion and regularity, Newton was looking for a mechanism by which God could periodically restore the quantity of motion and the regularity of movement among celestial bodies.

The discovery of the periodicity of comets led him to suggest, in the first edition of the *Principia* in 1687, that the mechanism probably owed something to the ability of comets' tails to dissipate slowly into space. Since comets' tails always point away from the sun, Newton imagined that they were endowed with a force that overcame the sun's force of gravity. (Early in his career, he sought the explanation of gravity in the permanent circulation of a diffused alchemical ether.) Thus this mechanism could replenish the Earth and the planets, but not the Sun. Only much later, at the time of the second edition of the *Principia* (1713), did Newton secure the replenishment of the sun as well. Struck by the astonishing proximity of the paths of certain comets to the sun, he adopted the theory that comets could fall from their regular orbits into the body of the sun and thereby replenish the sun itself.

The most elaborate version of the Newtonian cosmogony appears in John Conduitt's account of a conversation he had with Newton towards the end of his life (quoted in Kubrin 1967, p. 340):

[Newton repeated] what he had often hinted to me before, viz. that it was his *conjecture* (he would affirm nothing) that there was a sort of revolution in the heavenly bodies that the vapours and light emitted by the sun which had their sediment in water and other matter, had gathered themselves by degrees in to a body and attracted more matter from the planets and at last made a secondary planet (viz. one of those that go round another planet) and then by gathering to them and attracting more matter became a primary planet, and then by increasing still became a comet which after certain revolutions by coming nearer and nearer the sun had all its volatile parts condensed and became a matter set to recruit and replenish the sun . . . and that would probably be the effect of the comet in 1680 sooner or later.

There is no doubt that Newton is considering here the formation of a given body from its earlier stages—in an extended solar atmosphere—on to its shaping into a comet. This body passes through the stage of satellite of a planet, then becomes a planet itself, before finally turning into a comet apt to fall into the sun. Newton refers back to an argument invoked in the letters to Bentley, that only comets can be naturally formed out of brute matter just because they follow erratic paths. The fact that comets are *final* products of this natural evolution avoids undermining completely the sacrosanct idea, that planets as we observe them cannot be formed naturally, precisely in virtue of the highly organized pattern of their orbits. There is a *temporary* planet state only inasmuch as the original stuff is supposed to be emitted from the sun—it does not come from remote regions where comets travel. Accordingly, in his final cosmogenesis Newton does not seem to have touched upon a conception according to which planets are 'pre-existing'; the very process whereby comets apt to collide with the sun are formed already depends on the presence of certain planets, of which future comets are first satellites.

The real originality, of course, lies in the possibility of comets colliding with the Sun. Comets are sufficiently massive to collide with the Sun if they were planets first. Does this mean that *all* comets were planets first? Or may there exist other comets, the origin of which is not or cannot be accounted for by Newton's cosmogony? In the *Principia* we find a more explicit statement, which forms a fitting complement to Conduitt's 'Memorandum'. Newton explains at length that "comets are a sort of planet revolved in very eccentric orbits about the sun." Among the comets, and precisely because of the comparison, "those which in their perihelion approach nearer to the sun are generally of less magnitude", so that "they may not agitate the sun too much by their attraction". Furthermore, it is only incidentally that the mutual attractions between comets influence their ability to fall upon the body of the sun: by their mutual disturbances, "their eccentricities and the times of their revolutions will be sometimes a little increased, and sometimes diminished". About the comet that appeared in 1680, "in its aphelion, when it moves the slowest, it may sometimes happen to be yet further retarded by the attractions of other comets, and in consequence of this retardation descend to the sun" (1729, pp. 532, 540, 541). Attractions of other bodies are occasional; they simply add to the descent. This shows that collision with the Sun can be due only to 'intrinsic' properties of the comet as such. Any comet that is already due to fall into the Sun bears strictly incidental relations to other bodies. In the conversation with Conduitt, Newton endeavoured to relate these intrinsic properties to the history of the comet's existence.

Notwithstanding obscurities in Newton's handling of the origin of comets, it is therefore misleading to suggest that Newton's final cosmogony is concerned merely with "celestial transmigration," as David Kubrin calls it (p. 346). In this view, satellites and planets would be disturbed by comets and become comets themselves only because of new combinations of their motions. If Newton meant something like that, the propensity of comets to fall into the Sun would be totally independent of the possibility of new creation. But Newton speaks of this propensity as the very key to the possibility of new creation; therefore genuine transmutation, rather than transmigration, must be involved in the final theory: the theory does not merely implement what could only superficially be called here a dynamic (non-static) picture of the universe.

In fact, on this account, the 'non-static' picture of the world is already that of the *Principia*, with its intricate machinery of forces being activated by motions. What the unpublished sources reveal, like this conversation and some of the significant manuscript evidence, is something wholly different. Indeed, if transmigration serves to justify a continual cyclical re-creation of the systems of the world, genuine transmutation goes so far as to question the very possibility of cycles. Even if he nowhere asserts it explicitly, Newton's problem is whether the 'explosion' following a collision between the sun and a comet, which is fully exemplified in the phenomenon of novae, embraces the *whole* solar system—planets, satellites, and, possibly, remaining comets. The true alternative that Newton faces, which alone justifies his reserve when speculating about the nature of the universe, is not a static universe versus a dynamic one, but rather renewal versus final destruction.

In manuscripts contemporary with the letters to Bentley, Newton suggests with curiously Leibnizian accents both the infinity and the eternity of the world. He argues, however, that "no thing is by eternity and infinity made better or of a more perfect nature, but only of larger duration in its own kind, and either greater or more numerous, to the honor of God the creator" (in J. McGuire 1978, p. 121). The true originality of Newton's final theory is to implement a new picture of the relation of things "in their own kind" to God: the very duration of things might well be physically limited. This contention broached a number of theological issues that Newton was obviously reluctant to face. He had the example of his friend Edmund Halley, who, accused of unorthodoxy, defended himself by demonstrating the physical impossibility of theologically unacceptable facts like the eternity of the world. As a result, as Simon Schaffer has put it, "what was then objected to was not Halley's unorthodoxy, but his use of physical considerations in theology" (1977, p. 28). This kind of objection might certainly have restrained Newton from tackling directly the most delicate matter of all: reconsideration of the formation and stability of the entire universe in the light of what he had finally discovered about the origin and fate of comets on the scale of the solar system.

Significantly enough, even the post-Newtonian secularization of the mechanistic world view was not instrumental in changing the basic issue. For it was not only Newton, but the physicists of the nineteenth century and even the early Einstein himself who seem to have based their fundamental cosmological constructs on the attempt to ward off the universe running down. (See S. Jaki 1974, pp. 276–305 and J. Merleau-Ponty 1983, pp. 211–253.)*
Thus, Newton's views on the loss of quantity of motion in the universe were formulated more than a hundred and fifty years before the second law of thermodynamics. And at the other end of the seesaw, Newton's desire for a mechanism to replenish the sun is repeated almost verbatim by the early twentieth century geologist, Arthur Holmes. Holmes argued that some external force must replenish the sun because of "the cyclic processes of nature . . . in the universe nothing is lost . . . in its cyclic development [is] the secret of its eternity" (1913, p. 121.) Holmes then invoked the name of the nineteenth century philosopher Herbert Spencer, who had endorsed the idea of endless cycles of what he called "evolution and dissolution". The belief in the eternal cycles has this much in common with the belief in the static universe: both offer a way of accommodating the anathema on universal decay and death.

---

* A. Lightman (personal communication) has pointed my attention to the material of this and the following paragraph. See also A. Lightman *et al.* (1989).

In the early part of the nineteenth century, J. Fourier (1824) was probably the first to point out that thermic and mechanical processes cannot be reduced to one another, yet heat, like gravity, is everywhere present. Fourier made a series of conjectures on the energetic properties of a cosmic substratum associated with space, but it was not until the second half of the nineteenth century, when the global implications of a principle of heat engine formulated by Carnot in the 1820s were first perceived, that the issue of decay of energy became palpable. Lord Kelvin immediately recognized the disturbing consequences of Carnot's discovery and warned that if the universe was something like "a single finite mechanism", it would be "running down like a clock, stopping forever" (1862, p. 357). In an age of rampant positivism, opinions were fairly divided on how to make sense of the matter. Lord Kelvin, for instance, came out with a series of loosely cosmo-theological conjectures about the irreversible energetic degeneration of the universe, and Clausius, the father of the notion of entropy, thought much in the same way. On the other hand, a thermodynamicist like William Rankine suggested that vast concave ether walls in space collected and refocused the energy lost by stars, so that the universe was given "the means of reconcentrating its physical energies and renewing its activity and life" (1852, p. 202). Joule and Mayer ventured to speculate on the cosmic sources that generate heat.

But uncertainty over the true relationships of heat to light prevented anyone from settling the question. It is rather wonderful to think that this confused background of argument and counter-argument about thermodynamics may well have been one of Einstein's motivations to see the deficiencies of Newtonian cosmology in the *gravitational* context. Perhaps here we have some clue as to why Einstein clung to a static universe: in the memoir where he formulated his model for the first time, after pointing out the problems in Newton's theory and providing an alternative, he felt more secure with a system where the "behaviour of the gravitational potentials at infinity would not . . . run the risk of wasting away which was mooted . . . in connexion with the Newtonian theory" (1917b, p. 181). The words "wasting away" give us some inkling of Einstein's underlying concern. It may be that for Einstein, as for Newton, a dynamic universe could only be a decaying universe. As we will see in Chapter 5, the clash between this specific anathema on 'devolution' (far more than a simple fear of change) which underlay both the search for permanency and staticity, and the independent theoretical criteria of general relativity, may also have influenced the establishment of non-static metrics for the universe. For Eddington, the rediscoverer of these metrics, could at last recognize that no *necessary* connection holds between the two.

In the period of our immediate survey, just prior to this development, it is probably again Holmes who most clearly perceived the implications of this speculative debate. He articulated the radical alternatives of precisely the kind that was later resolved by the dynamic view of the universe: "If the development of the universe be everywhere towards equalisation of temperature implied by the laws of thermodynamics, the question arises—why, in the abundance of time past, has this melancholy state not already overtaken us? Either we must believe in a definite beginning . . . or else we must assume that the phenomena which we have studied simply reflect our limited experience" (1913, pp. 120-1). Holmes was ready to incline towards the latter solution. And oddly enough, an almost identical confession of ignorance came from those who had finally developed ways of investigating the sky which were not merely speculative—the astronomers of the latter part of the nineteenth century.

3. Astronomical ideas on the universe as a totality

It is a quite remarkable fact that the Newtonian theory of gravitation is in itself at odds with the conception of a completely uniform universe. A logical corollary of the theory is that matter should tend to concentrate towards a centre of universal attraction, and the density of matter should decrease as the distance from the centre is increased. However, the idea of uniformity is immensely appealing if we want to treat the whole universe as a self-consistent system. Ever since the time of Giordano Bruno's speculations on an infinite universe replete with similar systems everywhere, uniformity has been used as a powerful heuristic aid by astronomers who have investigated the universe outside the solar system.

At the very end of the nineteenth century, the great American astronomer Simon Newcomb wrote in one of his essays about the significance of a science of the whole universe:

We know that several thousand of these bodies [the stars] are visible to the naked eye; our giant telescopes of the present time . . . show a number past count . . . Are all these stars only those few which happen to be near us in a universe extending out without end, or do they form a collection of stars outside of which is empty infinite space? . . . Taken in its widest scope this question must always remain unanswered by us mortals . . . Far outside of what we call the universe might still exist other universes which we can never see (1906a, pp. 5-6).

Newcomb's words faithfully reflect the general tenets held by astronomers at that time. There seems to be a radical divorce between the *extent* of the universe in space and its physical *constitution* in the sense that all stellar systems are regarded as populating a non-physical and purely geometrical Euclidean space. As we have seen, any speculation about a higher unity inhering in these systems, independent of the means of observation, was to be found only in isolated conjectures scattered through the then very young science of thermodynamics. But just as the astronomers tended to give up any hope of ever finding observational evidence about the material structure of the universe, so the guiding theme in speculative thermodynamics with any applications for a theory of the universe, was that of the death of energy.

The whole tradition of empirical cosmology was inaugurated in the latter part of the eighteenth century, when William Herschel systematically directed his telescope to every part of the sky. His main achievement was to absorb an increasing number of nebulosities into an aggregation of stars. He was equally convinced, however, that not every nebulosity would be resolved by increased telescopic power; some nebulae could serve as a sort of seat of attraction for the formation of stars. But the distribution, as well as the constitution of these nebulae was keenly debated until the early 1900s. It was not really until the 1920s, thanks to Hubble's classifications (which followed H. Curtis's demonstration of the extra-galactic character of the Andromeda nebula) that some consensus was finally reached. (See H. Curtis 1917. For a historical perspective over the observational problem, see also R. Berendzen et al. 1976.)

Such uncertainties are explained, at least in part, by the way in which a stellar astronomy developed throughout the nineteenth century. The historian of astronomy Agnes M. Clerke, writing at the end of the century, said that astronomy had now certainly reached the zenith of its achievement: where the Ancients merely described, and the advent of Newtonian physics led to causal understanding, "the third and last division of celestial distance may properly be termed 'physical and descriptive astronomy'. It seeks to know what the heavenly bodies are in themselves, leaving the How? and the Wherefore? of their movements to be otherwise answered" (1893, p. 2). The one great advance after the invention of the telescope in the early part of the seventeenth century, was when observational cosmology moved from a descriptive to a physical stage as the spectrometer and the camera were finally applied to astronomy in the second half of the nineteenth century. The theory of gravitation had very little to disclose in this area for, as Clerke went on to argue, "it was virtually brought to a close when Laplace explained to the French Academy, 19 November 1787, the cause of the Moon's accelerated motion. As a mere machine, the solar system, so far as it was known, was found to be complete and intelligible in all its parts" (pp. 2-3). As to the world beyond the solar system, "the problems which demand a *practical* solution are all but infinite in number and extent" (p. 528, my italics), but their existence does not in any way invalidate the type of theoretical approach that had already been worked out in celestial mechanics. From this point of view, it appears that the paradoxical situation of nineteenth-century observational cosmology is that it tended to proffer a vision of the universe which was both well-founded *and* incomplete (see J. Merleau-Ponty and B. Morando 1976, p. 85).

The apparently autonomous character of observation was not without a degree of theoretical incidence, however. In 1847, the Earl of Rosse installed at Birr, Ireland a massive telescope (a 72-inch reflector) which remained the largest in the world until it was finally surpassed by that of Mount Wilson (Hubble's telescope) in the 1920s. This enabled astronomers to resolve an increasing number of nebulae into stars, but it was precisely this increased telescopic power that was to precipitate a fundamental problem of interpretation. The problem being: how far it is possible to match the results bearing on the *constitution* of these nebulae with conjectures over their *distribution?* As M.A. Hoskin has put it, "to equate all nebulae with star clusters would be to lend strength to the conviction that other island universes or galaxies exist beyond and outside our own" (1967, p. 80), for those milky patches that are too distant to be resolved and yet occupy a substantial portion of the sky must certainly be located at enormous distances from Earth. But sightings of great changes of brightness and of form among some nebulae were thought by other scientists to be incompatible with the existence of such huge and far-off star systems. The problem again posed radical alternatives: either *all* nebulae are star systems, or *all* of those that cannot be resolved are actually made up of some gaseous, diffuse matter. As Alexander von Humboldt had rightly emphasized in his celebrated *Kosmos* of the 1850s, the very possibility of an ever increasing telescopic power raised from the start the question of whether the either/or could be transcended once and for all; explorations of new and more remote regions of the sky would certainly reveal new nebulae, and the resolution of *all* of them into stars would require even greater telescopic power. Humboldt went on to consider the possibility of island universes separated from one another by such enormous distances that our most powerful telescopes would show only the dark depth of the sky.

In fact, the latter part of the nineteenth century saw the advent of entirely new investigative techniques besides the telescope: as we have seen these were the astronomical use of spectroscopy and photography. By 1864, Huggins had, by studying the spectra of some nebulae, established the reality of gaseous nebulae, and so did much to invalidate the basic assumption underpinning the island universe theory. In the 1880s, extreme revisions were necessary to the image of the Andromeda nebula after a nova was discovered in it; the nova was so bright that astronomers wondered how a single star could rival the light of an entire galactic system. Moreover, photographic plates of Andromeda (I. Roberts 1888) revealed its fine structure, and it was soon thought that the famous hypothesis regarding the formation of planets from nebular matter, which Laplace had put forward at the start of the nineteenth century in his *Exposition sur le Système du Monde*, was somehow "made visible"; observation seemed to directly illumine the reality of a planetary system in its stages of formation, with effect that by radically improving our knowledge of the constitution of nebulae, the perspective on their distribution was dramatically transformed. So much so that the island universe theory had fallen into all but total disrepute by the century's close. The discovery that our system of stars, the Milky Way, is a highly organized system came in turn to be associated with the fact that it was finite. To put it more accurately, the most widely accepted picture of what the universe is made up of and how its elements are distributed tended to be identified with the limits inherent in the empirical methods available. Simon Newcomb drew a puzzling parallel between the traditional philosophy, which vainly tried to grapple by using pure reason to cope with the problem of how far the universe extended, and modern science, which was limited by its observational methodology (1906b, p. 31). With the case of Kant's philosophy in mind, he wrote: "The fact is that the problem with which the philosopher of Königsberg vainly grappled is one which our science cannot solve any more than could its logic" (p. 32). Nevertheless, speaking of the universe as a totality, Newcomb could also argue that "it is a great encouragement to the astronomer that . . . he is gathering faint indications that it has a boundary" (p. 6). The indications amounted to little more than a few working assumptions that tallied with the subordination of empirical methods relating the actual distances of nebulae to those gathering evidence about their constitution:

Granting that, in any or every direction, there is a limit to the universe, and that the space beyond is therefore void, what is the form of the whole system and the distance of its boundaries? Preliminary in some sort to these questions are the more approachable ones: Of what sort of matter is the universe formed? And into what sort of bodies is this matter collected? (p. 33)

The fact that the views on distribution go along with those on constitution is important, for it was this which resulted in the previously mentioned alternative being thrown up earlier this century, with the revival of the island hypothesis.

Certainly it is in the connection between distribution and constitution that we find pure observation attaching itself to a generalized but highly active form of 'theory'. Long before the advent of the "third and last division" of astronomy, as Agnes Clerke called it, the idea of the stars having a perfectly uniform constitution was used in determining their distances. Thus, in the latter part of the seventeenth century, both Huyghens and Newton computed some distances by assuming that all stars have the same intrinsic brightness as the sun; Huyghens (1698) asked how far the sun should be in order to look like Sirius, and later on Newton (1729, p.596) used the inverse square law applicable to apparent luminosity to deduce a distance to Sirius which was actually 30 times more than Huyghens'.

But characteristic of the kind of problems raised by the connection of constitutive and distributive uniformity are the notes in Newton's own manuscripts, in which Newton appeared to be striving for a fully physical and natural explanation of stellar fixity in a universe of indefinitely large dimensions (see M.A. Hoskin 1977). In these pages Newton seemed to solve the very problem that Bentley formulated; namely, that a universe where each star is surrounded by equal and opposite gravitational forces is a universe in a state of natural repose. A high degree of symmetry is needful in order to prevent gravitational collapse. So he can prove that uniformity, Newton undertook to deduce stellar distribution from the number of stars that should belong to each degree of magnitude. His first assumption was that the scale of magnitudes used by astronomers provided the distances in a direct way. Using the inverse square law for brightness and making the assumption of equal absolute magnitudes, the distances of certain stars could be obtained directly by measuring their apparent brightnesses. In the attempt to reconcile data drawn from the observation of stellar magnitudes with the 'postulate' of uniform distribution, Newton soon realized that he was led to unacceptable conclusions: the increase in the number of stars was much too rapid when the distance from the sun increased. At this point, Newton had the choice either of dropping the hypothesis that all stars are uniformly bright or of dropping the belief in the linear relationship of magnitude to distance; he seems to have opted for a change in the latter so that the stellar distribution as computed from magnitudes could remain uniform. In actual fact, the nineteenth-century divorce between theory and observation found its primary vindication in the critical re-examination of this particular kind of correlation between the two uniformities. Thus, when Bessel measured stellar parallaxes for the first time (between 1837 and 1840), and proved in the process the validity of the Copernican theory by the use of an argument no longer confined in its application to the solar system, he took the proper motions of the stars instead of their brightness as his standard in measuring distance. His assumption was that the closer the object, the larger its proper motion would appear. Bessel drew from this the very important conclusion that the stars vary a great deal in their intrinsic brightness.

It was only in the early years of the twentieth century that an indirect method started to be used to measure larger distances. As Adams and Kohlschütter noted (1914), the dark bands of absorption lines in the spectra of the stars vary from star to star according to a definite proportion. The knowledge of this proportion, in addition to the absolute brightness of those stars whose parallax had been measured, enabled Adams and Kohlschütter to hit upon the absolute brightness of other, more remote stars. Thus, a more sophisticated form of constitutive uniformity led to better knowledge of the structure of our star system. At still further distances, where the spectra become far too tenuous to be measured, Miss H. Leavitt was able to establish an accurate relationship between the intrinsic brightness and the period of a particular type of variable star, the Cepheids (see E. Hertzsprung 1913). In consequence distances could be estimated on the basis of apparent luminosity alone; when this method was applied in 1913 to the smallest of the Magellanic Clouds, it was clear that this nebula was more distant than any known star. The postulate of constitutive uniformity was being applied with great success to a step-by-step evaluation of the distances, until the old idea of the island universe theory reared its head once more.

This revival was mainly due to the special attention paid by astronomers to one particular class of nebulae, the so-called spiral nebulae. Close investigation of their spectra revealed that they were, in fact, comparable to the sun's spectrum, i.e. they exhibited dark bands of absorption lines (J. Scheiner 1898). This seemed to settle the question of their affinity with star systems while also giving a serious blow to the Laplace nebular hypothesis (indeed an alternative hypothesis was suggested in 1904 by T.C. Chamberlin and F.R. Moulton in which these spirals were no longer viewed as proto-stars) (see Moulton 1905). Instead, it was suggested that the tidal effects between two neighbouring stars in this kind of spiral could give rise to some form of eruptions, and these eruptions would develop into planetary systems after a transitory period in which they presented themselves as spiral-like. According to this hypothesis, the spirals remained part of the Milky Way. But by around 1910, systematic measurements of the distances of spiral nebulae convinced an increasing number of astronomers that, as Eddington summed it up, although it is "highly speculative, the [island universe theory] may help us to a possible conception of our own system" (1912, p. 260). Then in 1912, V. Slipher became the first scientist to measure the radial velocity of a spiral. Purely in terms of theory, W.W. Campbell (1913) had speculated that the age of the stars was related to their radial velocities: the older a star, the greater its velocity would be. But the appealing idea of extending this relationship to the nebulae, in the framework of an evolutionary theory according to which these nebulae must be comparatively young systems, came into glaring conflict with the high velocities measured by Slipher for Andromeda and, later, for other nebulae as well. Strikingly enough, these velocities appeared to be both positive and negative. At first, Slipher sought for an explanation of these velocities in terms of rather sophisticated arguments about the implications of the Chamberlin–Moulton theory (Slipher 1915; see also N.S. Hetherington 1971). However, struck by an apparent asymmetry between the positive and the negative velocities, he went on to suggest the hypothesis of 'galactic drift', i.e. the possibility of detecting the motion of the Milky Way with respect to the nebulae just as the radial velocities of the stars made it possible to determine the sun's motion with respect to these stars. It was not until 1917 that Slipher announced his conversion to the island universe theory (see Slipher 1917).

True, the whole climate of opinion among astronomers made them very slow in realizing how the discovery of high velocities for the spiral nebulae was a foundation shaking event. In a letter of 14 March 1914 to Slipher, Hertzsprung explained that "the great question, if the spirals belong to the system of the Milky Way or not, is answered with great certainty to the end, that they do not" (quoted in R.W. Smith 1982, p. 22). According to Hertzsprung, it was these very velocities that would not allow the nebulae, if they were extra-galactic, to have any gravitational link with the Milky Way. In other words, the exceptional velocities seem to have initially only reinforced the old conviction that there was nothing but desperate fragmentation in the universe; on top of the fragmentation in space there was even the fragmentation of time, since the notorious theory of temporal evolution was found to be at variance with the observed velocities. And the theory that would match the observational evidence of extra-galactic nebulae with the rest of astronomy could simply not be a theory dealing with the temporal evolution of heavenly bodies. For by 1917 when the evidence finally came in (after the discovery of novae in spirals by H. Curtis and G.W. Ritchey) it was explicitly stated by Curtis himself that the spirals could not be proto-solar-systems; instead, the new scale of magnitudes which allowed the first distinction to be made between various types of nebulae contained within itself a proof that the seat of such early systems must lie within the Milky Way itself (see Curtis 1917).

Curtis's work of 1917 was based on the assumption that the absolute brightness of novae is the same within the Milky Way and Andromeda, and it certainly represents a high point of development in the articulation of the ideal of constitutive uniformity in the universe. From here on, it was a matter of clearing the ground for Hubble to settle the question of distances when he discovered a Cepheid variable in Andromeda. But until the late twenties, the real issues were blurred by H. Shapley's independent theory, based on the identification of yet another class of nebulae as island universes—the globular clusters. Because these clusters systematically avoid the galactic plane, Shapley was quick to assume that they are somehow subordinated to the Milky Way. He was of course right in thinking that the globular clusters are satellites of the Milky Way, but this prompted him to develop a model of the Milky Way which turned out to be incompatible with the existence of external galaxies; the Milky Way now appeared so big that it absorbed the spirals (see Shapley 1918). It was in 1921 that the celebrated dispute about the nebulae took place between Curtis and Shapley, the so-called 'Great Debate', which undeniably gives vivid expression to the problems and uncertainties of the time (see R.W. Smith 1982, pp. 77–90).

It is most important in order to understand cosmological science before Einstein's revolution to realize that, ever since the day when William Herschel pointed his telescope towards the nebulae, however powerful the heuristic value of constitutive uniformity as a postulate, it could not lead to a coherent picture of the distribution of the nebulae. William Herschel himself explained the feeling of everything being in disarray which was engendered by this situation. Place an observer at the centre of any aggregation of stars, and "the whole universe . . . will be comprised in a set of constellations, richly ornamented with scattered stars of all sizes. Or if the united brightness of a neighbouring cluster of stars should, in a remarkable clear night, reach his sight, it will put on the appearance of a small, faint, whitish, nebulous cloud" (in J.L.E. Dreyer 1912, p. 226). But if the aggregation is not perfectly symmetrical, then "the heavens will not only be richly scattered over with brilliant constellations, but a shining zone or milky way will be perceived to surround the whole sphere of the heavens, owing to the combined light of those stars which are too small, that is, too remote to be seen". As a result, and irrespective of the power of our telescopes, the appearance of the sky is always a function of "the confined situation in which we are placed". Almost one and a half centuries later, the situation had not changed in any fundamental respect. But the sudden convergence between the two ideals of uniformity, that of constitution and that of distribution, came to light at the end of the 1920s when a relationship of an apparently universal kind was established by Hubble between the distances and the velocities of extra-galactic nebulae (see Hubble 1929). This was made possible by the correct identification, in 1926, of the actual centre of the Milky Way—it had simply escaped attention until then because it is shrouded in an absorption area of dust particles. For instance, it was realized that if the Andromeda nebula is approaching the sun with a velocity of about 300 km/sec (as found by Slipher in 1913), this is mostly due to the fact that the sun takes part in the rotation of the Milky Way around its true centre. From the true centre, *all* extra-galactic nebulae recede from us and the negative velocities become mere kinematic illusion. But for this final and unambiguous step to be made, it was necessary to disentangle the problems of overall *structure* from those of *evolution* of the basic constituents of the universe. In fact, throughout the second decade of the twentieth century, three conjectures in all had managed to involve the radial velocities of nebulae in some systematic way: that of Campbell, which postulated a link between the velocities and the age of the nebulae and which rapidly fell out of favour; that of Slipher, who suggested that these velocities were an effect of the proper motion of our galaxy; and finally that of De Sitter, which was also (like that of Slipher) purely structural and which will be discussed at length in Chapter 3.

Significantly enough, the De Sitter conjecture seems to have been put forward in relative ignorance of the then recent results of Slipher (see Hetherington 1973). Hubble has suggested that a reason might be the lack of communication: "Slipher's list of 13 velocities, although published in 1914, had not reached De Sitter, probably as a result of the disruption of communications during the war" (1936, p. 109). It is all the more important that De Sitter, the astronomer, was much more of a theoretician when he proposed the relation, and a theoretician engaged in a formidable dispute with Einstein. That a relatively consistent picture of the universe, on a new and gigantic scale, could be finally attained by the astronomers came at least as much from the purely theoretical side of the whole endeavour. Ultimately it was the totally new view of structure brought into play by the relativity theory which allowed evolutionary considerations to recover their full legitimacy: this was, as we shall see in Chapter 5, the foundation of the 'big bang' model in the early 1930s.

4. Theoretical conceptions of the universe

Newton's law of gravitation owes its 'universality' to the fact that, for more than two centuries, it was the most powerful *heuristic* principle available. The law derives from Kepler's empirical laws of planetary motion in the solar system, and its universality first proved suasive when Newton managed to discover the equality between the force of gravity at the surface of the earth and the attraction of the Moon towards the Earth. A major triumph of the law, long after Newton's death, was William Herschel's demonstration, early in the nineteenth century, that the force uniting binary stars is none other than the same Newtonian gravitational attraction. This was certainly an example of the unity of the cosmos, yet the kind of unity offered made very little difference to the perception of a basically 'fragmented' universe since the straightforward application of the allegedly universal attraction to the *whole* of the universe still led to seemingly insuperable difficulties. In fact, the difficulties involve not so much an outright impossibility (perfectly consistent Newtonian models of the universe can be built today, as we shall see in Chapter 5) as the particular way in which absolute space and absolute time were held as the foundations of the law of attraction.

In his *Principia*, Newton provided justification for the possibility of a truly exact, or mathematical, science of nature by distinguishing between the experimental measurement of things and the measured quantities themselves. As it happens, kinematic measurements of relative positions and motions of material bodies prove inadequate in distinguishing between appearance and independent reality—between apparent and true motion. This distinction would be possible from the kinematic point of view only if there existed a body in absolute rest. Even if such a body could be presumed to exist, we have no empirical means at our disposal to recognize it. That is why Newton located it beyond the sphere of fixed stars (1729, pp. 9–11).

Only if one distinguishes between the causes and effects of motion can one distinguish between true (dynamic) and apparent (kinematic) motion. Once inertial forces are accepted as effects of an accelerated motion, rather than the reverse, the famous argument of the bucket experiment allows for a successive elimination of all 'ordinary' material systems as possible candidates for the inertial system: the bucket itself, the ground, the other planets, and so forth. All are eliminated except one, the system of fixed stars, which may indeed account for the inertial effects undergone by water in the rotating bucket. In order to eliminate this system, Newton introduces a new experiment, this time in the field of pure thought: the two rotating globes connected by a cord in an otherwise empty universe. (A more complete discussion is in Chapter 2, Section 2c.) In other words, to measure purely inertial effects in the Newtonian sense, the observer is obliged to go 'outside' the real world (that of the interplay of gravitational forces), even though the inertial effects occur only in response to the exerted forces (pp. 10–12).

The two globes do exert true and measurable forces on each other, but their reinsertion into a world of remote bodies keeping the same relative positions has geometrical meaning only. This world, of course, is quite similar to our starry universe, and "from the translation of the globes among the bodies, we should find the determination of their motions" (p. 10). This process makes it possible to curtail the necessarily indefinite regression towards a material inertial system. The stars themselves are only the best approximation in our sensible measures, but they never make us coincide with the proper "measured quantities" (p. 10). In this sense, inertial effects are measured in relation to a space that is totally ideal, that is, the gravitational effects of the universe of stars are a priori negligible. In principle, there is no continuity between the successive approximations towards the really existing inertial system and the proper "measured quantities". But Newtonian science succeeds in discarding the vicissitudes that arise in practical measurement; whatever the material system used in experiments for making measurements, it is sufficient that it be 'good' enough for the measurement to be 'true'.

This is the first far-reaching consequence of Newton's mathematical physics: the distinction between causes and effects can be accomplished 'locally', in the sense that physics is provided with a means to be indifferent to the concrete universe as a whole. To deal with the possibility of a purely local physics being at the same time 'true', Newton is obliged to conceive of systems that exhibit similar effects in a world quite different from ours. For instance, the bucket experiment would yield identical results in a universe devoid of any other matter. It is even primarily this similarity that makes dynamics 'true'.

It is strange to reflect that, if he could have dispensed with the requirement of absolute space, Newton would have had the appropriate tools for constructing a consistent and infinite universe. For, as W. Rindler has pointed out, "*without* absolute space, a general homogeneous contraction under gravity of the infinite distribution would be possible: no star would move in a preferred manner *relative to the rest*" (1977, p. 199). In fact, this question raises one of the fundamental issues of Newton's famous controversy with Leibniz. Accelerated motion can be considered a prototype of absolute motion, inasmuch as it is the observable effect of a non-observable absolute space. Clarke, defending Newton's theses, insisted upon the following argument: if it so happened that God was willing to impart to the whole universe a change in its state, an acceleration would follow, the effects of which would be perfectly felt and measured inside the universe. In order to accommodate this possibility to God's will, and notwithstanding Leibniz's metaphysical criticisms about the nonsense involved in such a displacement, it appears that it is actually necessary for the universe to be finite, to have an 'outside' where there should be nothing but empty space (Clarke often insisted upon the necessary movable feature of all that is finite; see Alexander 1956, p. 46 and p. 101). It cannot be the case that the same effects could be measured if the *infinite* universe were to be disturbed, for then the whole universe could never be identified with the kind of observation we have of particular bodies *inside* the universe. Leibniz's objection in this respect, translated into physical terms, implies that nothing effectively observable ensures that the whole accelerated universe would exhibit the same effects as a bucket does in the world. However, Clarke's argument is not really connected here with how Newtonian physics actually works, but rather with the conditions that make it possible when it is extended to an arbitrarily large system. It appears that absolute space preserves its identity at all levels, from the local to the global, only if the universe can always be regarded as finite in its material extension.

As to the distribution of matter in the universe, Newton seems to argue that *any* variation on uniform distribution shows God's handiwork. Suppose gravity is some kind of 'innate' force, that is, it is something left to itself, a force which is not controlled by God. In that case, Newton says, "the hypothesis of matter's being at first evenly spread through the heavens, is, in my opinion, inconsistent with the hypothesis of innate gravity, without a supernatural power to reconcile them" (Cohen 1978, p. 311). His argument runs as follows:

For if there be innate gravity, it is impossible now for the matter of the earth and all the planets and stars to fly up from them, and become evenly spread throughout all the heavens, without a supernatural power; and certainly that which can never be hereafter without a supernatural power, could never be heretofore without the same power.

The argument seems somewhat circular. The conclusions have meaning only if the nature of gravity is subordinated in advance to an absolutely free and almighty will, a hypothesis which is ruled out from the start since gravity was taken to be 'innate'; but without such subordination, any configuration of nature might *ipso facto* not be a real configuration. And even if it were real, there is apparently no reason why gravity should operate at any time so as to induce a return to an earlier configuration.

What Newton had in mind is that the true organization of physical systems is intelligible only if we assume God's absolutely free and almighty will. As a matter of fact, this intelligibility is brought within man's reach by way of the mathematical physics that governs our planetary system:

> To make this system therefore, will all its motions, required a cause which understood, and compared together, the quantities of matter in the several bodies of the sun and planets, and the gravitating powers resulting from thence . . .; and to compare and adjust all these things together, in so great a variety of bodies, argues that cause not to be blind and fortuitous, but very well skilled in mechanicks and geometry (Cohen 1978, pp. 286-7).

It is impossible to prove directly that the constant of gravitation governing the interaction between bodies is the same for *all* bodies. Yet in the *assumption* that the constant is truly universal, a remarkable harmony is formed between all the possible variables: the masses, the periods of revolution, and the distances to the sun for all the planets of the solar system. Thus, through the mathematization of physics, our understanding of such a system as the solar system is made harmonious with the grandeur of God's design; and if the final causes of this design remain inscrutable for us, it is because only God's skill in mechanics and geometry is at one with His capacity to make such things as a sun and planets.

But what does mathematical physics teach us about the universe as a whole? Newton begins by explaining that, were the extent of the material universe finite, there could not be any physical centre without motion. Indeed, a motionless centre would require the existence of a point placed so exactly at the centre that it was subjected to an equal gravitational attraction from all sides:

> That there should be a central particle . . . without motion, seems to me a supposition fully as hard as to make the sharpest needle stand upright on its point upon a looking-glass. For if the very mathematical centre of the particle be not accurately in the very mathematical centre of the attractive power of the whole mass, the particle will not be attracted equally on all sides (Cohen 1978 p. 292).

A hypothesis advanced by Newton in his *Principia* provides an enlightening parallel. This is a hypothesis which he demonstrates only through the consequences drawn therefrom: "*That the centre of the system of the world is immovable.*" He is undoubtedly dealing here with one of those basic hypotheses that are not "feigned" (1729, p. 547), that is, with a hypothesis of the sort that everyone knows is not directly testable, but that is necessary nevertheless just to get scientific investigation started in accordance with the methodology laid out in the "Rules of Reasoning in Philosophy". From that hypothesis, Newton concludes that "*the common centre of gravity of the earth, the sun, and all the planets, is immovable.*" Indeed, "that centre either is at rest, or moves uniformly forwards in a right line; but if that centre moved, the centre of the world would move also, against the Hypothesis" (pp. 419-20). There are two implications here: first, that the common centre must be considered the centre of the world; second, that for this centre rest and rectilinear uniform motion are not identical states. (This is contrary to the axioms, or laws of motion, pp. 19-20.) A state of absolute rest appears to differ from a state of uniform motion by virtue of the prior adoption of this hypothesis.

Similarly, absolute space is also at absolute rest. To prove this, Newton offers no physical argument but only an argument of intelligibility. It is absurd, he says, to imagine that space is able to move, since true motion is already displacement from one absolute place to another. The displacement of absolute space would involve the movement of absolute space into absolute space, that is, the ability of parts of absolute space to change their place in that space. Newton's argument here rests upon the very definition of true motion. Thus the physical centre of the world is a fixed point because of the immobility of absolute space itself: but this immobility is deduced, in turn, from its very intelligibility. Furthermore, this centre is "mathematical" only (as Newton calls it), for it does not quite coincide with the Sun's centre; by virtue of the law of action and reaction, the Sun's centre moves a little around the point that forms the common centre of mass of the Sun and the moving planets that pull on it. It thus follows that the possibility of a physical centre, which is consonant with the possibility of a point in absolute rest, is strictly mathematical.

Whereas the discrepancy between the mathematical and the physical centres seems to be of little importance in the solar system, Newton does find it a decisive argument with regard to the universe at large. He now has to compare a mass that is assumed to be the centre of the universe with masses that are no longer negligible, like those of the planets. The hypothesis of a central, motionless particle is therefore hard to accept, simply because the basic constituents of the world system are at variance with those of the solar system. Newton argues further that, were matter of infinite extent, accuracy in the equilibrium would be necessary, not for one particle but rather for an infinity of particles. Newton seems to extend his argument uncritically from finite to infinite universe, so that the infinite universe is simply an infinite set of centres and does not imply in any way a change in the nature of the concepts and structures (central forces) which we started with. Yet, in the case of infinity, Newton acknowledges the possibility of this accuracy of equilibrium on condition that it be realized by divine power.

It is just this divine power that, as Newton put it in the *Principia*, prevents the stars from falling on each other. He says there that God, in order to prevent collapse, "hath placed those systems at immense distances from one another". Newton's statement means either that the universe is materially finite or that the force of gravity itself is somehow limited in space. But he could well have in mind neither of these meanings, simply keeping to the obvious consequences that would result from an infinite accumulation of stars. In that case, however, the minimization of the effects of gravity in an infinite universe would not really preclude an infinite power of gravity at infinity. In an infinite universe, it would be more consistent to speak of gravity as being totally deprived of effects, that is, not only would gravity have zero result, but the very force itself should not exist as cause. On the other hand, if the law of gravity is considered to be purely 'local', the universe would be nothing more than a set of independent centres, and the solar system would have to be seen as distinct among them. A 'privilege' of this sort is patent in the *Principia*, where Newton appears deliberately to confuse the solar system with the real centre of the world. And indeed, the *Principia* itself contains no theory about the universe in its totality. In the *Opticks* Newton goes as far as to deny to the universe in its totality the status of an object that would be open to the kind of regularity expected in scientific reasoning: he admits that God can alter the laws of Nature as it pleases Him, not only in time but in space as well. Thus different laws in different places are perfectly possible, and there is no contradiction in these differences so long as God wills them (1730, pp. 403–4). It was in the second of his four "Rules of Reasoning in Philosophy" in *Principia* that Newton had forcefully stated that "to the same natural effects we must, as far as possible, assign the same causes" (1729, p. 400).

A physical cosmology requires that the universality involved in the very form of the law of gravity be treated as a concrete property of this law. But gravity's role is too small when it is acknowledged to have a limited effect (there is no connection between an indefinite number of centres) and too great when it applies immediately to the whole universe (the connection between centres precludes finding a single true one). In the face of this paradox, Newton seems fully committed to the minimization effect. Indeed, the immense distances from one star to another not only offset the gravitational paradox occurring when all stars are taken together, but also and surprisingly enough, they help overcome the concomitant effect that could be understood as the optical paradox: that God placed the stars at immense distances from one another is justified by the fact that "the fixed stars . . . must be all subject to the dominion of One; especially since the light of the fixed stars is of the same nature with the light of the sun, and from every system light passes into all the other systems" (1729, p. 544; this is the closest that Newton comes to evoking the premises of the so-called Olbers' paradox which will be discussed shortly).

Newton's hypothesis of an initial state (infinite matter generating the raw material of a multiplicity of stellar systems) is therefore 'fictive', in the sense that the 'innate' gravity it presupposes cannot alone secure its own existence. This is precisely the point at which any idea of evolution of the universe out of natural conditions comes at variance with the way of reasoning in physical theory. Newton's implicit metaphysics can even be seen as a subtle justification of the impossibility of such a natural evolution *in reality*. For the very statement that nature left adrift by itself cannot generate anything but chaos immediately testifies to our human point of view failing at some point to coincide with God's. When Newton declares in his fourth letter to Bentley that at no moment in time can nature itself return to its initial chaotic state without the intervention of a supernatural power, he may well mean that it never 'really' passed through such a state instead of simply reasserting God's all-mighty will. In fact, should we trace nature back to its origin in time, we would find that uniformity cannot prevail throughout the whole universe. It is in those terms that Newton's approach to the question of the world as a whole is closely associated with a 'static' picture. As a counterpart to the powers of human understanding giving out at some point, the human perception of 'change' in nature becomes nothing more than a 'downgraded' representation of what is a permanent and immutable identity of being in God, and consequently in God's products of creation as well. In his second reply to Leibniz, Clarke has gone some way towards explicating this metaphysical issue: "The present frame of the solar system (for instance,) according to the present laws of motion, will in time fall into confusion; and perhaps, after that, will be amended or put into a new form. But this amendment is only relative, with regard to our conceptions" (in Alexander 1956, pp. 22-3). The downgraded status of human perception of change parallels the non-reality of the spatial world as a whole for physical theory.

Nineteenth century stellar astronomy was to enshrine the widening gap between all possible astronomical applications of Newtonian physics and global considerations in the context of that physics. On the one hand, observation leads to a view of the universe which conforms with the practical limits of the activity, i.e. it is a finite system of stars and nebulae embedded in an otherwise empty, infinite space; the systems that could conceivably populate this space seem to have no connection with the Milky Way. On the other hand, the few purely speculative accounts to be found in this period almost uncritically assume that the universe is an infinite set of stars. We have just seen how Newton, apart from the gravitational paradox, might well have coped with an optical paradox of the universe: the addition of the lights of all the stars might create a problem analogous to that of the sum total of gravitational forces. The problem seems to have been first envisaged with some precision by Halley in 1721, but it is Olbers' name (1826) which is associated with the articulation of the paradoxical argument, according to which the stars distributed uniformly throughout an infinite universe should result in a sky ablaze with the equivalent of sunlight (see S.L. Jaki 1969). Olbers thought the paradox could be overcome by postulating the existence of some light-absorbing interstellar matter. Another solution was proposed by John

Herschel (1848, pp. 184-5), that a quite specific spacing between the stars or the galaxies could annihilate infinite light. This is a version of the 'hierarchical universe' hypothesis (first suggested by J.H. Lambert as early as 1761) according to which stars form galaxies, galaxies clusters, and clusters superclusters *ad infinitum*. It is significant that only this hypothesis does full justice to the spirit of Newtonian physics, since its sole reality is that of centres of force without any definite or ultimate periphery that would be imposed by these centres themselves—every periphery is always 'local', in the sense that it is produced by a suitable arrangement of dimensions. This hypothesis does not seem to have been ever seriously entertained by the vast majority of theoreticians beyond its virtues as way out of a disturbing paradox. The simple reason is that it is incompatible with complete homogeneity, i.e. no volume in the universe is large enough to be typical.

The hierarchical hypothesis was revived and presented as a model by the Swedish astronomer C.V.L. Charlier in 1908, this time in the context of the gravitational problem. It was shown that a suitable arrangement of dimensions may also enable us to construct a universe with zero average density, thereby overcoming the cumbersome occurrence of infinite gravity at infinity when the average density is finite. The speculation was to remain quite marginal. Already when Newcomb discussed attempts of this kind in his overall account of astronomy and cosmology (1902, p. 233), he showed that this was indeed the only plausible way of salvaging the possibility of an infinite universe but that there was no truly scientific evidence to support it. In fact, Olbers' paradox, as Newcomb argued elsewhere (1906a, p. 6), only says what the universe is *not* like; as a piece of theoretical speculation, it could only work to further reinforce the autonomy of observational methods.

As to the problems associated with gravitation alone, we find a similarly negative picture. By the end of the nineteenth century, Seeliger and Neumann proposed a modification of the Newtonian law of gravitation that would make no perceptible difference within the solar system but would dispense with the disturbing increase of attraction over larger distances. I will discuss this in more detail in the next chapter. Suffice to say here that this modification was designed to restore homogeneity at large, and that it was just as *ad hoc* a solution as the absorbing matter had been in the case of the optical paradox.

In the context of Newtonian mechanics' success stories, the famous hypotheses advanced by Kant (1755) and Laplace (1796) on the formation of celestial bodies from a primitive nebula are early instances of a *physical response* to the apparent necessity of supernatural forces. Before he went on to dismiss the absolute validity of any rational cosmology in his later, critical philosophy, Kant's early speculations were highly distinctive. His *Theory of the Heavens* is a very original answer indeed to the problem of how to make the existence of an indefinite number of centres of forces compatible with universal motion as predicted by the law of gravitation. The model is a dynamic one, derived from the idea of the simplest possible conditions prevailing at the moment of 'creation', when all things came into being. Kant offers a solution to the problem of how to combine centre and infinity by endowing the very first moment, when the material universe came into existence, with a sort of transgeometric function which it bestowed on a centre in an otherwise infinite universe. For Kant this centre is transgeometric in the sense that the act of 'creation' itself is the thoroughly *physical* and *temporal* process of 'organization' that works upon brute, primitive matter from such a centre. Kant thought he could derive the very occurrence of attractive force from an infinite diversity of specific densities among the primitive particles, and thus explain from the instability of the primeval chaos both the production of organized material entities and the existence of motion. His universe is an expanding sphere of ever more highly organized matter, contained within an infinite chaos. Newton's argument for a 'pre-existing order' is thus pushed to its ultimate limit, which Kant identifies with the continuing act of creation rather than its products. But the price to pay for this bold extension of the limits of mechanical philosophy was enormous, because Kant was compelled to argue that "the greatest geometrical precision and mathematical infallibility can never be demanded from an essay of this sort" (1755, p. 92).

As Newton had explained to Bentley, the new mechanics did not account for all the peculiarities of the solar system. The supernatural cause which was deemed responsible for these peculiarities was also taken by Newton to be concomitant with the overall *stability* of the solar system. In the first half of the eighteenth century, several attempts were made to combine Newtonian gravitation with the relevant elements of Cartesian physics, in which the structure of our solar system was also derived from the cosmogonical scheme known as the vortex theory of planetary motion (Descartes 1630). In his cosmogony Descartes limited his basic concepts to matter and motion; ignoring forces implied the world had to be characterized as plenum. The fusion between the two theories was never successfully achieved, for it was precisely the possibility of an interplanetary void that, in the Newtonian system, allowed the peaceful coexistence of interpenetrating orbits like those of the comets and planets. These attempts culminated in Daniel Bernouilli's model of the 1730s, where the production of planets and comets was accounted for in terms of an extended solar atmosphere so tenuous that it really bore little resemblance to Descartes's original model (D. Bernouilli 1752; see E.J. Aiton 1972 pp. 228ff.). It was not until the end of the eighteenth century that Laplace was able to show (working solely within the Newtonian framework) that if the planets are subjected to a new, external influence, they will *spontaneously* revert to a pattern of stability; and the peculiarities of the solar system are precisely the physical conditions of its stability. Guided by the apparent self-sufficiency of this Newtonian law and deploying probabilities to account for the emergence of particular physical structures, Laplace (1796) went on to develop a nebular hypothesis of his own, quite independently of Kant. According to this model, a rotating nebula gave rise to the planets and the comets by a process of gradually cooling and slowing down.

There are two major differences between Kant's and Laplace's model, which illustrate in a striking way the kind of cosmological problem involved in Newtonian physics. First, Laplace's nebula is limited to the solar system, and simply provides an explanatory and causal model for any conceivable system in the universe, while that of Kant is truly all-embracing and seeks to account for the existence of both planet and star systems. Secondly, Kant's critique of the mathematical perfectibility of cosmology tends to undermine the validity of his own model. The model appears to violate the law of the conservation of angular momentum, since the solar system (and indeed the whole universe, on Kant's account) would have to generate its own angular momentum from nothing. This inconsistency is eliminated by Laplace, who simply starts with a nebula that is *already* in motion. The law of angular momentum is a law of the 'before = after' form, to the extent that it can be applied to cases where it is not possible to use the 'instantaneous' change of motion. Newton wrote his force law in the instantaneous form, and it was not until Euler in 1775, well after the formulation of Kant's model, that the relationship between force and acceleration as we know it today was fully spelled out; and it is from this that the conservation of angular momentum follows as a matter of course (see C. Truesdell 1964). To summarize: it is certainly impossible for the solar system to generate its own angular momentum from nothing, as Kant thought, but Laplace's strategy is determined and limited by the idea that the laws which *preserve* stability are also those that quite simply *create* it.

These developments after Newton therefore re-assert the great stumbling-block of Newtonian physics in its original form: that the power of its own method of investigation robs it of a physical theory of origin (in time) and totality (in space). Nor can it resort to a theory explaining the cause of a pre-existent gravitation. That is, while gravity as an innate property of matter would have inconsistent implications at the global level, the way such gravity would actually operate seemed to Newton impossible to think. As he said to Bentley it would act "so that one Body may act upon another at a Distance thro' a *Vacuum* without the mediation of anything else" (Cohen 1978, pp. 302–3). But the fundamental fact is that the cause of gravity (as providing the mediation) remains out of reach even if we try to match some empirical object of experience with the idea of such a cause. Indeed, actions of contact do not make this cause more intelligible: the agent causing gravity might still as easily be either material or immaterial. There is little doubt that Newton is here applying to the force of gravity, which "propagates its virtue on all sides to immense distances", the action he evokes at the end of the *Principia*, that "most subtle spirit which pervades and lies hid in all gross bodies"

(1729, pp. 546-7). He even believes that the other physical phenomena like electricity and light operate in accordance with the laws of this same universal spirit. And it was certainly part of the whole programme set up by the *Principia*, that the final unity of physical science should be sought in the apprehension of this spirit; as he said in the preface: "I wish we could derive the rest of the phenomena of Nature by the same kind of reasoning from mechanical principles, for I am induced by many reasons to suspect that they may all depend upon certain forces . . . by some forces hitherto unknown" (p. xviii). In other words, if we can speak ultimately of the unity of the Newtonian universe, it is not via its concrete figure but rather via the mode of action that is common to its various forces.

In arguments of this kind Newton reveals the intrinsic connection between the cosmological problem and the problem of the cause of gravity, that is, between the global application of the laws of physics to the universe and the very nature of these laws. Probably because of the closeness of the parallel, Newtonian physics was not generally perceived in the eighteenth and in the nineteenth century (except perhaps by Kant and Lambert) as relating to *what* we may predicate of the entire universe. Rather, the separate but central issue was *how* the Newtonian law of attraction (and the other laws which would follow from the same kind of reasoning) actually works within the reach of our experience.

5. The concept of field: Physics and mathematics of boundary conditions

To be sure, the Newtonian law of gravitational force in its original form is much more a conceptualization of the notion of force than a true explanation of the phenomenon of gravitation (see G. Buchdahl 1970). The internal nature of the force is left quite independent from its mathematical expression, in the sense that the expression remains unchanged whether the force is action at a distance or action of contact. Now, towards the middle of the nineteenth century, Poisson's reformulation of the original force law was an attempt to get rid of this ambiguity, and this had major consequences for questions of cosmology.

In order to avoid the mysterious action at a distance, Poisson substitutes the *local* variation of a quantity, called potential, for the Newtonian force law. Let there be a material point of mass m and another point of unit mass; the attractive force F exerted by the former on the latter ($F = -Gm/r^2$) is a derivative of the function ϕ ($\phi = Gm/r$), and this function is the Newtonian potential of gravitation. Poisson's law, which is the expression of Newton's law in terms of potential, is:

$$\Delta\phi = -4\pi G\rho \qquad (1.1)$$

where ρ is the density of matter and Δ is the Laplacian of the function ϕ, i.e.

$\Delta = \partial^2/\partial x^2 + \partial^2/\partial y^2 + \partial^2/\partial z^2$. The local significance of the law comes from the fact that the function ϕ is expressed in terms of partial derivatives.

A *field of force* is defined by the specification of magnitude and direction of a force at each point; such a field is uniform when magnitude and direction do not change in space and time. As a result, all regions of space are indistinguishable from one another in the case of a uniform field of force, but the idea of potential allows distinction: it is primarily the variation from point to point of the potential that is associated with the existence of a force. A uniform variation of the potential corresponds to a uniform field of force, but an irregular variation indicates a non-uniform field. It is only when the potential is constant everywhere that no force can be said to exist at all. In the nineteenth century, the force law and the potential law were held to be strictly identical—there was no experimental evidence to discriminate between them. Only the interpretation differed: in a uniform field of force, the particles themselves are responsible for the distinction between regions of space, while the potential allows the distinction to be made irrespective of the discrete nature of particles. Thus, the gravitational force between the Sun and the Earth, for instance, could from then on be understood in terms of a *continuous* field produced by the Sun, a field that fills in all spaces except the origin itself (when $r = 0$). Calculation of the external potential from a mass shows that the inclusion of the origin is not possible, for that would lead to describing the Sun as acting on itself. Indeed, the function ϕ can be written explicitly as $\phi = G\rho\iiint_v dV/r$ (where V is the volume containing the matter with density ρ), and the rules of derivation under the integral signs yield the following value for the Laplacian:

$$\Delta\phi = G\rho\iiint_v \Delta(1/r)\,dV = 0, \qquad (1.2)$$

since the Laplacian of $1/r$ is zero. But the derivation itself can be carried only if the function, say F, under the integral signs is continuous throughout the domain of integration; for an interior point, $F(1/r)$ is infinite at the point $r = 0$, so that the rules of derivation can no longer be applied. As $\phi = F(1/r)$ is linear, calculation shows that $\Delta\phi$ indeed leads to Poisson's equation in the interior case; for the exterior case, the equation $\Delta\phi = 0$ is called Laplace's equation.

Because it is only the variation of the potential from point to point, and not its absolute value, which indicates the existence of a force, the exact numerical value of the potential is determined only up to an additive constant. This constant is just the constant of integration which results from the calculation of its numerical value. In the case of a uniform field of force, the constant remains utterly arbitrary. Its value can be fixed in the case of a non-uniform field, since it is then possible to say that in those regions of the field where the field vanishes, the potential itself vanishes; when a definite value is ascribed to the potential in one part of the field, the exact value can be derived in every other part—a typical case being that of the gravitational field which is at an infinite distance from a mass.

The very idea of space is transformed when the concept of potential is brought into play. Indeed, space becomes somehow 'materialized', it is endowed with some of the substantial properties of physical entities. Think of 'empty' space ($\rho = 0$), when Poisson's equation turns into Laplace's equation. The solution to this equation continues to represent the Newtonian potential of gravitation as a function of the density existing within matter. (This solution of $\Delta\phi = 0$ is expressed as $\phi = \int \rho \, dV/r$.) While, in the force law, gravitation was the external phenomenon of a cause which manifests itself only when at least two masses interact, the notion of potential identifies the 'reality' of a physical object with its sphere of action or influence. In other words, contrary to Newton's conjectures, what mediates the gravitational effects is not necessarily different in nature from the source of these effects. And the limiting condition at infinity (vanishing potential) allows us to think of the gravitational field as a medium in which material bodies are the *only* sources, since it is then that the *totality* of the numerical value of the potential at any finite distance is derivable solely from the existence of such sources.

I shall return in the next section to this attempt to view space as a substance where the nineteenth-century electromagnetic theory will be examined in terms of its impact on the theory of relativity. In the context of gravitation theory there is the problem of what status to give the limiting condition at infinity. Clearly, if the boundary condition had a straightforward physical significance, the whole material universe would have to be seen as a finite island within infinite empty space. In fact, it is not physical at all, because the solutions to Poisson's equation are meant to apply to *any* physical system, i.e. they vanish at infinity however large the system may be. As a 'loçal' equation, expressed in terms of partial derivatives, the potential law is consistent with the *strictly mathematical* possibility of there being no matter at infinity; in practice the situation is that, with respect to any conceivable physical system, the 'rest' of the universe is viewed *as if* it were nothing at all. And in principle there is no problem in regarding this idealization as *remaining* in force when the system is nothing less than the total universe, for even an actual infinite universe cannot be incompatible with the limiting condition of vanishing potential at 'infinity'. It is simply a consequence of the concept of potential that it should be applicable to all possible cases, to systems of unspecified dimensions (including the infinite), because it is a magnitude which is defined from point to point and can therefore be 're-gauged' in an ideal way without jeopardizing its physically local significance.

That is to say, the law of potential is undeniably concomitant with conditions which are generally thought of as 'cosmological', because it abolishes any sense of borderline between an 'interior' and an 'exterior'. Any sufficiently isolated system may be regarded as a sort of self-containing totality in the same way as the system which *is* in effect the sufficient and entire universe. Furthermore, because the potential is a linear function in $1/r$, the gravitational field can be said to derive from a potential only if the 'work' of the field vector moving between two points is path-independent; this means that the potential is a scalar, in contrast to the forces which are always exerted along some direction. Thus, because the potential law is quite independent of any privileged direction, it is allied to the property of *isotropy*, by which we mean equality in every direction from a given point. Note that isotropy from all points implies *homogeneity*, but that the converse fails to be generally true (see Rindler 1977, p. 202). The question then arises whether the potential law is actually compatible with the property of homogeneity in the universe. Suppose a uniform distribution of matter up till infinity: the potentials of gravitation will remain constant throughout space. Now, with $\phi =$ constant in Poisson's equation, we have $4\pi G\rho = 0$, or $\rho = 0$, and this result is quite independent of the ideal value of the potential at infinity. In other words, the only possible uniform distribution of matter would be that in which there is no matter at all, but such a universe would be also the only one in which the fixing of the boundary condition was not utterly arbitrary.

An infinite universe with constant potential throughout is a universe devoid of global gravitational effects. The difficulty is that it cannot be identified with a universe of pure inertia either. There is a well-known theorem of Newtonian mechanics, according to which a particle placed inside the cavity of a spherical surface of matter will be subject to no net gravitational force (this is due to the symmetry of the system, see Newton 1729, p. 193). But if the matter around the cavity comprises an infinite number of spherical mass shells, it becomes impossible to determine the field within because if we integrate the potential, the integral diverges to infinity—this is another way of saying that Poisson's equation allows for no solution where $\phi =$ constant. The assumption that the field inside the cavity remains zero is of no help at all, because if we reintroduce some matter into it, it will collapse under its own force. That is, isotropy from *all* points is lost.

By applying the concept of potential to the consideration of the cosmological problem, we see, therefore, that a sharp conflict, if not an outright antinomy, arises between the law of force and the law of potential. Starting from a homogeneous distribution of matter, it seems impossible for any matter to exist; and starting from a given point, in accordance with the point-to-point strategy implicit in the concept of potential, the resulting isotropy also excludes homogeneity. The nineteenth-century philosophers and scientists (whether they were advocating a finite or an infinite universe) did little to burden themselves with any reference to the paradoxes generated by the concept of potential in a stable universe. (See S. Jaki 1972, pp. 268–275.) True, the whole nineteenth-century idea of 'materialized space' in keeping with the potential theory of gravitation would seem to add much credibility to

Samuel Clarke's contention (alluded to in the previous section) that it should make as much sense to speak of a physics *of* the whole universe as it does to speak of a physics *within* the universe. But if the overlap were tenable in any sense of a consistent cosmology in its own right, then, at the very least, the limiting condition at infinity should be capable of thoroughly unambiguous physical interpretation. In fact, the positing of a vanishing potential at infinity is a way of allowing for every possible potential (for every system and every observer) to be re-gauged in a truly universal way; but it is precisely that definition which seems to be responsible for the conflict between a force-oriented and a potential-oriented interpretation at the cosmological level. What we find here is probably the clearest single instance of how the nineteenth century divorced cosmology from practical physics.

It was the theory of general relativity that led to the realization that there was a profound connection between the law governing a constant of integration like the potential at infinity and another limitation familiar to classical physics, namely the notion that the Cartesian system of coordinates is the only possible system of coordinates. Such a view implied, in turn, a drastically revised idea of how to tackle the cosmological problem. But we cannot hope to understand this shift before we look more closely at the way general relativity was first discovered and established; in fact, the way it was discovered is significantly entangled with some of the most profound foundations of the theory. It all came from another part of the nineteenth-century conceptualization of 'materialized space': the electromagnetic theory of light.

6. The cosmological significance of the path from the special to the general theory of relativity

In his famous treatise on electricity and magnetism, Maxwell (1873) defined a constant c as a relationship between the units of electrostatic and electrodynamic charges. This constant has the dimensions of a velocity, and can be determined experimentally. In addition, Maxwell's theory predicted the existence of electromagnetic waves that propagate in a vacuum with the velocity c. Now, it appeared that this velocity was also the measured velocity of light in vacuum, i.e. light had to be a phenomenon comparable with the electromagnetic wave. The question then arose as to what could be the physical support for the propagation of such waves, and Maxwell found his answer in proposing the existence of an 'ether' in absolute rest. In this sense, the idea of ether is simply the electromagnetic side of the nineteenth-century saga of 'materialized' space. While the fixed stars were regarded by Newton as only a fair approximation of what an inertial system must be, the ether was identified with the one absolute immovable system; the electromagnetic theory seemed to be a step on the way to a fully physical account of the foundations of physics itself.

But then a hiatus cropped up between electromagnetism and mechanics. Einstein explained it brilliantly enough (1917a, p. 14):

If the principle of relativity (in the restricted sense) does not hold, then the Galilean coordinate systems K, K', K'', etc., which are moving uniformly relative to each other, will not be *equivalent* for the description of natural phenomena. In this case we should be constrained to believe that natural laws are capable of being formulated in a particularly simple manner, and of course only on condition that, from amongst all possible Galilean coordinate systems, we should have chosen *one* (K_0) of a particular state of motion as our body of reference. We should then be justified . . . in calling this system 'absolutely at rest', and all other Galilean systems K 'in motion'.

In mechanics, the connection between a Galilean system K and another such system K' can be done either directly or via a system K_0 in absolute rest:

$$\begin{array}{ccc} K & \rightarrow & K' \\ & \searrow \quad \nearrow & \\ & K_0 & \end{array}$$

This is no longer the case in electromagnetism, where the ether is a sort of embodiment of the absolute space of mechanics, that is, the mediation of K_0 becomes unavoidable:

$$\begin{array}{ccc} K & & K' \\ & \searrow \quad \nearrow & \\ & K_0 & \end{array}$$

As Einstein emphasizes, the situation is a serious impediment to the principle of relativity.

In order to understand what this is all about, it will be useful to refer to an earlier critique of Newtonian mechanics, from which the conceptual development of those ideas which lead up to Einstein's special theory of relativity can be illuminatingly reconstituted. The critique is Kant's. Not only does it provide a conceptually interesting antidote for some of the excesses of the nineteenth (and early twentieth) century anathema against the metaphysical probes into the foundations of physics, but as we shall see it is especially relevant to the cosmological problem. In fact, as Michael Friedman has noted (1983, Introd.), reference to the Kantian framework was an almost necessary mediation in any discussion of the foundations of physical science at least up till the early 1930s.

In the work he devoted to the metaphysical foundations of Newton's science, Kant poses as an *axiom* the classical formulation of the principle of relativity. He does this by showing how the principle can be applied to the motion of a body in order for it to become an object of experience: "Every motion as object of a possible experience can be viewed at will either as motion of a body in a space that is at rest, or as rest of the body and motion of the space in the opposite direction with equal velocity" (1788, p. 28). What he regards here as space is something materialized under the form of a reference system. True, there are no mechanical means at our disposal for distinguishing the uniform rectilinear movement from the state of rest. At that kinematic level, there is only relative motion: absolute motion would be one which is referred to a non-empirical space, but such a postulated space has no reality as experience.

By contrast with kinematics (or phoronomy to use Kant's word), dynamics is defined by Kant as dealing, not with motion in general as an object of experience, but rather with the movable body itself. It is precisely the correct understanding of the effects of its motion with respect to an independent space that is the fundamental issue. Indeed, in the case of dynamics, it becomes much more delicate to assess the difference between space as a set of reference systems, and space as an entity which is a priori independent of material bodies and experience. Circular rotation is a typical example of such effects: rotation of the Earth about its axis can be established through a certain number of facts of experience, but the distinctive feature of these facts is that equivalence with a surrounding space (such as motion of the Earth relative to the stars) can no longer be complete. In fact, the celebrated Foucault pendulum or the observation of the Earth's equatorial bulge are proofs of the circular motion of the Earth about its axis; these proofs are quite independent of the relation of the Earth to the stars. With the help of these facts, true motion can be distinguished from mere appearance, i.e. from the illusion that it makes no difference whether the Earth stands still and the heavens revolve about it or whether the Earth rotates and the heavens stand still. Passing from kinematics to dynamics, it therefore appears that relativity is limited to the class of inertial states defined by the principle of relativity.

Now, this limitation of relativity can be established not only in relation to the definition of the reference system, but also in relation to the nature of the operational means by which the effects of motion (what Kant calls the possibility of experience) may be ascertained. From the latter point of view, dynamics is just the inverse of kinematics: while no kinematic means allow the inertial states of rest and uniform motion to be distinguished from one another, *only* dynamic means allow a given reference system or body to be determined in its non-inertial state (see L. Sklar 1974, p. 188). For Newton, and for all classical physicists, this kind of circularity is of purely empirical significance. There are simply no other means, i.e. it is basically accidental that nature does not provide us with a supreme kinematic system that would differentiate between inertial and non-inertial states. At any rate, it could not change one's reasons for asserting the absolute nature of a motion like rotation. However, Kant at this point raises the foundational aspect of the whole theory. His original account is that *true* motion, in particular the Earth's motion, cannot be viewed as *absolute* motion. True motion, in the dynamic sense, is declared to be relative, i.e. it is the relation to one another of the parts of the movable body. For instance, Foucault's pendulum determines the rotation of the Earth as a true motion for it is attached to the Earth as its part; the determination is relative neither to the stars nor to absolute space. Accordingly, the only thinkable absolute motion in a consistent Newtonian theory, Kant says, would be the motion of the whole universe in empty space. Indeed, in that case, true motion would occur irrespective of other matter. However, this motion can never become an object of experience: it is no longer by accident, but by principle, that all means are lacking that would enable us to distinguish the effects of such a motion from any other effects *within* the universe (1788, p. 131).

In criticizing the metaphysical basis of the Newtonian science of motion in this case, Kant has indicated clearly the *limits* of the classical principle of relativity. He does so without altering the definition of the reference system postulated according to this principle, but only in relation to the possibility of experience. Absolute space cannot be taken in any physical sense whatever. While the space in which we experience motion "must also be capable of being sensed" (p. 19), absolute space is non-empirical and serves only as a presupposition of empirical space.

The new mechanics embodied in Einstein's special theory of relativity, which completely revolutionized the principle of relativity at the beginning of the twentieth century, proceeds from a yet more radical interpretation of absolute space. Kant's statement that "absolute space is in itself nothing and is no object at all" (p. 20) is no longer taken as a limit of the principle of relativity, but is incorporated as a feature of the special theory. Einstein's early ideas arose from a particular fact of experience (1917a, pp. 17–20 and p. 51). The rise of electromagnetic theories in the nineteenth century appeared to prove the reality of absolute space, since these theories were based on the Newtonian model and postulated an ether at absolute rest, as a support for the propagation of electromagnetic waves. Among the data that seemed to uphold this was Fizeau's experiment on the velocity of propagation of light in a material medium and the phenomenon of aberration of light waves. (On the emergence of the theory of special relativity in connection with these observations, see A.I. Miller 1981.) This image of a stationary ether seemed to make it inevitable that the value of the speed of light measured by an observer in motion with respect to the ether would depend upon the magnitude and direction of his velocity. If c is the velocity of light and v is the colinear velocity of the observer with respect to the ether, then, according to classical kinematics, the velocity measured by the observer would be $(c - v)$ or $(c + v)$ according to whether he moves in the same or the opposite direction as the light. An observer ignorant of his velocity relative to the ether would be able to determine this experimentally by emitting light signals in all directions and measuring the time taken by these signals to reach the surface of a sphere centered at the source of light. Any motion with respect to the ether would create an 'ether wind' which would affect the signals in such a way that the first one to reach the sphere would have travelled in a direction opposite to that of the motion whereas the last one would have travelled in the same direction. This was the principle of the celebrated Michelson–Morley experiment (1887) which attempted to find the motion of the Earth with respect to the ether. But the result was negative, so that the experiment showed that the speed of light was independent of the observer's motion.

In order to explain the negative result of the Michelson–Morley experiment, Fitzgerald in 1889, and Lorentz in 1892 propounded a set of hypotheses that would make it possible to preserve both the *unobservability* and the *existence* of the ether (see H. Erlichson 1973). According to these hypotheses, the bodies which move in the ether with a certain velocity undergo a length contraction in the direction of their motion; this contraction neatly compensates for the lack of a time difference, that would otherwise have been measurable in the Michelson–Morley experiment. Setting about the task of formulating the transformation of coordinates which would leave Maxwell's equations invariant *and* include the effects of the length contraction, Larmor (in 1898) and, again independently, Lorentz (just before 1904) established an equation of transformation of *time*. But while both Larmor and Lorentz seem to have regarded this transformation as a purely mathematical device, Einstein, in his epoch-making special theory of relativity (1905), gave both the length contraction and the time dilatation a *physical* meaning.

How did this come about? Let us look again at Einstein's popular exposition. Suppose the speed of light to be variable, as it is according to classical mechanics. A result of this nature "comes into conflict with the principle of relativity... For, like every other general law of nature, the law of the transmission of light *in vacuo* must, according to the principle of relativity, be the same" in any inertial system (1917a, p. 18). It is absolutely crucial to note that Einstein takes an *observed fact*, the result of a measurement, to be a true *law*. This leads to a contradiction within the theory, not a simple clash between a fact and a theory that is as yet insufficiently refined: should we give up the law of the constant velocity of light and maintain the principle of relativity (which is to say modify electromagnetism), or should it be the other way around? Einstein urges on us the simplicity of the principle of relativity which, as he says, "appeals so convincingly to the intellect because it is so natural" (p. 19). At the same time, he shows that a more complicated law of light propagation would not be compatible with the observed phenomena:

The epoch-making theoretical investigations of H.A. Lorentz on the electrodynamical and optical phenomena connected with moving bodies show that experience in this domain leads conclusively to a theory of electromagnetic phenomena, of which the law of the constancy of the velocity of light *in vacuo* is a necessary consequence. Prominent theoretical physicists were therefore more inclined to reject the principle of relativity, in spite of the fact that no empirical data had been found which were contradictory to this principle.

In Lorentz's eyes, the constancy of the velocity of light was plain proof of the existence of the ether, for then, and only then, could length contraction be regarded as something absolute: because this property is imposed on bodies by the ether, it cannot be directly observed. According to this interpretation, length contraction is akin to a fictive force in the Newtonian sense (it can be transformed away by an appropriate choice of the reference system, i.e., the ether in absolute rest), since it can be observed only in a *non*-inertial system of reference like the Earth. As Minkowski has said, the hypothesis of length contraction comes to something "extremely fantastical, for the contraction is not to be looked upon as a consequence of resistances in the ether, or anything of that kind, but simply as a gift from above—as an accompanying circumstance of the circumstance of motion" (1908, p. 81). In other words, the contraction is both 'fictive' and 'absolute', since no experience can make manifest the ether. Einstein observed that light as a universal constant is already present in Lorentz's transformations and that, furthermore, the form of these equations is not modified by taking *any* inertial system as reference—the choice is not limited to the ether. Combining these two statements, Einstein reinterpreted length contraction and time dilatation as effects of the chosen inertial system, and thus obviated the need for any physically unverifiable hypotheses regarding the properties of matter immersed in the ether.

Thus, in contrast to the earlier and mainly speculative critique formulated by Kant, the negative result of the Michelson–Morley experiment does much more than simply exemplify the limits of the classical principle of relativity with respect to Newtonian dynamics: it actually shows them to be *contradictory*, at least if the laws of electromagnetic theory are admitted as true. Einstein's new dynamics was the product of this contradiction. It maintained the classical idea of relativity in opposition to the assumption of those proposed internal properties of bodies that would mask the expected effects of absolute space. These properties are reinterpreted in Einstein's theory as effects of the chosen reference system. In consequence, absolute space in the physical sense is merely the class of inertial systems; the possibility of further restriction to absolute rest (which defined absolute space in the classical theory) is explicitly rejected on the basis of the Michelson–Morley experiment. Just as Kant had realized that empty space could not be said to exist materially, for this would be a hindrance to the kinematic relativity of motion

(1788, p. 132), so absolute space in the special theory simply remains as a pure form. Not surprisingly, it then appears that rigid rods and standard clocks do provide the observer with none other than those kinematic tools, impossible in Newtonian dynamics, that allow us to make the distinction between inertial and non-inertial reference systems.

The combination of space and time coordinates in the special theory has profound implications for the causal structure of the world. The classical distinction between past and future, as well as the very notion of simultaneity, are changed in quite substantial ways. According to Poisson's equation, a material support makes it possible to conceive of the propagation of phenomena (via the potential) with a finite velocity, but propagation in a vacuum remains instantaneous. Whereas with the special theory, we have in place of potentials governed by Poisson's equation the so-called 'retarded potentials' expressed in terms of what has come to be known as d'Alembert's equation,

$$\phi = \int \frac{\rho(t - v/c)}{r} \, dV.$$

This means that the equality between two potentials at two points, *in vacuo*, can no longer be instantaneous: between two points A and B separated by a distance r, the potential at B, at time t, is identical to the potential that existed at A, at the earlier time $(t - r/c)$. Because the special theory postulates no physical support for the propagation of light, and modifies the laws of classical kinematics, the velocity of light becomes a limit that no other form of propagation can overcome. In terms of the problems of cosmology as I have outlined them, it is now clear that physics has embarked upon the task of defining, in its own terms and in a more-than-ever radical way, how the world of material and electromagnetic phenomena sustains itself, without the aid of external mediating agents. The nineteenth-century notion of field had gone only part of the way towards the achievement of this task, as its emphasis upon the materiality of ether made it still captive of causality as absolute mediation. The special-relativistic way of thinking moves away from the nineteenth century to the extent that the abolishment of either calls for reconsidering the intelligibility of nature's self-sustaining process. To the classical distinction between past and future, special relativity adds that of 'absolutely elsewhere'. This conceptual breakthough was due to Hermann Minkowski who, in his celebrated lecture, *Space and Time* (1908), spelled out what the implications were of the absence of observer-independent, transitive simultaneity between events for our ontological conception of the external world. In his 'Autobiographical Notes', Einstein (1949, p. 59) gave to Minkowski the credit for turning the relation between mathematics and physics, as it had been classically conceived, completely around, and indeed for initiating an entirely new approach to the fundamental notion of 'the world'. Central to Minkowski's approach is the priority he accords to invariance, or symmetry properties of the field equations over their 'physical' properties. Thus, Lorentz discovered the first important symmetry principle in fundamental physics, which was found as a *mathematical* property of Maxwell's equations which, in turn, were based on the *experimental* laws of electromagnetism. But the symmetry itself was a secondary discovery; Minkowski started with the Lorentz invariance and required the field equations to have a particular form (known as covariance) that would tally with that invariance. Minkowski's strategy has dominated the scene of physics ever since, with the result that the unification of *all* fundamental forces of nature is conceived of today in terms of 'supersymmetry'.

Minkowski denotes by the word 'world' the totality of values for all the spatial (x, y, z) and the temporal (t) coordinates that are *thinkable* (1908, p. 76). To the field of pure thought he argues that there corresponds a field of *observability* in which "not to leave a yawning void anywhere, we will imagine that everywhere and everywhen there is someting perceptible". Minkowski then goes on to designate as *substance* all that may, like matter and electricity, be observable. A substantial point is located at the world-point x, y, z, and t; and the world-line of such a point is a curve in the world which determines what Minkowski calls the "everlasting career" of the point. World-lines are the 'absolute' of physical science in the sense that "physical laws might find their most perfect expression in reciprocal relations between these world-lines". In other words, only the intersections between world-lines have a meaning in Minkowski's physics; the behaviour of a single world-line anywhere between these intersections is regarded as irrelevant. It is no overstatement to say that this idea was the essential prolegomenon that made possible Einstein's further researches, in particular the development of general relativity. For the preservation of the existing intersections is expressed mathematically as general covariance; and it was Einstein's aim to write down generally covariant equations when the field included gravitational phenomena.

If we go further into Minkowski's ideas, we may even discover their connection with what formed one of the bases of Einstein's way to cosmology. Consider the group of linear, homogeneous transformations of the expression $x^2 + y^2 + z^2$ (rotation of the axes, x, y, z at time $t = 0$ about the origin) and the group of linear translations (in which x, y, z, t can be replaced by $x - At, y - Bt, z - Ct, t$, with A, B, and C arbitrary constants). The fundamental problem of how to combine these two groups is expressed by Minkowski in the following terms: "Hence we may give to the time axis whatever direction we choose towards the upper half of the world, $t > 0$. Now what has the requirement of orthogonality in space to do with this perfect freedom of the time axis in an upward direction?" (p. 77). In classical physics, the plane $t = 0$ passing through some point divides in an absolute way the future from the past events which may have had an interaction with the point. But if we want to combine time and space, in accordance with the spirit of the special theory of relativity, it is the double-cone $x^2 + y^2 + z^2 - c^2t^2 = 0$ that operates the division, with the effect of designating as 'absolutely elsewhere' all that lies outside the cone (see Fig. 1.2). In order to understand how this comes about, let us first consider rotation about the origin. The equation $c^2t^2 - x^2 = 1$ represents hyperbola branches, with the light cone defining their asymptotes. A rotation from x to x' changes t into t' if we follow the parallelogram A'B'OC'. In the new axes (x', t'), the equation $c^2t^2 - x^2 = 1$ will keep the same form, i.e. it will become $c^2t'^2 - x'^2 = 1$, if OA' is taken to be the inverse of OC', i.e. if we take OC' = 1 and OA' = $1/c$. The group of transformation G_c results from the addition to these rotations of the group of arbitrary displacements of the origin. It appears that it is the group G_c that fixes the direction of time with respect to the choice of a spatial frame of reference. Indeed, if we made c tend to infinity (G_∞), $1/c$ tends to zero and the angle between the asymptotes becomes larger and larger; when c is actually infinite, x' coincides with x, so that t' can take *any* (positive) value. This case G_∞ is exactly that of pre-relativistic space and time, in which there is no connection between these two entities. When c is taken to be a finite parameter, the connection is achieved and the parameter can readily be identified with the finite speed of light in Einstein's theory.

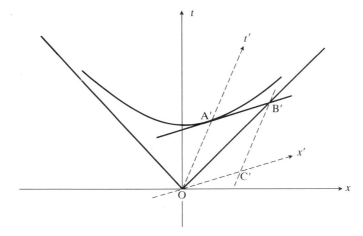

FIG. 1.2. Minkowski's representation of space–time axes in special relativity (adapted from 1908, p. 78).

Our understanding of every sphere of action and of reality itself has been affected by Minkowski's achievement. The Minkowski world is one in which it does not make sense to speak of a three-dimensional Euclidean space associated with linear time. Instead, the world becomes simply a four-dimensional continuum in which space and time can no longer be distinguished from one another; they are linked together indissolubly, and the ordinary distinction between space and time (the 'natural' measurements of the world by means of rods and clocks) becomes a matter of "projection", to use Minkowski's word for it (p. 83). Accordingly, the primary indissolubility of the four-dimensional world becomes a matter of *postulate*, i.e. the "postulate of the absolute world". In Minkowski's eyes, this formulation is justified because the universal validity of the postulate is consonant with the equally universal coordination between the thinkable and the observable: the coordination bestows objectivity upon "the idea of pre-established harmony between pure mathematics and physics" (p. 91).

The four-dimensional continuum, as it was recognized soon after Minkowski's paper by such prominent scientists as Weyl and Eddington, is neither time nor space, so that the classical dichotomy between the processes as they take place in space and as they occur in 'history' (their forward motion in time) is abolished as well. In particular, any concept of an external world evolving *in* time has no formal counterpart in the laws of special relativity. As Weyl has said, "The objective world simply *is*, it does not *happen* (1949, p. 116); or again in Einstein's own words: four-dimensional *existence* supersedes three-dimensional *evolution* (1954, p. 150). This does not mean, of course, that the fact of motion and the reality of change are denied. Rather, there is no becoming in the world only to this extent—that nothing in the formal aspect of the theory embodies such a notion. And if there is harmony between the formal and the factual, then the only consistent thing that can be said about becoming is that it is purely a matter of the individuation of four-dimensional entities. Here too, Weyl has expressed this thought in its most profound form: he argued that "only the consciousness that passes on in one portion of this [four-dimensional] world experiences the detached piece which comes to meet it and passes behind it, as *history*" (1918, p. 217). True, the harmony in question is decided as a postulate; it is not something that we may hope to fully grasp conceptually, for it is the very starting point, the precondition of whatever it is we may grasp. Physically, this relative absence of history is concomitant with a view of what is 'objective' that renders the world basically *static*: objective time is simply proper time, that is (to refer to Fig. 1.2) it is always allowed "to introduce OA' as a new axis of time, and with the new concepts of space and time thus given, the substance at the world-point concerned appears as at rest" (Minkowski 1908, p. 80). Nature allows systems of coordinates in which the proper time of a clock shall remain the same when it has come to rest in one of those systems; this property is even defined for those clocks that go correctly. The conservation of the "static measure", as it may be called after Weyl (1918, pp. 176-7), applies equally to the rigid rod used for measuring because "after coming to rest in an allowable system of reference, it shall always remain exactly the same as before". As to the old sense of simultaneity between events, the impossibility of attaching a definite physical meaning to the comparison of two clocks in terms of some independent time implies a degree of arbitrariness in what is meant by 'simultaneous'. Arbitrariness means in this context that simultaneity is no longer a primitive notion, but is defined in terms of the theory itself. That is, simultaneity is not invariant and depends on the inertial state of the observer. There is no physical process that intrinsically defines simultaneity, i.e. there is no 'natural' simultaneity—only one built out of clocks and light signals.

The crucial point is that this new form of static vision of the world (with its denial of time as an independent dimension in the world) was closely associated with the new relationship between mathematics and physics. Some time before Minkowski set about formalizing it (and in fact just before the construction of special relativity by Einstein), Poincaré could already identify something very important in the outcome of the attempt to express the laws of physics in terms of differential equations. Indeed, Poincaré described any process of resolving complex phenomena into a larger number of elementary phenomena as hypothetical (a matter of postulate), and he went on to evaluate the consequences for our view of time by saying that "instead of embracing in its entirety the progressive development of a phenomenon, we simply try to connect each moment with the one immediately preceding. We admit that the present state of the world only depends on the immediate past, without being directly influenced, so to speak, by the recollection of a more distant past" (1905, p. 154). Minkowski's view of the absolute world certainly constitutes the most radical expression possible of these foundations of field physics. And however critical Einstein became of Poincaré's conventionalist position with regard to space, he never really felt the need to re-articulate the implications of this philosophy of time.

It is indeed against such a background of static world (understood in Minkowski's hypothetical sense) that the theory of general relativity was first erected. For the basis of the transition from the special to the general theory lies in the fact that it remained meaningful to speak in that sense of rigid rods and standard clocks when *gravitational* phenomena were included in the schema. The inclusion endorses Einstein's plea for seeing the general theory as an extension, not a radical revision, of the special theory. Clearly, the plea implies something of a paradox in relation to the projected unification of physics. As Einstein himself has remarked in his 'Autobiographical Notes', "one is struck by the fact that the theory . . . introduces two kinds of physical things, i.e. (1) measuring rods and clocks, (2) all other things, e.g. the electromagnetic field, the material point, etc." (1949, p. 59). Einstein called this an inconsistency, for space and time appear, on this account, as "theoretically self-sufficient entities", that is, they are not on the footing of the other entities. The point is that such inconsistency would be temporary only, and it would be the obligation of the future unified field physics "of eliminating it at a later stage" (p. 61). And indeed with regard to space in particular, the arbitrary character of the independence from all other things was an assumption that Weyl, as early as 1918 when he addressed the unification of gravitation and electromagnetism, was unable to find any justification for. And even before the advent of general relativity, the viability of the assumption had already been put into question by Poincaré on epistemological grounds. (Later chapters have a detailed discussion of this.) Most important among the "other things" was, for Einstein, the status of the material point. For this touched on the very nature of the whole of field physics and its attempt to get rid of some of the most delicate problems of interpretation involved in Newton's conception of forces. Indeed, as Einstein again explains in his late 'Notes', "if one had the field equations of the total field, one would be compelled to demand that the particles themselves would *everywhere* be describable as singularity-free solutions of the completed field equations. Only then would the general theory of relativity be a *complete* theory" (p. 81). That is, field physics was designed so as to implement the view of continuity in nature, or, in other words, to overcome any distinction between the field and the sources of the field; but obviously this view clashes with the statistical interpretation of the quantum theory that had been established a little more than ten years after general relativity. Overcoming any dual source-field was a matter of re-working the very *postulates* that Einstein had originally laid down in accordance with Minkowski's programme. Before quantum theory would come and completely disturb the terms of this programme, *cosmology* was an important development because it reflected the implications of Minkowski's postulate of the absolute world at the level of gravitation. In the terms of the unified view, the radical step of sketching a cosmological model was taken by Einstein in order to provide the most natural *distribution* of the sources that could be compatible with general relativity. So, before we turn to cosmology, we must take a close look at the nature of the postulates of this new theory of gravitation which was conceived of as an extension of the special theory.

7. Postulates, principles, and assumptions of general relativity

Much work has been done in recent times on Einstein's actual path from the special to the general theory. The reappraisal of older views on this subject is due mainly to John Stachel (1979) and John Norton (1984 and 1985. Compare, for instance, C. Lanczos 1972). This re-evaluation brings about a certain number of qualifications in the traditional view of what Einstein's disciples took to be the foundation principles of the theory. In fact, Einstein's intellectual journey during these decisive years, from 1907 to 1915, was quite a solitary one, and the other leading physicists of the time working on the problem of gravitation (such eminent figures as Mie, Nordstrom, and Planck) found his line of thought impossible to accept. During the eight-year period during which Einstein was struggling in almost complete isolation to develop a consistent theory of gravitation, he had to cope with various conceptual difficulties, all of which stemmed from an early intuition he had had in 1907. To appreciate how much Einstein's own intuition departed even from authoritative presentations of his theory by others, we need only compare two statements. In a letter of September 1950 to Max von Laue, Einstein said that "what characterizes the existence of a gravitational field from the empirical standpoint is the non-vanishing of the components of the affine connection, not the non-vanishing of the Riemann tensor. If one does not think in such intuitive ways, one cannot grasp why something like curvature should have anything to do with gravitation at all' (quoted in J. Stachel 1986); at the other extreme, there is John Synge's famous textbook (1960, p. ix) formulation that, in order to do justice to Einstein's theory, we must concede that "either there is a gravitational field or there is none according as the Riemann tensor does not or does vanish".

The fundamental intuition which underlies everything else has been called the principle of equivalence, and it has to be spelled out at some length if we want to understand the nature of the general-relativistic field equations. This breakthrough is the first act of the story, which covers the period 1907–12; another act, probably the most disturbing and germinative in the whole of Einstein's career, goes from 1912 to 1915; and the third act of the drama (as Stachel rightly calls it) begins at the end of 1915, when the final formulation of the field equations was effected. This final formulation is the following generalization of Poisson's equation:

$$R_{\mu\nu} - \tfrac{1}{2} g_{\mu\nu} R = - \kappa T_{\mu\nu} \tag{1.3}$$

in which $R_{\mu\nu}$ is the Riemann curvature tensor, the $g_{\mu\nu}$ are the geometrical analogues of the potential ϕ in the classical theory, R is the curvature invariant, κ is some suitable 'coupling' constant, and $T_{\mu\nu}$ is the energy–momentum tensor. These field equations generalize Poisson's equation but (because Einstein's actual aim was to extend the special theory) it can be said that they describe the dynamics of space–time, much as Maxwell's equations describe the dynamics of the electromagnetic field. In these equations, the Riemann tensor is important because it is the genuinely dynamic quantity.

In other words, what we have can be stated like this:

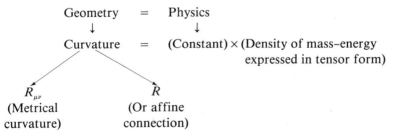

In short, any measurement, anywhere in the world, produces a ds^2, an infinitesimal interval of space-time, that Riemannian geometry expresses in terms of functions ($g_{\mu\nu}$) of the chosen coordinates. The tensor $R_{\mu\nu}$ is, at any point, completely determined by the value of $g_{\mu\nu}$ throughout any neighbourhood of the point; it bestows meaning upon the possibility of comparing the same distances at different points in space. R is a scalar determined locally by $R_{\mu\nu}$, that is, the affine connection is inherent in metrical space in the sense that the transference of *distances* carries with it some rule governing the transference of *directions*. The combination of the metrical structure and the affine connection goes to constitute the geometry of the world. (It must be noted that Einstein first attacked the problem of the structure of space-time from the side of the Riemannian metric. The geometric notion of affinity was brought in physics later by Weyl, in 1918, but this historical fact will be of importance to us only in Chapter 4. To be sure, there is a certain degree of independence between the metrical and the affine structures on a space-time manifold, and this raises the question of what is physically prior to what.) The energy-momentum tensor $T_{\mu\nu}$ represents all energy and momentum sources at a point, save for gravitational energy whose potentials are represented by the $g_{\mu\nu}$ in the geometric part of the equation.

In the case of the vacuum, the analogue of Laplace's equation is

$$R_{\mu\nu} = 0 \qquad (1.4)$$

and this means that the addition of matter implies something *more* than the mere substitution of $T_{\mu\nu}$ for 0 in the right-hand side of this equation. In fact, Eqn (1.3) can be re-written in the form

$$R_{\mu\nu} = -\kappa \, (T_{\mu\nu} - \tfrac{1}{2} g_{\mu\nu} \, T). \qquad (1.5)$$

The 'correction' from the more obvious field equations,

$$R_{\mu\nu} = -\kappa \, T_{\mu\nu}, \qquad (1.6)$$

to the actual equations (1.3) rests upon some fundamental considerations about the kinds of things we want to actually represent when we write down field equations in the first place. (In fact, Eqn (1.6) can be rewritten $R = -\kappa \, T$, which means that the two types of curvature are blurred. See W. Rindler 1977, pp. 178-9.) Two requirements are quite 'classical': in a vacuum, $T_{\mu\nu} = 0$, and the field equations are reduced to the desired form (1.4); the correction also complies with the classical demand of energy and momentum conservation (on all this, see Einstein 1921, Ch. 4). But two other criteria are connected with Einstein's attempt to embody the original spirit of his theory, and hence its superiority over the classical one; by an extraordinary combination of circumstances, he failed to realize at the time that a sufficiently 'patched up' classical theory can meet the same demands. These are general covariance and non-linearity, which will be discussed in turn.

Einstein knew all too well, throughout his life, the necessity of having geometrical structures that would give rise to non-linear equations. As he said in his 'Autobiographical Notes' (1949, p. 89), "The true laws cannot be linear nor can they be derived from such." This property is related to the fact that Eqns. (1.3) are doing two things at once: not only do they describe the force field but they also function as equations of motion for a test body. (This point was first spelled out clearly by Eddington in 1920, Ch. 6.) Indeed the affine connection has a geometrical structure that defines the privileged class of 'straight lines' (i.e. the direction) of that geometry—these are the geodesics. Since the free particles (particles acted on by no external forces) follow geodesics of the postulated affine connection, they have an equation of motion that is defined in terms of this connection. On the other hand, and by contrast, both the Newtonian theory of gravitation and the electromagnetic field of the Faraday–Maxwell theory require additional equations in order to predict motion. They start with the description of matter in an undisturbed state—that is, the paths of particles in the Newtonian theory are independent of one another, just as the fields are in the Faraday–Maxwell theory. Interaction is then described in terms of a perturbation that adds up to an undisturbed state, and this addition is treated mathematically as the linearity of the resulting equations; thus, the motion of a test charge in the electromagnetic field has no effect in itself on the field which is tested. But the field equations of general relativity are non-linear in the $g_{\mu\nu}$ and their derivatives, which means that the introduction of a test body in a field is not compatible with the linear addition of the field created by the body and by a pre-existent field; this feature certainly reflects the interactive nature of the resulting laws, but it is by no means clear that departure from the classical form also implies that, because the gravitational field is not included among the source terms represented by $T_{\mu\nu}$, this field may also act as its own source. (Authorities differ: compare J.A. Wheeler 1962 and R. Adler *et al*. 1965, p. 275.) The new concept of gravitational field is an attempt at changing the old ideas of the field as a mediation between bodies. The fact is mathematically translated as non-linearity, but the conceptualization itself derives from Einstein's wish to incorporate what he called a principle of equivalence into general relativity.

Ever since Galileo's formulation of the laws of free fall in a constant gravitational field, it has been remarked that the mass of a body as an inertial (intrinsic) property and the mass in its response to gravitational forces are equal within the minutest empirical limits. Gravitation therefore acquires a special relationship to inertia, but this fact is simply accidental in Newton's theory. Einstein struggled to make them appear equal by virtue of their very essence—as he said, a satisfactory theory of gravitation must account for the equality in such a way that it is *"wesensgleich"* (equal by virtue of essence) (1921, p. 64). In 1907, Einstein compared two systems of reference, K and K', the first at rest in a homogeneous gravitational field, and the second uniformly accelerated in a region where gravitation was negligible; if the acceleration of K and K' is the same, then the two systems are equivalent for the description of all physical phenomena. And in 1911, he made the further assumption that a reference system in free fall in a gravitational field can be considered to be 'at rest' in this field, at least as far as the form of the laws of nature is concerned. While we tend to think today of the special theory of relativity as a gravitation-free case (i.e. a theory in which the gravitational field is itself made felt via the intrusion of some kind of perturbation into the Minkowski metric) Einstein seems to have conceived of the special theory as the theory of a special type of gravitational field (the uniform one), and it is just this recognition that enabled him to go further and require generalization.* In laying down a principle of equivalence, it was never Einstein's intention to endorse the idea that, in certain cases, gravitational fields have only a 'relative' existence, dependent on the choice of the reference system. Of course it is true that Einstein would have agreed that gravitational fields can be equated with inertial fields arising when reference systems are subject to arbitrary accelerations, but this is not how he first formulated the principle of equivalence (which he applied only to uniform acceleration). And his particular formulation proved crucial in promoting the search for the field equations.

As he made unambiguously clear in the first paper which had reference to the principle, Einstein's purpose was indeed to formulate an appropriate *extension* of the principle of relativity to the case of accelerated motions (1907, pp. 414 and 454). In the process, the principle of equivalence took on an extraordinarily powerful heuristic role in the course of the transition from the special to the general theory. Because the principle was at the furthest remote from any possibility of transforming away *arbitrary* gravitational fields, Einstein began to develop a theory of static gravitational fields which constituted an approach to those gravitational fields associated with what is today identified as Minkowski space–time. This occupied him from 1907 to 1912, only to discover in the end that the new field equations admitted the desired association only in small regions of space–time. Even so, the later identification of the principle of equivalence with the assumption that one can always transform away an arbitrary gravitational field within an *infinitesimal* portion of space–time (see this standard formulation in W. Pauli 1921, p. 145) has little to do with Einstein's original insight. For Einstein's trouble was that this restriction could not be reconciled with his simple case involving association of uniform acceleration with homogeneous fields. It was by thinking through this problem that he concluded in late 1912 and early 1913, getting the mathematical assistance of Marcel Grossmann, that no scalar generalization of Newton's theory could be possible. What was needed

---

* In this and the following paragraph, I mainly summarize Norton's thesis (1985).

GENERAL RELATIVITY 73

was a four-dimensional generalization of classical gravitational fields; adopting a non-flat metrical tensor with Minkowski signature, Einstein and Grossman realized that their model could represent not only the space–time structure but also a generalization of the classical notion of potential (see Einstein and Grossmann 1913). This enabled Einstein to conceive of the generalized properties of the Minkowski metric as a way of accommodating the treatment of arbitrary gravitational fields. But until November 1915, he could not be swayed of his conviction that the static space–times of his earlier theory ought to be solutions of the new field equations as well. This was the real battleground, for, in the Einstein–Grossman theory, the coordinate systems adapted to uniformly accelerating frames of reference could not be regarded as being 'at rest'. A misleading confusion between coordinates and structure (in Chapter 2 I shall discuss it in greater detail) led Einstein to believe for more than two years that no generally covariant equations for the gravitational field were possible. In the complete solution presented by Einstein in his 1915b paper, general covariance was finally acknowledged to be the natural counterpart of the earlier insight about the principle of equivalence, so that the limitation to infinitely small regions could be dispensed with as well. In a 1916 formulation of the theory, Einstein went on to describe the principle of equivalence virtually as a particular case of the principle of general covariance: while the latter assumes that "the laws of physics must be of such a nature that they apply to systems of reference in any kind of motion", the former allows *uniformly* accelerated reference systems to be just as 'privileged' or 'stationary' as the unaccelerated ones (1916, p. 113).

This final relationship between the two principles raises a certain number of fundamental issues about what precisely is meant by Einstein's proposal that the general theory should be conceived of as an *extension* of the special theory. Einstein seems quite rapidly to have arrived at a comprehensive, geometric representation of gravitation in the case of inhomogeneous fields. From his early treatment of the rigidly rotating disc, it became clear that the space–time coordinates can have no direct metrological significance: in a system of reference 'at rest' in a rotating disc, the ratio of the circumference to the diameter must be greater than π (see J. Stachel 1980). Similarly, it was already clear from the 1911 version of the principle of equivalence that gravitation drastically affects the comparison of clocks at the different places of the field (Einstein 1911, p. 106). Therefore, in a case like that of uniform rigid rotation in Minkowski space–time there was an imposition of non-Euclidean spatial geometry, and Einstein then used arbitrary space–time coordinates, much in the manner of nineteenth-century methods for dealing with non-Euclidean spatial geometries. But rotation as treated here is regarded as a local phenomenon, in which the connection between various geometries for various physical systems in different places is not at all spelled out. The connection is made when an appropriate understanding of general covariance can be reached. Einstein arrived at this in his 1915b paper, after a long and tortuous transitory conception according to which the energy–momentum tensor would not uniquely determine the gravitational field if general covariance was maintained; that is, it would be necessary to limit the choice of possible coordinate systems—and thus of possible coordinate transformations—in such a way that the field is completely determined by the source term of the relativistic equations. But the final form of thinking overcame this unsatisfactory limitation. According to Einstein's correct field equations, it became only an appearance (derived from physics that pre-dates general relativity) that two distinct fields centered at two different spatial points will somehow enter into conflict with each other when they interpenetrate. In fact (as J. Stachel 1986 has pointed out), general covariance enables us to view the metric field as dictating the individuation of events (that is, coincidences in space–time), instead of the other way around, so that two space–times associated with two such events do correspond to a single gravitational field.

As a result, the postulated extension can be conceived of as a twofold process (see Weyl 1918, pp. 219-20). In the first place, we have a purely mathematical process: starting from the fundamental metrical form, $ds^2 = \Sigma g_{\mu\nu} dx_\mu dx_\nu$, it is possible to formulate the physical laws in such a way that they be invariant under arbitrary transformations of coordinates. Eqn (1.3) is written in the form of a tensor equation because the concept of tensor just reflects this property of invariance. Then the situation becomes truly physical, when it is realized that the metrical structure of the world cannot be given a priori; on the contrary the above-mentioned ds^2 is annexed to *matter* via generally covariant laws. According to Einstein, were it argued that the ds^2 is to possess an immediately physical meaning, without relation to the actual distribution of matter, it would not even describe an empty space–time—it would describe nothing at all. This is not, however, a picture that could be arrived at spontaneously, nor has it ever compelled universal assent. Thus the current attempts to quantize the general theory of relativity by means of a special-relativistic theory with some sophisticated symmetry superimposed on it are founded on a return to an independent rigid space (defined by the Lorentz invariance). Already in his day Einstein was not without some good arguments of his own that were critical of all quantum theories that start from a special-relativistic basis. In fact, the basic formalism laid bare here does not lend itself to an unambiguously coherent *interpretation*. Problems of interpretation throw into relief the fact that the postulated extension cannot work without some damage being done to the attempt at capturing, in accordance with Minkowski's programme, a pre-established harmony between the mathematical and the physical. With these problems of interpretation, we pass beyond the domain of the laws, as they were established by Einstein in November 1915, to that of their solutions; or, to put it differently, from the understanding of how the field operates, to the study of physical systems as they actually exist in nature.

Equations (1.3) cannot be read either way, from left to right or right to left. In the former case, the equations would say nothing new about nature. At best they could be used simply to extend the class of possible worlds: we start from a given geometry (that is, we fix the $g_{\mu\nu}$), which can be that of any possible world, and infer the consequences for the material content of that world ($T_{\mu\nu}$). Only in the latter case can the equations teach us something about the real: having determined some properties of matter, we discover the kind of metrical structure that reflects them (see H. Bondi 1965, pp. 410ff.) By so doing, solutions may be derived from the equations. The fundamental problem, however, is that the left-hand side of the equations *already* incorporates something which traditionally belonged to the physical, right-hand side, i.e. the forces of gravitation. It is just because this incorporation is successfully achieved by Einstein's equations that *not* any kind of space-time is possible in Nature. There is a restriction on the a priori indefinite number of possible geometries in the world that the physics of $T_{\mu\nu}$ might impose, and it is at this level that the wish to conceive of the general theory in terms of an extension of the special theory becomes critical. Einstein's hypothesis was that it still makes sense to speak of 'free' variables in the classical way, when the variables represent the field at infinity: there is no field of gravitation at an infinite distance from all masses (on all this, see Eddington's remarkable exposition 1920, pp. 77-92). Once this restriction is introduced, it is possible to draw up a sort of inventory of all the $g_{\mu\nu}$ that could reflect a real property of the world (Einstein 1921, pp. 82ff). One must then distinguish between:

a) those $g_{\mu\nu}$ which correspond to the type of *space-time* met with in a particular physical system under consideration;
b) those $g_{\mu\nu}$ which correspond to the type of *coordinate* system chosen to describe that space-time.

The actual values of the $g_{\mu\nu}$ derive from this procedure.

It would be tempting to see Einstein's preliminary characterization of the field at infinity as a mathematical procedure that has no real counterpart in the physical world, much after the pre-relativistic conception. This, however, will not quite work. Some hiatus between the mathematical and the physical crops up in relation to this remnant of classicism. It is worth looking at Weyl's presentation of the problem, for it leads us straight to the heart of the cosmological question in general relativity.

So far, as Weyl says (1918, p. 273),

the general theory of relativity leaves it quite undecided whether the world-points may be represented by the values of the four coordinates x_i in a singly reversible continuous manner or not. It merely assumes that the *neighbourhood* of every world-point admits of a singly reversible continuous representation in a region of the four-dimensional 'number-space' (whereby 'point of the four-dimensional number-space' is to signify any number-quadruple).

This is the level at which there is a complete parallel between general relativity and differential geometry: the properties of uniqueness and continuity refer only to a piece of surface or to a portion of space-time, not to the total surface or the total space-time. The properties of continuity over the whole are not the province of metrical or affine determinations. They belong to an analysis which is of the highest degree of generality: the topological analysis, i.e. the study of those properties that remain unchanged by continuous deformation. That is why, as Weyl goes on to say, general relativity "makes no assumptions at the outset about the inter-connection of the world". However—and this is the real crux of the matter—this does not mean that general relativity is not in need of a certain number of such assumptions. So much so that the ultimate significance of the cosmological question in general relativity is the problem of whether these extra assumptions do or do not form part of the original conception of the theory. Hence, also, whether some other physical theory might fill the gap in case general relativity does not.

Weyl's quote above points out a fact we will have to bear in mind throughout the remainder of this book, namely, that general relativity (at least in its original form) specifies only local connections. All we have said so far about its structure and its objects is true only from that point of view. The space-times of general relativity are constrained by the field equations connecting local features of space-time's metric structure with the distribution of mass-energy. But many different space-times are consistent with the field equations to the extent that they may have different topologies. In this way, a specific mass-energy distribution is compatible with many different space-time worlds. Even though the metric is the primitive reality in general relativity (together with its inherent affine connection), topology as a global property might change the metrical determinations. So at this level it would be misleading to maintain without some qualification what we said above, namely, that Eqns (1.3) are meaningful only when we read them from right to left. For this way of reading amounts to saying that the structure of space-time is univocally *caused* by the distribution of mass-energy in the world. Rather, there is a sort of interaction between space-time structure and mass-energy distribution: the equations tell us that they are joint features of the world.

Precisely by virtue of this interaction, the understanding of the exact nature of the field equations, i.e. what these equations actually say about the world, is a controversial issue. We shall take this feature as the starting-point of our enquiry into relativistic cosmology, since it is this very problem that is associated with the way to cosmology. Cosmology does not 'solve' it but expresses it in an explicit way. Constructing cosmological models is of relevance to general relativity as a theory, because this is where it is forced to exhibit the very way it says something about the world. The residual 'pliability' in the metrical determinations is this point of conceptual importance that early history of relativistic cosmology illustrates in a remarkable way. For the absence of complete correspondence between mass–energy distributions and particular space–time structures can be appreciated from two different aspects, each of them defining the sequence of discovery of relativistic cosmology. First, the solution of differential equations such as the field equations (1.3) requires boundary conditions, very much as any pre-relativistic differential equations would; the cosmological context of this specification was De Sitter's discovery in 1916, as we shall see in Chapter 2. Second, the relation of metric to topology discussed by Weyl in 1918, which sets cosmology to its properly relativistic character (Chapters 4 and 5). We begin with the far-reaching implications of this discussion because it virtually contains the key to what cosmology can tell us about the essence of general relativity.

The possibility of assigning metrical and affine characterization to the whole of space–time appears to be the consequence of attempting to reconcile the facts and the theory. The theory goes like this:

Every world-point is the origin of the double-cone of the active future and the passive past. Whereas in the special theory of relativity these two portions are separated by an intervening region, it is certainly possible in the present case for the cone of the active future to overlap with that of the passive past; so that, in principle, it is possible to experience events now that will in part be an effect of my future resolves and actions (Weyl 1918, p. 274).

But it is a matter of fact that

the very considerable fluctuations of the $g_{\mu\nu}$'s that would be necessary to produce this effect do not occur in the region of the world in which we live.

Astronomers have indeed found, until this day, that the actual metrical structure of space–time in the observable portion of the universe does not diverge in any significant way from Euclideanity. These facts relieve us from being visited by fearful spectral images of the world. But there is more in this relief than a simple "gift from above", to use Minkowski's words. As Weyl makes clear

. . . there is a certain amount of interest in speculating on these possibilities inasmuch as they shed light on the philosophical problem of cosmic and phenomenal time. Although paradoxes of this kind appear, nowhere do we find any real contradiction to the facts directly presented to us in experience.

Over and above the facts, which might prompt us to ignore the problem of the universe's structure, general relativity embodies, by rights, a *philosophical* perspective on the problem; and this perspective leads us, in turn, to the consideration of *time* which differs from that of special relativity in that it no longer simply works as a dimension of consciousness. Tracing this connection will preoccupy us in the following chapters. It should be remembered that general relativity could not be derived, and indeed was not derived, from global considerations, because as a local theory it does not in itself provide the topology needed for such considerations. But if we start from the metrical and affine determinations accessible from within the observable world, then the extension of these determinations to the whole does not quite remove us from the philosophical foundations of the theory.

8. The problem of the planets

Our last analysis has shown that the new tension which general relativity created between the mathematical and the physical, extending beyond the Minkowskian programme for a geometrical physics, finds its natural expression in the distinction between the local and the global. The first solution of the field equations in which the distinction plays an explicit role was provided in early 1916 by Schwarzschild (see his 1916a); this is the solution of the gravitational field produced by the Sun, from which the relativistic equations of planetary paths could be derived.

In his formal developments, Schwarzschild makes use of a certain number of hypotheses, which are in fact quite similar to those that Einstein used a short while earlier when he found that his field equations correctly explained a small phenomenon that no nineteenth-century physics could account for satisfactorily—the anomalous precession of the perihelion of Mercury (Einstein 1915a). According to general relativity, the gravitational field of the Sun is not exactly of the inverse square law form, except at very large distances from the Sun; whenever gravity fails to obey exactly the inverse square law, a precession of the planet's elliptical orbit occurs. Einstein's solution was based on a method of approximation. In order to find the ds^2 appropriate to gravitational orbits, four conditions were imposed on its form:
(1) all spatial components are time-independent;
(2) the metric is not only time-independent, but no term contains mixed space and time, that is, $g_{i4} = g_{4i}$ for $i = 1,2,3$ (space terms);
(3) the solution is symmetric with respect to space, i.e. an orthogonal transformation of the spatial coordinates does not change the solution;
(4) at infinity, we have the Minkowski metric, i.e. motion at infinity is uniform and rectilinear.

The first two conditions express the fact that the motions and the masses of the planets are negligible against the Sun's mass; we deal with a *static* space-time. Of course, the separation of space and time can have no absolute meaning in relativity; a curvature of space alone or time alone only expresses a relationship between the intrinsic curvature of space-time and the frame of reference of an observer using rigid rods and standard clocks as 'natural' measures. Condition (3) states that the field produced by the Sun is spherical. Condition (4) is the most delicate to understand, but also the crucial one as Einstein was to acknowledge later. In fact, Einstein's 1915 derivation was an approximation in the sense that only the time curvature was considered. In polar coordinates, the spatial part of the metric, $d\sigma^2$, can be written $d\sigma^2 = dr^2 + r^2(d\theta^2 + \sin^2\theta \, d\psi^2)$, where r is the radius from the Sun (centre of the symmetry) to any point of the solar system, and the angular coordinates θ and ψ measure the meridian of the point. In principle, the gravitational effect of the sun should yield a modification of this ordinary spherical geometry. We should have something like $d\sigma^2 = F(r)dr^2 + r^2(d\theta^2 + \sin^2\theta d\psi^2)$, where the function $F(r)$ governs the gravitational effects. But the assumed approximation implies $F(r)$ to be very little different from 1 for slow motions, since in the Minkowski metric the main part of the ds^2 is contributed by dt^2 (which is multiplied by c^2). Suppose we try to synchronize all clocks in this static field. A clock at r_1 runs faster than an identical clock at r_2 ($r_2 < r_1$) according to

$$1 + \frac{Gm}{c^2}\left(\frac{1}{r_2} - \frac{1}{r_1}\right).$$

If all clocks are synchronized with respect to an undisturbed clock at infinity, it is found that the proper time of a clock at distance r will be accelerated by a factor $1 + \phi/c^2$, where ϕ is the Newtonian potential. Einstein could then write down the desired metric: $ds^2 = (1 + 2\phi/c^2)c^2dt^2 - d\sigma^2$. And what Schwarzschild did was to provide a more complete, *exact* solution in which the same four conditions were used. We need not be concerned here with the details of his derivation. What will be relevant is the more thorough use of condition (4).

It must be noted that the physical meaning of this boundary condition in the corresponding Schwarzschild metric remains a very controversial topic. *Supposition* of an asymptotically Minkowskian continuum by no means leads to the necessary *existence* of this continuum at infinity (see the discussion in A. Grünbaum 1973, pp. 420 and 840). Some authors have tried to demonstrate that such a symmetrical solution implies the Minkowskian continuum at infinity (S. Weinberg 1972, p. 337). Also, it was not until 1923 that G.D. Birkhoff (1923, pp. 255-6) could prove that, while remaining the most general solution of the field equations with spherical symmetry, the Schwarzschild solution does not imply that the field is *static*; this static condition is quite superfluous. Birkhoff's result is certainly of great generality for it proves that *any* spherically symmetric field *in vacuo* is static. But in our present context, suffice it to bear in mind that the Schwarzschild solution with its 'classical' boundary condition was first of all greeted by Einstein as conveying a further proof of the validity of the field equations. On 16 January 1916, he read before the Prussian Academy Schwarzschild's paper on his behalf, since Schwarzschild was in the German army at the Russian front at that time (he died shortly after). A week earlier, he wrote to Schwarzschild about his enthusiasm and only noted a problem of *interpretation* in regard to the boundary condition. I shall discuss it after expounding briefly Schwarzschild's reasoning.

It is useful to follow the reconstruction of this reasoning offered by Tolman (1934, pp. 202–205) as this has become the standard presentation. The general line of the argument goes as follows. A priori, Schwarzschild's metric can be written

$$ds^2 = -e^\lambda dr^2 - r^2 d\theta^2 - r^2 \sin^2\theta d\psi^2 + e^\mu dt^2$$

where λ, μ, θ, and ψ are functions of r only (ϕ and ψ are also independent of λ and μ), that is, they are time-independent. The purpose of introducing the functions λ and μ by means of the exponential is that the ds^2 is reducible to the Minkowski metric when both λ and μ are zero, so that the problem is to determine λ and μ in relation to the actual distribution of matter. The calculation of the potentials is done with the aid of the field equations (1.3) for the empty space around the central mass. The following system of equations is derived:

$$-e^{-\lambda}\left(\frac{\mu'}{r} + \frac{1}{r^2}\right) + \frac{1}{r^2} = 0 \qquad (a)$$

$$-e^{-\lambda}\left(\frac{\mu''}{2} + \frac{\mu'^2}{4} - \frac{\lambda'\mu'}{4} + \frac{\mu' - \lambda'}{2r}\right) = 0 \qquad (b)(1.7)$$

$$-e^{-\lambda}\left(\frac{1}{r^2} - \frac{\lambda}{r}\right) + \frac{1}{r^2} = 0 \qquad (c)$$

in which μ' and λ' mean the derivatives of these functions with respect to r. This system provides the following relationship:

$$\mu' + \lambda' = 0, \qquad (1.8)$$

from which follows $\lambda = -\mu + $ constant. The constant, say k, can be taken to be equal to zero, since for $r = \infty$, the metric becomes that of Minkowski in which $e^\lambda = e^\mu = 1$. As a result, we have

$$\lambda = -\mu = 0. \qquad (1.9)$$

In other words, the mathematical condition used by Einstein to limit the number of possible equations for the gravitational field finds here a most remarkable physical significance. For once the condition (1.9) is applied, the Schwarzschild metric for the motion of a planet around the sun can be derived:

$$ds^2 = \frac{dr^2}{1 - 2m/r} - r^2 d\theta^2 - r^2 \sin^2\theta d\psi^2 + (1 - 2m/r)dt^2 \qquad (1.10)$$

where m is defined by the ratio GM/c^2 (M = solar mass, G = Newtonian constant of gravitation); this the the 'relativistic mass', which has the dimension of length.

The metric (1.10) is known as Schwarzschild's *exterior* metric, for in the same year 1916 he went on to write down the metric corresponding to the *interior* of a spherical, uniformly distributed matter (Schwarzschild 1916b). This interior metric is obtained from the field equations (1.3) with non-vanishing components for the energy–momentum tensor. With a co-mobile system of coordinates in a perfect fluid, this tensor becomes $T_1^1 = T_2^2 = T_3^3 = p$, and $T_4^4 = \rho$ (with $T = 0$ if $\mu \neq \nu$), where p stands for the pressure. If we now take the values $-8\pi p$ for the right-hand side of Eqns (1.7a) and (1.7b), $8\pi\rho$ for the right-hand side of Eqn (1.7c), we get after some transformations the fundamental relation of perfect fluids as it must be used in any static metric:

$$\frac{dp}{dr} + (p + \rho)\frac{\mu'}{2} = 0. \quad (1.11)$$

A decreasing pressure from centre to periphery of the sphere is here a definite option for the boundary condition of the pressure. A null pressure at the periphery (when $r = R$) leads to the general solution of Eqn (1.7c):

$$p + \rho = Ae^{\mu/2} \quad (1.12)$$

(where A is a constant of integration), the solution of which provides the desired metric:

$$ds^2 = -\frac{dr^2}{1 - r^2/R^2} - r^2 d\theta^2 - r^2\sin^2\theta d\psi^2 + (A - B\sqrt{1 - r^2/R^2})dt^2, \quad (1.13)$$

where $R^2 = 3/8\pi\rho$; $A = -(3/2)\sqrt{1 - r^2/R^2}$; $B = -1/2$. It is essentially the boundary condition of null pressure at the periphery of the sphere that allows to be found the numerical values for the two constants of integration, A and B.

These first elementary considerations on boundary conditions in the case of local solutions of the field equations open up a vast area of questions, which Einstein will explicitly develop when he turns to cosmology. Particularly relevant is the problem already alluded to above of the physical meaning of the boundary condition in the case of Schwarzschild's exterior metric. Significantly enough, when Tolman turns to the presentation of the cosmological arguments first used by Einstein to describe the whole universe, he points out the strong analogy they have with the exterior metric (static symmetry) and then, without the slightest critical comment, he goes on to assume that in passing from (1.8) to (1.9) it is sufficient to assume that for *small* values of r the line element will reduce to flat space–time, "owing to the known validity of the special theory of relativity for a limited space–time region" (p. 334). This suggests that the kind of boundary condition to be selected in the relevant cases remains very dependent on the kind of *object* one is interested to describe physically. Now, Rindler (1977, pp. 137-8) has shown very elegantly that there is even no need to *assume* flat Minkowski space-time at infinity in the whole process of deriving the Schwarzschild exterior metric; that is, the value of the constant in the equation which comes between (1.8) and (1.9) can be taken as zero by simply rescaling the unit of time from t to $e^{-k/2}t$ (which has the effect of adding a constant to λ). And Synge (1960, p. 276) has aptly observed that absorbing the k by such an operation enables us to recover a precise definition of the coordinate t as proper time at the centre of the sphere of matter on the axis $r = 0$; what we have by letting r tend to infinity is that "dt is the element of proper time for a particle which is fixed in the sense that r, θ, and ψ are constant".

Einstein may well have ignored this general significance of the Schwarzschild metric, but the way he presses Schwarzschild on interpretation of the metric clearly reveals his fundamental problem. The issue turns around the question of how far a parallel can be drawn between the classical and the relativistic representations of a solution to the respective equations. In the classical case, as we know, the strategy is that the 'rest' of the universe around any finite physical system is conceived as nothing. Now, clearly, this cannot be maintained in the relativistic case without some crucial qualification being made. As both the calculation and the final expression of the exterior metric suggest, the flat Minkowski metric at infinity *is* the relativistic counterpart of the *classical* requirement of there being zero potential at infinity; that is, all observers will agree on what 'infinity' is without having to bother about the actual appearance of the whole universe. But while this agreement can be utterly 'ideal' in the classical case, it is certainly no longer so in the relativistic one, for even if the Minkowski continuum at infinity is not associated with any particular *object*, at least it can be identified with a definite *structure*. And if the strict partition between the objectual and the structural is indeed loosened in general relativity, in accordance with how its *laws* have been erected, then there seems to be no way of escaping that the Schwarzschild exterior solution offers the picture of a locally distorted, but pre-existent, pseudo-Euclidean continuum.

This is just what bothered Einstein, and he said it sharply in the objection he expressed to Schwarzschild in his letter of 9 January 1916. He asked what would happen if we were to remove all objects from the world, and answered: "According to Newton the Galilean space of inertia remains, while according to my conception *nothing* at all". In other words, the view by which the rest of the universe is feigned to be nothing at all no longer works: for now either the rest *is* nothing or, if it is not nothing, it *is* something. No mere mathematical expedient could now save us from tackling the classical and arbitrary lack of explicit distinction between the local and the global as a problem of its own, because if what separates the two is translated in terms of its relativistic counterpart, the correct representation of even the smallest physical system necessitates an awareness of the larger and larger portions of space–time of which it is a part. The key word for Einstein, the clue to an understanding of inertial structure, is *Wechselwirkung*, which combines the ideas of interaction and reciprocity between every existing mass in the universe.

This emphasis on the idea of interaction is Einstein's paid tribute to the overwhelming influence of Ernst Mach, one of the most eminent physicists and philosophers of science in the latter part of the nineteenth century. It seems that as a young man, Einstein was a rather strict empiricist in his attitude to the basis of physical science, and in this respect he was very much Mach's disciple. His view of theoretical physics was quite positivistic—physics amounted to the more-or-less accurate depiction of observations and experiments. No systematic principle of any sort should dictate the process of description, and Mach himself identified the only possible 'all-encompassing' principle of physics with *nature as a totality* (1883, pp.287–8):

Nature does not begin with elements, as we are obliged to begin with them. It is certainly fortunate for us, that we can, from time to time, turn aside our eyes from the over-powering unity of the All, and allow them to rest on individual details. But we should not omit, ultimately, to complete and correct our views by a thorough consideration of the things which for the time being we left out of account.

A good number of these empirical ground-rules were blown wide open by formulation of the general theory of relativity, because it became clear that mathematical expression could no longer be regarded as a straightforward and rather simple translation of the facts of experience. However, it is essential to realize that, if Einstein did undoubtedly move on to the recognition of principles of rational significance as forming the root of all physical science, he did not depart, at least in the first instance, from the conviction that these principles somehow communicate with the 'All' as it was envisaged by Mach.

Of course, Mach's reverence for the All has nothing to do with what we understand today by cosmological science, but Einstein's early attempt to implement the idea of cosmology into general relativity *has* a lot to do with it. From the historical/conceptual point of view, Mach's ideas can now be seen as a crossroads between pre-relativistic theories and relativity itself. He arrived at them by taking a fresh look at the implications of *rotation* in Newtonian physics.

As we know from the previous section, Kant had already argued for the impossibility of effects of absolute space, but he did so on transcendental, not physical, grounds (such as the special theory of relativity might prompt), for he identified this impossibility with one particular object, the entire universe. That is why he dismissed out of hand both the reality of absolute space and the possibility of physical cosmology. However, close attention to his argument shows quite clearly how the new refutation of absolute space achieved in the special theory of relativity could already lead to something very different, even something opposed to Kant's arguments against cosmology as a science. When he refers to the motion of the entire universe, Kant speaks of the *rectilinear* motion only as a proof of the impossibility of absolute space. What he seems to have in mind is Newton's attempt to demonstrate *ad absurdum* the validity of the third axiom of motion, that is, the law of action and reaction: if the attraction between the various parts of the Earth were not reciprocal according to this law, then the motion of the Earth would be in one direction. Kant extends the idea to the parts of the whole universe and concludes that "every divergence from this law would move the common center of gravity of all matter". But he adds that "this would not happen if one were to represent the universe as rotated on its axis; therefore, it is always possible to think of such rotation, although to assume it, as far as one can see, would be quite without any conceivable use" (1788, p. 131). Now Mach's critique of the non-empirical presuppositions of Newton's theory pointed out quite precisely the relevance of conjectures about the rotation of the whole universe (1883, pp. 284–90). Mach makes reference to Newton's interpretation of the notorious rotating bucket, where Newton denied that the occurrence of centrifugal forces resulting from the concavity of the water inside the bucket was relative to any possible material system. Mach remarked that the only empirical test that could be imagined for disproving the hypothesis that rotational motion was relative (with respect to the universe as a whole), would be to compare Newton's experiment, as he performed it, with one in which the bucket is left undisturbed and the universe is made to rotate around it. Precisely *because* the test is impossible to carry out we are not obliged to accept Newton's interpretation of his experiment. The consequence is that his empirical case in favour of absolute space collapses as well. Nothing can stop us ascribing the inertia of a material body in the universe to the influence of all other bodies in the universe.

This testifies to what is, by implication, the new relation of the special theory of relativity to the physical problem of the entire universe, since the theory is already grounded on the impossibility of accounting for the effects of absolute space in the case of an object *within* the universe. In fact, the problem becomes explicit in the general theory, as Einstein was to understand almost in the very act of establishing the significance of Schwarzschild's first exact solution. Just as the classical type of relativity is *limited* when it comes to the question of the existence of a possible connection between two systems as far apart as the Earth and the stars, so the extended principle has been perceived by Einstein, under the influence of Mach's critique, to *imply* the reality of such a connection.

However, Einstein's evaluation of Mach's critique of the Newtonian conception of rotation was not in itself sufficient for a necessary drastic revision of classical cosmology. Equally decisive were the speculative contributions about the possible extension of non-Euclidean geometry to the entire universe.

9. Variations on a non-Euclidean theme

Here too, were it not for Einstein himself, very little seemed to prepare physical science of the early twentieth century to make a new start with cosmology, however much general relativity itself took its bearings from the earlier developments of non-Euclidean geometry.

Parallel lines in the Euclidean plane are lines which never meet. Ever since the Greeks, the so-called parallel postulate of Euclidean geometry was known to assert that for any one given line and any one point outside it in the plane, there is exactly one line that may pass through that point and be parallel to the other given line. But it gradually became clear that this particular postulate, in contrast to the others of Euclidean geometry, does not carry with it any convincing or immediate evidence. In the early eighteenth century, Saccheri made the epoch-making attempt to show that no alternative was possible (several parallel lines to the given point, or perhaps none) by demonstrating that any such alternative would lead to a variety of utterly strange conclusions. Although the conclusions were strange, they did not involve outright contradictions. Not until near the middle of the nineteenth century did other mathematicians (Bolyai, Lobachevski) finally show that an effective geometry of the plane could indeed be constructed by using an axiom that contradicted the Euclidean postulate. (For a history of the subject see R. Bonola 1955.) Several other attempts at axiomatic approaches revealed that the parallel axiom was neither evident nor necessary, and this had the effect of making room for different models of space. Ideas started to shift from mathematics as a description of our particular and unique universe to a description of *various possible universes* by way of the axioms that would satisfy them all. By the end of the nineteenth century, Poincaré among others showed that the non-Euclidean geometries were at least as consistent as ordinary Euclidean geometry, by demonstrating that models could be constructed for them within the terms of Euclidean geometry.

In the 1870s, W.K. Clifford was one of the first to speculate that the geometry of space as conceived by physical laws might not be Euclidean. Basically, he proposed four ideas: that space in the small portions of the world is like so many hills on a surface which is, on average, flat; that this 'distortion' or curvature is continually being transmitted from one portion of space to the other rather in the manner of a wave; that motion of matter (whether ponderable or ethereal) has just this variation; and that all phenomena of the world are reducible to a variation of this kind. These speculations were indeed prophetic of the general theory, even though the concept of space-*time* was of course totally absent. And in fact, in the wake of Clifford, non-Euclidean geometry came to be often linked with arguments borrowed from theoretical physics long before the advent of the general theory. Thus, non-Euclidean forms of hydrodynamics, of elasticity, of Maxwell's electrodynamics, existed in some proliferation by the end of the nineteenth and the early twentieth centuries. (John North 1965, pp. 73-4, mentions the existence of no less than eighty papers on the subject over this period.) But because all the relevant predictions were far beyond the threshold of measurement, these speculations were not taken seriously; they were simply the toys of mathematical curiosity. By contrast, the theory of general relativity was the first that could offer predictions verifiable within the limits of measurement. In 1919, Eddington detected the predicted deflection of starlight in the Sun's gravitational field (that is, the deflection of light rays grazing the Sun's disc) during his famous and spectacular solar eclipse expedition. A second test was the precession—or slow drift—of the planetary orbit of Mercury. A third test, which occupied astronomers in the early 1920s, was the gravitational redshift: radiation escaping from a body like the Sun loses energy because of the pull of gravity; as the energy decreases, the frequency of light rays also decreases, so that the wavelength of these rays increases as they travel away from a gravitating body—the wavelength is shifted towards the red. (On the story of this latter test, see J. Earman and C. Glymour 1980.)

Prior to the advent of general relativity, it was not the micro-physical but (even at this stage) the astronomical scale of phenomena that seemed to offer something more reliable with respect to the observational evidence of non-Euclideanity (see J. North 1965, pp. 72ff, and M. Jammer 1969, p. 147). In the early nineteenth century, C.F. Gauss had pioneered the method of actually testing the reality of Euclidean geometry by empirical means, in that he was credited with being the first to verify the Euclideanity of ordinary space by carefully establishing triangulation between the tops of three mountains. Later, F.W. Schweikart extended this method to celestial distances. Three positions can be taken as triangulation points: by definition, the annual parallax of a star is the angle at the vertex of a Euclidean triangle, the base of which is constituted by two points, E and E', which are positions of the Earth with respect to the Sun separated by a six-month interval (this is the axis of the annual revolution of the Earth around the Sun). The value of the parallax in terms of the two measurable angles A and B (see Fig. 1.3) is $P_E = \pi - (A + B)$. If the triangle is not Euclidean, the angle P is given by $P = \pi - (A + B) + KS$, where K is the constant of curvature and S is the area of the triangle (see R. Bonola 1955, pp. 135-6). Clearly, for $K = 0$, the geometry is Euclidean. The problem is thus to find the *sign* of K from the measured values of the angles A and B. In the case of a *very remote* star, the angle at the vertex is almost null for every possible geometry. This reduces the number of unknowns, since the last equation becomes $\pi - (A + B) = -KS$. Were space Euclidean, observations should give $A + B = \pi$; if $A + B < \pi$, then $K < 0$ and space is hyperbolic (negative curvature); if $A + B > \pi$, then $K > 0$ and space is spherical (positive curvature).

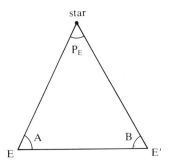

FIG. 1.3. Testing the reality of Euclidean geometry by using the annual parallax of a star.

In 1900, the same K. Schwarzschild, who was to provide the first exact solution of the field equations of general relativity, was the first to refine this technique in terms of actual star counts. However, all sorts of conceptual difficulties were involved in the calculation: most particularly the fact that the method was regarded as being based on a purely geometrical hypothesis, irrelevant to physical consideration. First of all, how could one know that a star was indeed very remote? The danger of circularity arose because the distances were being computed with the aid of the usual, Euclidean formulas of trigonometry. The only way out was to extend the method so as to take in a whole range of stars; the problem being to find whether the sums of the angles A and B *tend* towards π or not. Secondly, and most importantly, these methods implicitly identified a Euclidean straight line with the path of a ray of light. This was what led Poincaré to formulate his famous critique (1902, pp. 72–3):

. . .what we call a straight line in astronomy is simply the path of a ray of light. If, therefore, we were to discover negative parallaxes, or to prove that all parallaxes are higher than a certain limit, we should have a choice between two conclusions: we could give up Euclidean geometry, or modify the laws of optics, and suppose that light is not rigorously propagated in a straight line. . .

The passage is open to at least three kinds of interpretation (see J. Giedymin 1982, pp. 16ff). For our purposes, however, it is enough to take it on board at a broader epistemological level, as Einstein himself did (1921, pp. 35-6). Einstein argued that Poincaré's real implication was that any theory of nature could be meaningful only to the extent that it related to the *sum* of geometry + physics. Thus, there was no crucial experiment which could unequivocally decide between the types of metric geometry alluded to since only the sum geometry + physics could be meaningfully tested. As Einstein put it: "Envisaged in this way, axiomatic geometry and the part of natural law which has been given a conventional status appear as epistemologically equivalent." Whatever new things experience may teach us must simply not be allowed to contradict that sum. Einstein went on to argue that, in his opinion, Poincaré was right *sub specie aeterni*. But he offered a non-conventionalist interpretation of Poincaré's thesis by appealing to the stage of development at which physical science had arrived: space and time remain independent ideas, "for we are still far from possessing such certain knowledge of theoretical principles as to be able to give exact theoretical constructions of solid bodies and clocks". In conclusion, and for *practical* reasons (like the evidence of practically-rigid bodies over large portions of space), Einstein admitted the *absolute necessity* of certain hypotheses, that is to say, their relative independence from the sum geometry + physics. So the reality of non-Euclidean geometry remained by and large an empirical question in Einstein's eyes.

This is, of course, the basis of general relativity which claims to have definitively incorporated Riemannian geometry, at least as far as the understanding of intermediate levels between the sub-molecular and the cosmic order of magnitude is concerned. At the time of these Princeton lectures (1921), the extension to the sub-molecular order remained a matter of pure conjecture; Einstein was to struggle for a good part of the rest of his life trying to find an interpretation of atomic physics in terms of fields. But it was towards any attempt at extending things to the cosmic order of magnitude that Poincaré's critique was the most scathing. He rejected absolute space and absolute time very much in the manner of Mach: what experiments teach is only the relation of bodies to one another (1905, pp. 79 and 90); and furthermore any possible law must be interactive in nature since "no system exists which is abstracted from all external action; every part of the universe is subject, more or less, to the action of the other parts" (p. 103). Along these lines, he defined what he called a law of relativity (p. 76), asserting that the absolute initial position of a physical system, as well as its absolute initial orientation, could have no influence on the state of motion and the mutual distances of the bodies composing the system. This law is, of course, another way of asserting the 'passivity' of space: space cannot be a physical entity like the material bodies directly experienced by us. Poincaré went on to ask if the kind of flexibility in the interpretation of distances, allowed for in the ready-made hypothesis of taking distance in either a Euclidean or non-Euclidean way, could undermine the universal validity of the postulated law of relativity. Perhaps we could imagine some sort of crucial experiment which would prove the falsity of either of the possible geometries of the world. From Newtonian physics, we know that the Euclidean interpretation of distances *is* in agreement with the law, so perhaps there is an experimental way (like the parallax argument) of proving the falsity of non-Euclidean geometry. Poincaré's point is that this cannot be so because "to apply the law of relativity in all its rigour, it must be applied to the entire universe" (p. 77). Suppose we consider any finite physical system which is a *part* of the universe, then and only then will it be meaningful to speak of it as having an absolute position and, hence, of its geometry as a property fixed by nature. However, the geometry of the *whole* universe remains meaningless, it cannot be determined by way of some empirical and piecemeal procedure, just because position and absolute orientation in space cannot be defined in such a case by means of any conceivable experiment. In other words: our sense of the causal connection within the entire universe yields one truth about the nature of our physical laws (namely, that the law of relativity can be universally applied) but, at the same time, all our attempts to know something about the universe as a whole are doomed to failure.

It is not even overstating things to say that, in Poincaré's view, our definitive ignorance of the whole guarantees our grasp of the part. This is a powerful expression of the rejection of cosmology as a science in the very name of Mach's "overpowering All'. When he turns his attention to the large-scale implications of general relativity, Einstein will have to struggle all the way against this identification. To clear up our own ideas we must see Einstein's problem in terms of his own critical reckoning with Poincaré's arguments. In fact, the general-relativistic approach to an 'essential' relationship between gravitation and inertia can be thought of in *two ways*. As we have seen in Section 7: following the incorporation of the principle of equivalence into consistent field equations, gravitational forces could now be conceived of as part of the geometry of the world. In the terms of Poincaré's sum, G (geometry) + P (physics), this amounts to reducing gravitation to inertia, i.e. we have a comprehensive description in which both phenomena belong to G. But in view of the alleged interchangeability of what counts as geometry and what counts as physics, it may still make more sense to speak of an inverse reduction of inertia to gravitation, because both phenomena belong to what was *classically* designated as P. If so, then the two types of reduction are *not* indistinguishable from one another—as they should be in a straightforwardly general-relativistic picture—so long as we do not hit against a corresponding geometrical interpretation which reduces inertia to gravitation. Now it was primarily this tension that put Einstein on the track of a cosmological problem. As Max Planck had observed as early as 1887 (p. xiii, see also pp. 274–5), a theory based on action at a distance is always more general and global than a theory based on action of contact, because the former implies instantaneous action. With the success of general relativity, it remained necessary to remove the obvious contradiction of a more general theory based on action of contact. In retrospect, it is clear that the contradiction arose simply because the theory of general relativity could not sustain the classical framework. But Einstein's cosmology *derived* from the sheer impossibility of doing away totally and immediately with such classical image of the world. In Chapter 5, I shall examine the question of how far post-Einsteinian cosmology has moved from this image.

In our historical/conceptual analysis of the transition from Newtonian to Einsteinian conceptions about the nature of physical science, we had to take notice of an increasing concern about how to reconcile ideally global considerations with an ever more refined understanding of the physical properties of the world to which we have, as it were, ready access. We have also seen how the interpretation of the relationships that hold between mathematics and physics came to be profoundly affected by this move, because something like Minkowski's ideal of capturing a kind of pre-established harmony was associated with a conception of the 'world' which seemed altogether new.

The trouble was already there way back with Newton, when the discrepancy between a mathematical and a physical centre of forces first raised its head in a universe otherwise ruled (as he put it to Bentley) by a God "very well skilled in mechanics and geometry". Clearly, the absolute centre of forces of the solar system could only be an abstract centre of mass. And when further precision was brought to bear our knowledge of the mutual attractions between the planets by the celestial mechanics of the nineteenth century, the abstract centre of mass appeared even further away from the Sun. The upshot was no doubt the product of nineteenth-century Newtonianism at its most fundamental: the 'absolute' centre of forces of the solar system had always to be a function of our partial knowledge of its internal interactions and constituents. Moreover, if no finite system is in principle perfectly isolable from the rest of the universe, then the very notion of an absolute centre is null and void. With the coming of Mach's critique of Newton, the situation changed radically: Einstein could play upon the extension to the stars themselves of the problem of how to link a definite periphery to a definite centre. In particular, the physical use of non-Euclidean geometry effected a radical shift in what needed to be defined in the first place: the periphery came before the centre.

This transformation, this widening of scope, has affected our perception of scales of magnitudes so much that, unlike the earlier vague speculations about a non-Euclidean physics, our current theories about practically unreachable distances are taken very seriously indeed. The situation today is that the experimental and observational evidence is entirely out of kilter with the theory it is supposed to test. Let us take an example: one of the most important predictions of the grand unified theories, that of proton decay, would imply the observation of a distance of about 10^{-29} cm—it is like the detection of a speck of dust in the solar system. It really is true that, generally speaking, the energy levels at which serious experiments are made in particle physics are some 14 orders of magnitude below those at which the theories disclose their most interesting predictions. But the crux of the matter is clear: the higher the energy, the more 'comprehensive' and 'unifying' the force that corresponds with it. So, to catch a glimpse of the superforce thought ultimately responsible for generating all forces and structures, we should have to build a particle accelerator as long as the Milky Way. In other words, what we have today in atomic research is a need to probe the universe within the tiniest unit of its constituent matter. So much so that in order to compensate for the limits of human technology which can build no such huge accelerators, theoretical physics has made the universe itself into a natural laboratory. Thus, black holes and certain stars or galaxies in the final stages of their evolution seem to provide us with the desired levels of energy. Fundamental is the claim that the big bang was just strong enough to unleash nothing less than the superforce itself. Why should we believe that the discovery of any of the actual stage along the evolutionary sequence of the entire universe neutralizes the traditional hiatus between experiment and theory? One theoretical feature of the models of unification that have been proposed successively (electroweak interaction, hyperweak interaction, supergravity) is offered in support of these conjectures: the fact that each higher level of unification 'comprehends' the others so that the arbitrary constants of integration of previous theory become a structural part of the new advanced. Instead of purely arbitrary constants whose numerical values should be fixed independently, by way of experimentation, the values are determined by the theory itself—the higher theory comprehends them and therefore 'explains' them. In the end, the most comprehensive theory should be free of arbitrary constants.

Compare this with Einstein bewailing his lot when he wrote to Besso, just after he had gone through his first relativistic model of the universe in February 1917, that the model could never be verified empirically. (A. Einstein and M. Besso 1972, p. 104). Think of the calculation of the actual radius of curvature of the universe in this model. This could be obtained through an estimation of the mass density in space. Relying on the method of star counts, the rough value of 10^{-22} g/cm^3 for the density would yield a radius of curvature of 10^7 light years. The distance we are actually able to see, Einstein added, is of the order of 10^4 light years. (Current data indicate a still lower density, at most 10^{-30} g/cm^3, which would yield a corresponding radius of 10^{10} light years in Einstein's model. The discovery of receding galaxies has also altered ideas of distance: galaxies form clusters which in turn are grouped into superclusters, and only one such supercluster could have a size of up to 6×10^7 light years.) But Einstein quickly passed from despair to pure speculative assurance. By 1921, he found that the value of the radius could be estimated at some 100 million light years, and he made haste to assert that anyway "the exact figure is a minor question" (in A. Moszkowski 1921, p. 127). Empirical confirmation became almost irrelevant beside the power of theory. And here already, as he reflected in the latter part of 1916 on the nature of constants of integration in the new theory of gravitation, Einstein was precisely at the point of embracing cosmology.

References

Adams, W.S. and Kohlschütter, A. (1914). The radial velocities of one hundred stars with measured parallaxes. *Astroph. J.*, **39**, 341–9.
Adler, R., Bazin, M., and Schiffer, M. (1965). *Introduction to general relativity*, McGraw-Hill, New York.
Aiton, E.J. (1972). *The vortex theory of planetary motions*. Macdonald, London.
Alexander, H.G. (1956). *The Leibniz–Clarke correspondence*. Manchester University Press.
Aristotle (1934). *Physics*, (trans. P.H. Wickstead and F.M. Cornford). Harvard University Press, Cambridge, Mass.
Berendzen, R., Hart, R., and Seeley, D. (1976). *Man discovers the galaxies*. Science History Publications, New York.
Bernouilli, D. (1752). Recherches physiques et astronomiques. In *Recueil des pièces qui ont remporté les prix de l'Académie Royale des Sciences*. Paris, Vol. 3.
Birkhoff, G.D. (1923). *Relativity and modern physics*, Harvard University Press, Cambridge, Mass.
Bondi, H. (1965). Some special solutions of the Einstein equations. In *Lectures on general relativity*. Brandeis Summer Institute in Theoretical Physics, Vol. 1, Prentice-Hall, Englewood Cliffs. pp. 375–459.
Bonola, R. (1955). *Non-Euclidean geometry*. Transl. H.S. Carslaw, Dover, New York.
Buchdahl, D. (1970). Gravity and intelligibility: Newton to Kant. In *The methodological heritage of Newton* (eds. R. Butts and J. Davis) pp. 74–102. Blackwell, Oxford.
Campbell, W.W. (1913). *Stellar motions*. Yale University Press, New Haven.
Charlier, C.V.L. (1908). Wie eine unendliche Welt aufgebaut sein kann. *Arkiv for Math. Astr. och Fys.*, **4**, No. 24.
Clerke, A. (1893). *A popular history of astronomy during the nineteenth century*, 3rd edn., Adam and Charles Black, London.
Cohen, I.B. (ed.) (1978). *Newton's papers and letters on natural philosophy*. 2nd ed., Harvard University Press, Cambridge, Mass.
Curtis, H.D. (1917). Novae in spiral nebulae and the island universe theory. *Publ. Astron. Soc. Pacific*, **29**, 206–7.
Descartes, R. (1630). *Le Monde ou Traité de la Lumière*. In C. Adam and P. Tannery, *Oeuvres de Descartes*. J. Vrin, Paris, 1967, vol. XI.

Dreyer, J.L.E. (1912). *The scientific papers of Sir William Herschel.* Vol.1, The Royal Society and the Royal Astronomical Society, London.
Earman, J. and Glymour, C. (1980). The gravitational redshift as test of general relativity. *Studies in history and philosophy of science.* **11**, 175–214.
Eddington, A.S. (1912). Stellar distribution and movements. In *Report of the 81st meeting of the British Association for the Advancement of Science.* John Murray, London. pp. 246–60.
Eddington, A.S. (1920). *Space, time and gravitation.* Cambridge University Press.
Einstein, A. (1905). Elektrodynamik bewegter Körper, *Annalen der Physik*, **17**, 891–921.
Einstein, A. (1907). Ueber des Relativitätsprinzip und die ausdemselben gezogenen Folgerungen. *Jahrbuch der Radioaktivität und Elektronik.* **4**, 411–62.
Einstein, A. (1911). On the influence of gravitation on the propagation of light. (transl. W. Perrett and G.B. Jeffery) In *The principle of relativity*, A collection of original memoirs on the special and general theory of relativity (ed. 1952) pp. 97–108. Dover, New York.
Einstein, A. (1915a). Erklärung der Perihelbewegung des Merkur aus der allgemeinen Relativitätstheorie. *Sitz. Ber. Preuss. Ak. Wiss.*, 831–9.
Einstein, A. (1915b). Feldgleichungen der Gravitation, *Sitz. Ber. Preuss. Ak. Wiss.*, 844–7.
Einstein, A. (1916). The foundation of the general theory of relativity. (transl. W. Perrett and G.B. Jeffery) In *The principle of relativity.* A collection of original memoirs on the special and general theory of relativity (ed. 1952), pp. 111–64. Dover, New York.
Einstein, A. (1917a). *Relativity. The special and the general theory.* (transl. R.W. Lawson), Methuen, London, 1920.
Einstein, A. (1917b). Cosmological considerations on the general theory of relativity. In *The principle of relativity* (transl. W. Perrett and G.B. Jeffery) A collection of original memoirs on the special and general theory of relativity (ed. 1952), pp. 177–88. Dover, New York.
Einstein, A. (1921). *The meaning of relativity.* (transl. E.P. Adams), Methuen, London, 1922.
Einstein, A. (1949). Autobiographical notes. In *Albert Einstein, philosopher–scientist* (ed. P.A. Schilpp), Open Court, Evanston.
Einstein, A. (1954). *Relativity. The special and the general theory.* Fifteenth edition of (1917a).
Einstein, A. and Besso, M. (1972). *Correspondance 1903–1955* (ed. P. Speziali), Hermann, Paris.
Einstein, A. and Grossmann, M. (1913). Entwurf einer verallgemeinerten Relativitätstheorie und einer Theorie der Gravitation. *Zeitschr. für Mathematik und Physik*, **62**, 225–61.
Erlichson, H. (1973). The rod contraction–clock retardation ether theory and the special theory of relativity. *Amer. J. Physics*, **41**, 1068–77.
Fourier, J. (1824). Sur les températures du globe terrestre et des espaces planétaires. In *Oeuvres* (ed. G. Darboux), Vol. II, 1888–1890, pp. 97–125. Gauthier-Villars, Paris.
Friedman, M. (1983). *Foundations of space-time theories.* Princeton University Press.

Giedymin, J. (1982). *Science and convention*. Pergamon, Oxford.
Grünbaum, A. (1973). *Philosophical problems of space and time*. 2nd ed., Reidel, Dordrecht.
Herschel, J. (1848). Review of Humboldt's *Kosmos*. In *The Edinburgh Review*, **87**, 170-229.
Hertzsprung, E. (1913). Uber die räumliche Verteilung der Veränderlichen vom δ-Cephei-Typus. *Astron. Nachr.*, **196**, 201-10.
Hetherington, N.S. (1971). The measurement of radial velocities of spiral nebulae. *Isis*, **62**, 309-13.
Hetherington, N.S. (1973). The delayed response to suggestions of an expanding universe. *J. Brit. Astr. Assoc.*, **84**, 22-8.
Holmes, A. (1913). *The age of the earth*. Harper & Brothers, London and New York.
Hoskin, M.A. (1967). Apparatus and ideas in mid-nineteenth century cosmology. *Vistas in astronomy*, **9**, 79-85.
Hoskin, M.A. (1977). Newton, providence and the universe of stars. *Stellar astronomy: historical studies* (ed. M.A. Hoskin), Science History Publications, Chalfont St. Giles, Bucks, 1982.
Hoskin, M.A. (ed.) (1982). *Stellar astronomy: historical studies*. Science History Publications, Chalfont St. Giles, Bucks.
Hubble, E. (1929). A relation between distance and radial velocity among extragalactic nebulae. *Proc. Nat. Acad. Sciences*, **15**, 168-73.
Hubble, E. (1936). *The realm of the nebulae*. Yale University Press, New Haven.
Huggins, W. (1864). From an account of his researches on the prismatic analysis of nebulae given to the Royal Astronomical Society on 10 March 1865. *Monthly Not. Roy. Astr. Soc.* **25**, 1865, 155-7.
Humboldt, H. von (1845-1858). *Kosmos. Entwurf einer physischen Weltbeschreibung*. 4 vols, J.G. Gotta'scher, Stuttgart and Augsburg.
Huyghens, C. (1698). *Cosmotheoros*. In *Oeuvres Complètes*. M. Nijhoff, The Hague. 1888-1950.
Jaki, S.L. (1969). *The paradox of Olbers' paradox*. Herder and Herder, New York.
Jaki, S.L. (1972). *The Milky Way*. Science History Publications, New York.
Jaki, S.L. (1974). *Science and creation*, Scottish Academic Press, Edinburgh.
Jammer, M. (1969). *Concepts of space*. Harvard University Press, Cambridge Mass.
Kant, I. (1755). *Universal natural history and theory of the heavens* (transl. S.L. Jaki) 1981, Scottish Academic Press, Edinburgh.
Kant, I. (1788). *Metaphysical foundations of natural science* (trans. J. Ellington), Bobbs-Merrill, New York, 1970.
Kelvin, Lord (1862). On the age of the sun's heat. In *Popular lectures and addresses* Vol.1. 1891, Macmillan, London.
Koyré, A. (1957). *From the closed world to the infinite universe*, The Johns Hopkins University Press, Baltimore.
Kubrin, D. (1967). Newton and the cyclical cosmos: providence and the mechanical philosophy, *Journal for the History of Ideas*, **28**, 324-46.
Lambert, J.H. (1761). *Cosmologische Briefe über die Einrichtung des Weltbaues*. Eberhart Kletts Wittib, Augsburg.
Lanczos, C. (1972). Einstein's path from special to general relativity, in *General Relativity*. Papers in Honour of J.L. Synge. (ed. L. O'Raifertaigh), pp. 5-19. Clarendon Press, Oxford.

Laplace, P.S. (1796). *Exposition sur le système du monde*, Paris.
Lightman, A.P. and Miller, J.D. (1989). Contemporary cosmological beliefs. *Social Studies of Science*, **19**, 127-36.
Mach, E. (1883). *The Science of Mechanics*, (trans. T.J. McCormack), 4th edn, Open Court, Chicago, 1907.
Maxwell, J.C. (1873). *A treatise on electricity and magnetism*, Oxford.
McGuire, J. (1978). Newton on place, time and God: an unpublished source. *British Journal for the History of Science*, **11**, 114-29.
Merleau-Ponty, J. (1983). *La science de l'univers à l'âge du positivisme*, J. Vrin, Paris.
Merleau-Ponty, J. and Morando, B. (1976). *The rebirth of cosmology*, (trans. H. Weaver), A. Knopf, New York.
Michelson, A.A. and Morley, E.W. (1887). On the relative motion of the earth and the luminiferous ether, *Philosophical Magazine*, **24**, 449-63.
Miller, A.I. (1981). *Albert Einstein's special theory of relativity*. Emergence (1905) and Early Interpretation (1905-11), Addison-Wesley, Massachusetts.
Minkowski, H. (1908). Space and Time. In *The principle of relativity*, (trans. W. Perrett and G.B. Jeffery), pp. 75-91. Dover, New York, 1952.
Moszkowski, A. (1921). *Einstein the searcher: his work explained from dialogues*. (trans. H.L. Brose), Methuen, London.
Moulton, F. (1905). On the evolution of the solar system. *Astrophysical Journal*, **22**, pp. 165-81.
Newcomb, S. (1902). *The stars: a study of the universe*. J. Murray, London.
Newcomb, S. (1906a). The unsolved problems of astronomy. In *Side-Lights on Astronomy. Essays and Addresses*. Harper and Brothers, New York.
Newcomb, S. (1906b). The Structure of the Universe, in *Side-Lights on Astronomy. Essays and Addresses*. Harper and Brothers, New York.
Newton, I. (1729). *Mathematical principles of natural philosophy*, (trans. A. Motte, rev. F. Cajori), Univ. of California Press, Berkeley, 1934.
Newton, I. (1730). *Opticks*, 4th edn, Dover, New York, 1952.
North, J. (1965). *The measure of the universe. A history of modern cosmology*. Clarendon Press, Oxford.
Norton, J. (1984). 'How Einstein found his field equations: 1912-1915, *Historical Studies in the Physical Sciences*, **14**, 253-316.
Norton, J. (1985). What was Einstein's principle of equivalence? *Studies in History and Philosophy of Science*, **16**, 203-46.
Olbers, H.W.M. (1826). Über die Durchsichtigkeit des Weltraumes. In *Astronomisches Jahrbuch für das Jahr 1826*, **51**, 110-21.
Pauli, W. (1921). *Theory of relativity*, 2nd edn, Pergamon, Oxford, 1958.
Planck, M. (1887). *Das Prinzip der Erhaltung der Energie*, 2nd edn, Teubner, Leipzig, 1908.
Poincaré, H. (1902). *The value of science*, (trans. G.B. Halsted), Dover, New York, 1958.
Poincaré, H. (1905). *Science and hypothesis*, (transl. W.J. Greenstreet) 1952. Dover, New York.
Rankine, W. (1852). On the reconcentration of the mechanical energy of the universe. In *Miscellaneous scientific papers* (ed. W.J. Millar). 1881, Charles Griffin and Company, London.

Rindler, W. (1977). *Essential relativity: Special, general and cosmological.* 2nd edn, J. Springer, Berlin.
Roberts, I. (1888). Photographs of the nebulae M31, h44 and h51 Andromedae, and M27 Vulpeculae. *Monthly Not. of the Roy. Astr. Soc.* **49**, 65.
Schaffer, S. (1977). Halley's atheism and the end of the world. *Notes and Records Roy. Soc. of London* **32**, 17-40.
Scheiner, J. (1898). On the spectrum of the great nebula in Andromeda. *Astrophysical journal* **9**, 149-50.
Schwarzschild, K. (1900). Ueber das zulässige Krümmungsmass des Raumes. *Vierteljahreschrift der Astron. Gesellsch.* **35**, 337.
Schwarzschild, K. (1916a). Über das Gravitationsfeld eines Massenpunktes nach der Einsteinschen Theorie. *Sitz. Ber. Preuss. Akad. Wiss.* Berlin, 189-96.
Schwarzschild, K. (1916b). Über das Gravitationsfeld einer Kugel aus inkompressibler Flüssigkeit nach der Einsteinschen Theorie, *Sitz. Ber. Preuss. Akad. Wiss.* Berlin, 424-34.
Shapley, H. (1918). Globular clusters and the structure of the galactic system. *Publ. Astr. Soc. Pacific.* **30**, 42-52.
Sklar, L. (1974). *Space, time and spacetime.* University of California Press, Berkeley.
Slipher, V.M. (1915). Spectroscopic observations of nebulae. *Popular astronomy.* **23**, 21-4.
Slipher, V.M. (1917). Nebulae. *Proc. Amer. Phil. Soc.*, **56**, 403-9.
Smith, R.W. (1982). *The expanding universe.* Cambridge University Press.
Stachel, J. (1979). The genesis of general relativity. In *Einstein symposium Berlin* (eds. H. Nelkowski *et al.*) pp. 428-42. J. Springer, Berlin.
Stachel, J. (1980). Einstein and the rigidly rotating disk. In *General relativity and gravitation: a hundred years after the birth of Einstein.* (ed. A. Held) Vol.1, pp.1-15. Plenum, New York.
Stachel, J. (1986). What a physicist can learn from the discovery of general relativity, *Proceedings of the fourth Marcel Grossmann meeting on general relativity* (ed. R. Ruffini *et al.*). Elsevier Publications, pp. 1857-62.
Synge, J. (1960). *Relativity: the general theory.* North Holland, Amsterdam.
Tolman, R.C. (1934). *Relativity, thermodynamics, and cosmology.* Clarendon Press, Oxford.
Truesdell, C. (1964). Whence the law of momentum? In *L'Aventure de la science: Mélanges Alexandre Koyré.* Vol. 1, pp. 588-612. Hermann, Paris.
Voltaire (1738). *The elements of Sir Isaac Newton's philosophy.* (trans. J. Hanna) Stephen Austen, London.
Weinberg, S. (1972). *Gravitation and cosmology.* Wiley, New York.
Weyl. H. (1918). *Space-Time-Matter.* (transl. H. Brose of the 4th German edition) 1952. Dover, New York.
Weyl. H. (1949). *Philosophy of mathematics and natural science.* (transl. O. Helmer) Princeton University Press.
Wheeler, J.A. (1962). Discussion of McVittie, Cosmology and the Interpretation of Astronomical Data. In *Les théories relativistes de la gravitation.* (eds. A. Lichnérowicz and M.A. Tonnelat), pp. 269-73. CNRS, Paris.

2
From a universal physics to a physics of the universe

1. Introduction

In textbooks of modern cosmology, and histories of the subject, the name of Willem De Sitter occurs in relatively few contexts. He is usually associated with the discovery, in 1917, of a very strange and unexpected solution of Einstein's field equations: an *empty* universe need not have the metric of Minkowski flat space-time (see for instance P.J.E. Peebles 1980a, p. 4, and G.J. Whitrow 1980, p. 284). This result is itself a reaction to Einstein's announcement, earlier in the same year, of a metric for the first relativistic model of the whole universe—the so-called cylindrical universe (Einstein 1917b). Current literature also mentions an Einstein–De Sitter model of the universe, which has nothing to do with either of these solutions. It refers to a model which De Sitter constructed together with Einstein some fifteen years later, when the rediscovery of non-statical solutions made it clear that the statical solutions of 1917 were particular cases of a much wider class of solutions. In fact, the Einstein–De Sitter model is known for yielding the simplest of all non-statical cases: its geometry is an expanding Euclidean space, relieved of complicated relations among available variables. This simplicity has made it appropriate to the study of these more complicated relations, such as the occurrence of terms representing presssure, Interestingly enough, the De Sitter model of 1917 is also renowned for its remarkable simplicity in terms of symmetry; thus, in the words of Synge (1960, p. 266), "the flat space-time of Minkowski and the De Sitter universe are as symmetric as space-time can be". This feature makes it even possible to start with the De Sitter universe as the paradigmatic case of fields with spherical symmetry, and so a metric like Schwarzschild's may be derived from it.

Historical studies of contemporary cosmology also draw attention to the fact that the role of De Sitter was not restricted to these two contributions. A very decisive role was played by De Sitter's early critique of some of the most fundamental philosophical arguments Einstein had put forward when he constructed the theory of general relativity, prior to the emergence of cosmology. These studies suggest, indeed, that Einstein developed his cosmological ideas as a response to De Sitter, who had expressed objections to Einstein's views on the problem of relativity of rotation and origin of inertia. Thus, after having commented on Einstein's first cosmological memoir, J. Merleau-Ponty writes: "It was, in part, to take up De Sitter's challenge that Einstein had sought with so much obstinacy to find solutions to the field-equations which would be compatible with the principle of the relativity of inertia". Merleau-Ponty says further (1965, p. 52) that De Sitter was convinced right from the outset that such solutions were an "impossibility", a conclusion which Einstein eventually reached. Similarly, albeit already within an analysis of Einstein's later cosmological memoir, J. North hints that it was from De Sitter that Einstein took the idea "that one must. . .refrain from asserting boundary conditions of general validity" (1965, p. 72). Neither historian, however, dwells at length upon the alleged influence: they do not demonstrate its existence, nor do they attend to the substance of De Sitter's critique. On the whole, the impression gained from these histories of contemporary cosmology is that De Sitter's on-going objections to the cosmology were merely repetitions of his earlier criticisms of some of Einstein's philosophical positions. But nothing really is said about the substance of these criticisms in their original form.

At the other extreme, the literature dealing directly with these early philosophical contentions of Einstein portrays them against a background much broader than the cosmological dispute. This has the effect of minimizing De Sitter's role—for he is seen as contributing exclusively to the cosmological aspects of the story. (The most exhaustive studies are A. Pais 1982, pp. 281-8, and R. Torretti 1983, pp. 194-202.)

In the present chapter, I wish to re-examine De Sitter's place in the history of early general relativity, looking more closely than has previously been done at De Sitter's criticisms. The validity of the above as yet little-supported claims can thus be tested, and the crucial nature of De Sitter's influence on Einstein's path to cosmology can be definitively assessed. In the first instance, this will help complete the historical record. More importantly, the remarkable aspect of De Sitter's early criticisms which led to Einstein's cosmology, is that they emphasize a decisive epistemological feature of general relativity; while the latter ones which he directed against the already constituted cosmology, tended to shift the debate to more technical points. Neglect of the former has obscured the radicality and profundity of De Sitter's thought. We pass over in silence what was in effect the source of an invention. Thus, attention to these early criticisms, through a more careful historical appraisal, should provide a shrewder perception of the philosophical nature of the cosmological problem in these first, decisive years: cosmology will itself appear as part of a philosophical endeavour, a response to a tension perceived to pervade the newly established theory of general relativity.

INTRODUCTION

This historical discussion will focus on a series of published papers by De Sitter which have up till now been largely overlooked or simply ignored. These are papers which take on great significance when related to parts of the large body of the as yet unpublished correspondence between Einstein and De Sitter. This correspondence, which deals with both administrative and scientific matters, has only recently been discovered. (The story is narrated by C. Kahn and F. Kahn 1975. On De Sitter's life and thought, see C.H. Hins 1935, J.H. Oort 1935, and H. Spencer-Jones 1935.) Though gaps and references in this correspondence suggest it is seriously incomplete, enough remains to shed significant light on our problem. We will concentrate here on the short period which goes from the fall of 1916 up to the publication of Einstein's model of the universe on 8 February 1917.

This begins with De Sitter's first critical remarks on the ill-understood problem of the relativity of rotation. During 1916, Einstein and De Sitter met on several occasions in Leiden, together with P. Ehrenfest and H.A. Lorentz. (This phase of Einstein's career, in relation to the usually little known influence of Ehrenfest which I will develop later, is discussed by M.J. Klein 1970, pp. 193-216 and 293-323.) Because of the war, one of the reasons behind Einstein's making contact with De Sitter was to get his new theory known in England (The Netherlands maintained neutrality). De Sitter finally published two long articles on the theory in *Monthly Notices of the Royal Astronomical Society* (1916a and 1916c). Eddington, who was then the Secretary of the Society, had some difficulties in accepting the papers because of anti-war sentiments and was therefore "...interested to hear [from De Sitter] that so fine a thinker as Einstein is anti-Prussian" (Eddington to De Sitter, 13 October 1916). In a postcard of early January 1917, Einstein begins by praising De Sitter for his capacity "to throw a bridge over the abyss of misunderstanding". Apart from the wish to present a faithful exposition of Einstein's new theory, De Sitter began to delineate in September 1916 a critical account of his own, in which a highly original analysis of the problem of the relativity of rotation was offered (W. De Sitter 1916b; this was communicated on September 30, while 1916c was completed in September/October and published in the December issue of the *Monthly Notices*).

This critical problem, and with it that of the relativity of inertia, was fundamental to what Einstein has called his "indirect" and "bumpy" road to cosmology. Einstein had submitted to the *Annalen der Physik*, on 20 March, his memoir where the aims, methods and major results of general relativity were synthesized for the first time. The memoir was published in the October issue of this journal (A. Einstein, 1916). Einstein tackled there the whole issue of rotation after he had made numerous groping attempts in this direction in the two years preceding this publication. This he did on epistemological grounds in order to justify the very need for a relativity principle larger than that employed in the special theory. He came up, in the introductory part, with a thought experiment which he believed to be a fairly direct illustration of Mach's reflections on rotation. J. North rightly discusses the paramount importance of this epistemological justification, and he does so by referring to Einstein's concern with developing a satisfactory version of the general covariance principle (1965, p. 57). Making rotation relative may be thought to be identical with the problem of extending the idea of relativity to cover all motions, and not just the linear ones. In fact, Einstein derived the principle of general covariance (which he wrote in 1916 as asserting that the form of physical laws must be such that these laws only concern the intersection of world-lines, that is, the space–time coincidences) as a sufficient condition of the principle of general relativity. De Sitter, however, perceived how questionable it is to ascribe the possibility of this derivation to the relativity of rotation.

A little clarification with hindsight is certainly desirable, before we can embark upon a detailed analysis of the arguments. The most exhaustive commentary on this thought experiment is to be found in a more recent book which is, in fact, a most remarkable account of all the principles put forward by Einstein in the crucial years 1905–1916 (M. Friedman 1983, pp. 66-8 and 205-8). According to this account, most of Einstein's original scientific ambitions were enmeshed with philosophical motives, and this is the reason why these ambitions have generally failed, at least in their original form: "The present case is unique because of the extent to which philosophical motivations became entangled with both the new mathematics and the new physics—so thoroughly as to obscure the true relationship between the mathematics and the physics for half a century" (p. 212). Only today are we in a position to proceed to a fairly definitive distinction between the objects, the laws, and the formulations of general relativity. The peculiarity of this theory, as opposed to other views on space–time, is that these three notions are not interchangeable. For instance, the requirement of general covariance pertains merely to the formulation, and has nothing to do with the laws or the objects, contrary to what Einstein seems to believe in 1916 (pp. 213-14). The thought experiment under consideration is certainly a good example of his philosophical motivation; such various requirements as the general relativity principle, the general covariance principle, and what has become known as Mach's principle are interwoven. Be this as it may, it was precisely in trying to unravel this particular philosophical puzzle that De Sitter gave Einstein the push which made him point out the need for a cosmological picture in general relativity. For this reason, it certainly makes sense to look at the thought experiment independently of Einstein's own, later reservations which focus on the status of 'Mach's principle'. It appears that Einstein came to regard this principle as a particular form of the boundary conditions for the differential equations of gravitation. But the physical meaning of boundary conditions opens up those cosmological questions which—as De Sitter was to

INTRODUCTION

understand—reflect the philosophical foundations of general relativity.

True, Einstein's capacity to turn his attention to boundary conditions as a *physical* problem should not be taken for granted. It could be said that Einstein's establishment of his general relativity field equations represents the solution to a *mathematical* problem, in which the positing of some well-defined boundary conditions was not to be seriously questioned. In fact, the whole idea of reducing fields of force to a geometrical theory involved a formidable problem because an unprecedented amount of interaction occurred between the physical and the mathematical. But Einstein's laborious way out of it (with the help of the mathematician Grossmann, in the crucial years 1912–1915) also seemed to be dictated by the search for a somewhat classical type of resolution.

In a lecture which he delivered much later (1934, p. 83) Einstein recalled the twofold nature of his difficulties:

1. If a field-law is given in the terminology of the special theory of relativity, how can it be transferred to the case of a Riemannian metric?
2. What are the differential laws which determine the Riemannian metric (i.e. the $g_{\mu\nu}$) itself?

The two questions are determined by one problem whose concise form is: What are the $g_{\mu\nu}$ which represent a given field of gravitation, i.e. what is the physical meaning of the Riemannian geometry? Geometry being intrinsic, the choice of $g_{\mu\nu}$ is not arbitrary. We know that Einstein used the classical idea, that no field of gravitation can exist at an infinite distance of all masses, in order to answer the basic question of which type of space–time is at all possible in nature. This forms the basis of Einstein's answer to the first problem. General relativity is an extension (as he says, a sort of transfer) from special relativity. The second problem is the formulation of differential laws themselves: it is asserted that this pseudo-Euclidean space–time also exists in infinitesimal portions of the world, that is, in the sense of the principle of equivalence which warrants the local identity between a field of gravitation and a field of acceleration. Accordingly, the intrinsic geometry of nature is seen as Riemannian, firstly because only that geometry is compatible with a local reduction to pseudo-Euclidean space–time, and secondly because the presence of matter can be associated, *via* the appropriate spelling of the material tensor, with the warping of this pseudo-Euclidean continuum. But the idea of general covariance impeded Einstein's desire to conceive the new laws as an extension of the old ones. As he confesses, Grossmann and he had discovered the correct field equations quite early on but "we were unable to see how they could be used in physics". Indeed, Einstein adds, "I felt sure that they could not do justice to experience" (1934, pp. 83–4. See an exhaustive analysis of the problem in J. Earman and C. Glymour 1978). Once the extension was seen as complying with physics, general covariance no longer had a dominant role to play. In 1918, Einstein explained at length how general covariance must be seen as a natural expression (*"Ausdruck"*) of the principle of general relativity; only the latter is related to the objective aspect of the laws of Nature, namely, the space–time coincidences (1918b, pp. 241–2).

At this point, before autumn 1916, Einstein may have been satisfied with the new picture, since it fulfilled his hope of *enlarging* Newton's physics rather than *replacing* it with an entirely novel one; the Newtonian laws had to be seen as approximations of the new ones, just as the general principle of relativity looked like a mere extension of the special principle. However classical they may have been, boundary conditions could not really pose a physical problem for the new theory, since their application not only provided the definitive form of the field equations, but also an exact solution of these equations (the static isotropic gravitational field of a mass point like the sun). At this stage of research the classical form of boundary conditions was found happily to have no connection with the genuine originality of general relativity. So much the better for Einstein's comfort in his use of classical precedents. But things quickly changed. In early February of the following year, Einstein could open his new memoir by saying that, by *analogy* with Newton's theory, the *physical* nature of boundary conditions was a "fundamentally important question" (1917b, p. 177).

This is quite simply a complete reversal of the earlier attitude which had dictated discovery of the field equations. The progression, which led from a formally universal physics (a new physics, which none the less embraced the old theory as a particular case) to the revolutionary concept of a physics which would comprehend the entire universe, brought with it a fair share of traps for Einstein to fall into. His insight was determined by De Sitter's systematic opposition to the last residue of classicism in general relativity; this revealed the existence of something very new.

2. Einstein's 1916 thought experiment and the problem of relative rotation

(a) The philosophical background

In his recollections of how he actually established the laws of general relativity, Einstein mentions an additional difficulty among his "errors of thought which cost [him] two years of excessively hard work". The error is the following: "I believed that I could show on general considerations that a law of gravitation invariant in relation to any transformation of coordinates whatever was inconsistent with the principle of causation" (1934, p. 84). Thus, apart from the problem of connecting general covariance with the principle of general relativity, which Einstein sought to solve in terms of an appropriate understanding of the relativity of rotation, the new law of gravitation seemed to violate causality. It was in his 1918 review article that

Einstein clearly stated what is a fundamental consequence of causality in general relativity: this is what he calls 'Mach's principle', to the effect that "the R-field is *completely* determined (*'restlos bestimmt'*) by the masses of bodies" (1918b, p. 241). The word 'completely' here refers to the prescription that gravitation should be a mechanism by which the *instantaneous* distribution of matter determines the inertial properties of every test particle *everywhere*. The inconsistency Einstein has in mind seems to be that the existence of light-cones, in accordance with the special theory of relativity, is hardly compatible with the idea of 'everywhere' on a spacelike hypersurface. (It was only later that Einstein began to abandon the Machian demand: the pressure came as much from cosmology as from a growing faith in a unified field theory, according to which matter was to be interpreted as a particular configuration of the field rather than as the cause of the field.)

Thus, it appears that Einstein's early considerations on both the form of the new law of gravitation and its causal implications arise from a reflection on the work of Mach. There is no doubt that Mach's classic *Science of Mechanics*, published late in the nineteenth century, greatly influenced Einstein on the way from the special to the general theory. (For a full proof of this influence, see G. Holton 1973, pp. 219ff.) It was not Mach's programme to build a new physics altogether, but only to spell out some of the fundamental presuppositions of Newton's physics, and to lay bare a correct interpretation of its foundations. In order to understand how Einstein could perceive this critique as the conceptual basis of his new physics *as well*, I shall develop the interactions between Mach and Einstein in such a way that the two problems, relative rotation and causality, will appear closely related to one another.

Given Mach's influence on Einstein with respect to the problem of causality, we need to take a broad historical perspective, completing what was outlined in Chapter 1, if we want to come to terms with the reason why the problem of relativity of rotation was so fundamentally important for the transition from general relativity to cosmology. The problem can be readily traced back to thought experiments cited by Newton early in the *Principia* to support his theory of absolute space, and the attack on this theory mounted by Leibniz in his celebrated correspondence with Newton's disciple, Samuel Clarke. Leibniz there advocated a relational theory of space–time, though he hardly answered the specific arguments in favour of the reality of absolute rotation which Newton had invented. Instead he relied on general metaphysical insights, particularly the principles of sufficient reason and of the identity of indiscernibles, borrowed from a theological framework. For various reasons, some of them very revealing, his objections failed to halt the spread of the Newtonian concepts, and the latter became the commonly accepted view until the era of Mach and Einstein. There are some indications, however, that Einstein himself drew on Leibnizian concepts to aid his break with the Newtonian tradition. In the address which he delivered in 1918 at the

Physical Society in Berlin, on the occasion of Max Planck's sixtieth birthday, Einstein stressed his tribute to Leibniz by recalling that "no logical path" exists "between phenomena and their theoretical principles". This absence of bridge, as he said in a way which was also clear reference to Minkowski's programme, was so happily described by Leibniz "as a 'pre-established harmony'" (Einstein 1918a, p.4). In his critical approach to the problem of relative rotation, as we will see, Einstein also cites the principle of sufficient reason as an explicit way out of some apparently insuperable difficulties with the Newtonian theory.

Of all the problems that philosophical debates on classical physics have bequeathed to the foundations of relativity theories, that over the status of rotation has certainly remained essential. Rotation is, in fact, the paradigm of absolute motion in Newton's mechanics. The whole picture of the universe which rests upon this mechanics depends heavily on the validity of arguments for the reality of absolute rotation. (This is particularly patent in *Clarke's Fifth Reply*: See the Leibniz-Clarke correspondence edited by H.G. Alexander 1956, §§ 26-32, pp. 100-102.) Though Newton promoted his arguments as detecting absolute motions in general, they really served only to distinguish relative from absolute rotation. If a body is rotating with respect to a reference system where both the law of inertia and the fundamental equations of dynamics hold, centrifugal forces then appear; if a body is at rest with respect to this system, these forces disappear. The presence of such forces is thus a criterion for the state of rotation of a body. This argument provides no grounds at all for distinguishing a linear absolute velocity from a relative one, nor for distinguishing a linear absolute acceleration from a relative one. But they do mount a serious challenge for any claim that *all* motions are relative.

That Leibniz felt this challenge is clear. When Clarke puts forward this Newtonian argument (*Clarke's Fourth Reply*, § 13, p. 48), Leibniz is obliged to acknowledge that absolute motion is indeed dynamically possible. But if Leibniz grants that "there is difference between an absolute true motion of a body, and a mere relative change of its situation with respect to another body" (*Leibniz's Fifth Paper*, §53, p. 74), he invokes anti-Newtonian reasons in order to establish the difference: "For when the immediate cause of the change is in the body, that body is truly in motion; and then the situation of other bodies, with respect to it, will be changed consequently, though the cause of that change be not in them" (p. 74). Leibniz's counter-argument is thus still rooted in the distinction between absolute and relative motion. For example, Leibniz could well have concluded from the flattening at the Earth's poles that the immediate cause of change is intrinsic to the earth and has nothing to do with the 'fixed' stars. In this sense, such rotation ought to be called 'true absolute motion', whereas that of the stars is a 'simple relative change in situation'.

But Leibniz's point of view actually implements a dynamics very different from the Newtonian. A true motion is indeed a dynamic process, but this process is *not spatial*—every process in space is relative. In particular, inertial forces (such as centrifugal forces) are the *cause* rather than the effect of motion in space (see L. Sklar (1974, pp. 191-2). Two different kinds of motion are therefore to be distinguished: one is motion as a process in space, the other as a process interior to things. It is only the latter that should enable the observer to discriminate between kinematically equivalent motions. True, the whole ambiguity of Leibniz's dynamics is demonstrably pointed out by Clarke. He puts forward the objection that the fact of motion cannot be accounted for by sole observation: this is the example of a man shut up in the cabin of a ship (*Clarke's Fourth Reply*, § 13, p. 48). To this, Leibniz can do no more than reply: "motion does not indeed depend upon being observed; but it does depend upon being possible to be observed. There is no motion, when there is no change that can be observed. And when there is no change that can be observed, there is no change at all" (*Leibniz's Fifth Paper*, § 52, p. 74). In other words, the two conceptions of motion are far from being immediately distinguishable: in the case of motion, the possibility of being an internal process already depends on its possibility of being observed. This holds for the *possibility*, but the *nature* of the internal process cannot be understood in terms of this space in which all observations take place.

The crucial question is then to be found in *Clarke's Fifth Reply* (§ § 26-32, p. 101): what about this apparently absurd consequence derived from Leibniz's ideas, namely, the disappearance of all centrifugal force in the rotating sun when all matter around the Sun ceases to exist? From the internal point of view, there is no drawback in conceiving a unique, rotating sun with centrifugal forces. But the observability criterion does not allow this: it clearly signifies that there is no space devoid of material bodies, so that the motion of an isolated body is a mere fiction. For this single body, motion as process in space loses all meaning, yet no conclusion can be drawn as to its dynamic state. This particular difficulty is removed in Newton's dynamics, where space is not affected by the fiction of the isolated body: it is by "definition" that "absolute space, in its own nature, without relation to anything external, remains always similar and immovable" (Newton 1729, p. 6).

Throughout the eighteenth and nineteenth centuries, numerous attempts were made to understand the foundations of Newton's absolute space in terms that would make it meet with some of the (appropriately reinterpreted) objections raised by Leibniz. Significant episodes are the attempts of Berkeley and Kant. But it seems that only Mach succeeded in defining the nature of an internal dynamic state precisely in the rigorous sense of Leibniz's observability criterion. This is the substance of his arguments in the *Science of Mechanics*. Mach disregards all intrinsic differences between internal and external forces, granted that the origin of centrifugal force can be attributed to bodies other than those subjected to motion. For instance, in the case of flattening at the poles, fixed stars can be deemed the cause of centrifugal forces on the Earth (Mach 1883, pp. 229 ff). This tends to have the double effect of ruling out all non-spatial processes from the scope of physical science *and* of fulfilling Leibniz's desire to understand all spatial processes in terms of relative concepts. Having suppressed the non-spatial processes, the problem which crops up is to know whether the observability criterion can also entirely free the forces from a relation *to* space. How can this criterion ascertain that only relational space is meaningful?

(b) A first version of Einstein's philosophical argument

This, in fact, is the problem taken up by Einstein. The problematical status of the 'Leibnizian' relativity versus Mach's contribution which somehow 'solves' it, remain the two poles of Einstein's ideas on rotation. Two years before the synthesis of 1916, the issue is tackled in a palpably different text, dealing with some philosophical implications of general relativity. In a patently Leibnizian vein, Einstein aims at showing that any non-Machian theory "gives up a sufficient reason" (Einstein 1914, p. 346). By Machian theory, Einstein understands something he has had in mind ever since 1912, that "the *total* inertia of a mass point is an effect due to the presence of all other masses [in the universe] due to a sort of interaction with the latter" (1912, p. 37). In 1913, he calls it the "hypothesis of the relativity of inertia" (1913, p. 1261). In fact, Einstein wrote an important letter to Mach on the 25th June 1913, wherein he spoke of Mach's "happy investigations on the foundations of mechanics" as leading to the idea "that *inertia* originates in a kind of *interaction* between bodies" (this letter is quoted and translated in Misner, Thorne, and Wheeler 1973, pp. 544-5). This interaction, "*Wechselwirkung*", is a word which all often reappear when Einstein sets pen to paper throughout these years. It is not until his 1918 paper, quoted above, that Einstein realizes this concept is not a priori encompassed by the relativity principle: indeed, it is only then that he denotes it by the name "Mach's principle". There is yet a third principle which, he finds, cannot be derived from these two, and that is the principle of equivalence. He thus seems to have finally conceived the principle of relativity as the formal basis which, through its alleged universality, brings about the connection between a local and a global requirement. Note also that, apart from not saying anything about the nature of the interaction postulated by Mach's principle, Einstein does not specify whether the influence of all matter of the universe is exerted in the total history of the universe or just at the present time, whether it occurs at all times, at one time, or at all past times. This separation of space from time seems to be yet another instance of Einstein's assiduity in incorporating a rather classical world view into general relativity.

In his text of 1914, Einstein begins by stressing the demarcation between the special and the general theory of relativity (1914, p. 337). Until now, he claims, no physical experiment has illustrated the latter. In fact, Einstein was awaiting confirmation from a forthcoming eclipse, as he wrote explicitly to Mach. This situation made it desirable to strengthen the plausibility of general relativity on the basis of purely philosophical argumentation. The aim of the paper is to show that general relativity is not led to abandon the previous theory of relativity; quite the contrary, the new theory is a development of the old one, "a development which seems to me necessary from the philosophical standpoint" (p. 348). We know today that this hope is for the most part illusory: the principle of general relativity cannot be regarded as a straightforward extension of the principle of special relativity, unless a minimum of conditions are spelled out. What we now know is that general relativity is not a fully relationist theory: the space-time structure, for instance, is responsible for a certain number of effects that are not primarily due to the relations between concrete physical objects and events. This calls for a distinction between different aspects of the theory, which Einstein has conflated in his early treatment (see M. Friedman 1983, pp. 204-14). Thus, if the space-time structure is interpreted as being non-rigid, i.e. if the theory is said to have no 'absolute' object, then the new principle can be seen as a natural generalization of the former. But this statement on the objects of the theory has nothing to do with the demand of relativity of motion (that is, with the groups of transformations which characterize the laws of this theory), and even less with its formulation (general covariance). Conflating them leads to these confusions and obscurities which are seen as resulting from Einstein's overarching philosophical commitment. But part of the reason that led Einstein to believe his new theory was all-encompassing, that it realized a fully relationist theory, was his ardour for developing the very first cosmological model. Because cosmology has survived this later disillusionment, it thus remains to assess almost inversely Einstein's actual philosophical commitments *from within* his apparently 'wrong' interpretation of the formalism. In the remainder of this section, I shall attempt to reconstitute the various principles underlying the move to general relativity in the light of some of Einstein's profound motivations.

Let us thus see how the point of view of an extension is expounded. Einstein tackles the example of any two masses, very remote from all other celestial bodies. These masses are sufficiently close to each other so as to exert mutual interactions: "An observer follows the motion of the two bodies by pointing constantly to the sky in the direction of the line joining the two masses. He will see the line of sight to draw in the sky a closed line which will not change place with respect to the visible stars" (1914, pp. 344-5). From this observation, a theoretician steeped in Newtonian mechanics would conclude that the two masses define an inertial system. Nevertheless, a naive observer who knows nothing of geometry and mechanics might well arrive at a different conclusion: the motions of the masses are caused (at least partly) by the fixed stars, so that it is also the stars that govern these motions. The naive observer ascribes the causes of the phenomena of motion to that which is directly observable. To him, the 'scientist' would reply that not only the motions, but the laws of mechanics themselves as they apply to the masses are quite independent of the fixed stars. According to these laws, some given space (R) must exist in which the masses can move within a plane. Einstein goes on: "But the systems of fixed stars cannot have a rotation in that space, since it would be disrupted by enormous centrifugal forces. That is why it must necessarily be (almost) at rest if it is to endure". Observation shows that the relative velocities of the stars are indeed quite small. Moreover, neither the naive observer nor the scientist would ever deny that the two masses are truly in motion with respect to the stars—that is an indisputable fact of observation. Yet, the Newtonian argument is concerned with the search for a necessity underlying the capacity of fixed stars to form an inertial system. It is only if the system of fixed stars *is to endure* that this system can be given a purely geometrical function, deprived of mechanical effects. But the fact that the system endures has a merely accidental, not a necessary, status. The naive observer objects that the supposedly universal and necessary form of the laws of nature cannot be dependent upon any mere accident. What the Newtonian means by space, he says, is something "that I cannot see nor can I figure it". What would happen if it were to rotate? Would this rotation have to be distinguished from the rotation of the stars in space? What if space is represented by "a very subtle network of bodies" and the relation of other bodies to space by a relation to that network:

Then I will be in a position to figure, besides R, another network R' with any kind of motion relative to R (for instance, a rotating motion). Will, then, your equations be valid also relative to R'? To this question the learned man replies with assurance in the negative. So the ignorant man asks anew: How come that your masses know relative to which "spaces" R, R', etc., they are to move in accordance with your laws?... The learned man is now quite perplexed. He insists on the reality of such privileged spaces, but he is unable to give a reason why these spaces would be privileged (p.345).

The bodily networks R, R', . . . are bodies whose motions lack the ordinary effects of motion relative to space. This subtlety makes it possible to isolate what is supposedly the purely geometrical role of an inertial system. And what the argument tends to show is that no purely geometrical role of that kind is consistent with the possibility of space being *seen* (or *figured*). And when space is seen, the privilege of R comes to be at variance with the supposed existence of spaces such as R'. Newton himself argued that the *whole* of space could not be seen. In the famous Scholium following Definition VIII of the *Principia*, he was speaking of all things being placed *in* space, and it was at that point that he offered an argument of pure intelligibility: "... that the primary places of things should be movable, is absurd". Indeed, "suppose those parts to be moved out of their places, and they will be moved (if the expression may be allowed) out of themselves". Not only the whole of space, but also "the parts of space cannot be seen". That is why, in Newton's eyes, space in one of those "things themselves, distinct from what are only sensible measures of them" (1729, p. 8). A sensible measure of space is not to be confused with any kind of 'visible' space.

Einstein's example is intended both to enlarge the Newtonian principle of relativity and to preclude recourse to pure intelligibility. The equations of gravitation ought to be formulated in such a way that their form is always identical in all spaces $R, R', \ldots$ This is the strictly mathematical problem whose solution is concomitant with final establishment of the field equations of general relativity. According to the new equations, no physical quality belongs to certain privileged spaces. Yet, pure intelligibility (in the Newtonian sense) cannot be dispensed with quite so easily. The fact is that the multiplicity of spaces $R, R', \ldots$ remains, as it were, a conceptual possibility. The reality of space as a "subtle network of bodies" is not made explicit in terms of the physical relations between systems (such as the fixed stars and the two masses); it is all too obvious that this network is itself *ideally* immune to large-scale physical effects. In fact, Einstein has replaced a conception of space as object with one of space as pure structure. But on this account equally possible structures may exist. (Note also that the question of why the whole endures is not answered because it simply does not arise once Einstein's network plays its part in superseding Newton's absolute space.) In this respect, Newton is consistently speaking of a single, non-tangible space inasmuch as physics describes a single, given universe. Einstein's geometrical physics, on the other hand, implies a formidable epistemological task: how are we to secure the uniqueness (and by doing so the actual existence) of whatever materialized space happens to constitute our universe, without jeopardizing the new level of generality involved in the laws? When it comes to a consideration of the 'whole', physics, contrary to geometry, is not concerned with the problem of why other facts might just as easily be the case. That was the challenge to be taken up in the introduction to the 1916 memoir.

(c) The 1916 version

A clear statement of what Einstein really understands by 'space of reference' is to be found in the first of his four lectures delivered at Princeton. The bodily network is now the possibility to "form new bodies by bringing bodies B, C, . . . up to body A; we say that we *continue* body A. We can continue body A in such a way that it comes into contact with any other body, X. The ensemble of all continuations of body A we can designate as the 'space of the body A'... In this sense, we cannot speak of space in the abstract, but only of the 'space belonging to a body A'" (1921, p. 3). We have to bear this statement in mind when we turn to the new thought experiment of 1916. The idea of continuation does not imply here the sense of 'motion', except perhaps at an intuitive level.

The thought experiment sounds curious and really baffling (1916, pp. 112–13. Some comments on the thought experiment can be found in Kopff 1921; Born 1962, pp. 309–10; Pauli 1958, p. 165; Mehra 1973, p. 123; M. Friedman 1983, pp. 66–8 and 205–8):

Two fluid bodies of the same size and nature hover freely in space at so great a distance from each other and from all other masses that only those gravitational forces need be taken into account which arise from the interaction of different parts of the same body.

The situation described here is very different from the previous one, since no interaction between bodies seems to occur. Einstein posits further that the distance between the two bodies is invariable, and

in neither of the bodies let there be any relative movements of the parts with respect to one another.

Every mass is thus in equilibrium under the action of gravitation of each of its parts on the other and under the action of other physical forces as well. But

let either mass, as judged by an observer at rest relatively to the other mass, rotate with constant angular velocity about the line joining the masses... Now let us imagine that each of the bodies has been surveyed by means of measuring instruments at rest relatively to itself, and let the surface of S_1 prove to be a sphere, and that of S_2 an ellipsoid of revolution.

Einstein speaks here of a *verifiable* ('*konstatierbar*') relative motion (see Fig. 2.1), even though the experiment itself and its results are totally imaginary. The difference being observable, a reason for it may be sought. Non-inertial motions are precisely those which are concomitant with the non-symmetrical, differential effects produced by accelerations and rotations. Note that the spaces of reference in the example under consideration can be identified with the measuring instruments which come into contact with the bodies.

The propounder of Newtonian mechanics would answer that the laws of mechanics apply to the space of reference of the first body (R_1) but not to that of the second body (R_2), even though S_1 and S_2 are both at rest relative to their own space of reference. Were this the case, indeed, only the space of refer-

FIG. 2.1. Einstein's 1916 thought experiment with two rotating fluid bodies S_1 and S_2.

ence of the first body would be identified with the Galilean continuum of space and time, that is, with the privileged continuum in which the form of Newtonian laws remains invariant. S_1 is at rest in absolute space, whereas S_2 is rotating in that space. In other words, the only possible explanation in the framework of Newtonian laws implies that the observer will not regard himself at rest *only* with respect to the body he has been surveying by means of measuring instruments. For instance, the observer lies on the line joining S_1 and S_2, and then records the shapes of the bodies. A situation such as represented in Fig. 2.2 is impossible for him, because it has been verified that there is rotating, non-inertial motion and thus forces. *At least one* of the bodies, therefore, has to become ellipsoidal. On the other hand, *at most one* may be ellipsoidal, the interaction between the two masses having been excluded in the hypothesis. There remains a space both independent of the spaces of reference and capable of correlation with the causes of physical effects.

FIG. 2.2. An impossible alternative to Einstein's 1916 thought experiment.

Ever since the time of Mach's critique, the nature of forces and of absolute space in Newton's mechanics have been viewed in a way which is certainly instrumental in obfuscating Newton's original ideas. It is, of course, on the basis of this critique that Einstein perceives his own problem. But the Newtonian conception in the strict sense seems to run as follows. In contrast with kinematic properties, impressed forces like the gravitational forces upon bodies are the true cause of motion and generate it. However, Newton is cautious enough not to *simply* identify the force of gravitation with a true cause: rather, the force is our name for the true cause, whatever that may be. That is the reason why, according to Newton, the force of gravitation has already by itself much of a purely 'mathematical' quantity. This is even more manifest when we pass to the effects of motion. These effects enable the *vis insita* (the force inherent in a body) to occur, such as "the forces of receding from the axis of circular motion" (the centrifugal forces) (1729, p. 9). The *vis insita* is to be measured relative to absolute space. In this sense, the sensible measure of this force is of a particular kind, because absolute space seems to be directly responsible for its occurrence. This fact forms the basis of Mach's guiding idea: as we have no other indication of the existence of Newton's absolute space than centrifugal forces, Newton is supporting the hypothesis of absolute space simply by the fact that it was introduced as an explanation. Thus, if absolute space is taken in a truly physical sense, it comes into conflict with the law of action and reaction, because it exerts actions upon matter and suffers none. This type of opinion is strongly emphasized by the authority of H. Weyl, for instance (see 1949, p. 105). Or Einstein himself (1921, p. 62): "it is contrary to the mode of thinking in science to conceive of a thing...which acts itself, but which cannot be acted upon". However, if we are to follow Newton himself, space cannot be said to exert any effect and has certainly nothing to do with a *vis impressa*. For when its effects are measured, "the measured quantities themselves are meant", as Newton is painstaking in claiming. They have something of a 'purely mathematical' character, as does space itself.

Now, Einstein's thought experiment involves such constraints that it makes unavoidable the appeal to absolute space as cause. Indeed, in the case considered, nothing enables us to speak of motion for this or that body, unless the two spaces of reference are beforehand discriminated in respect to their effects on the shape of the bodies. Without such discrimination, S_2 could well be an *ellipsoid of revolution in absolute rest*. That is why Einstein is entitled to name the space of reference, R_1, the "merely *factitious* cause" of the difference between S_1 and S_2 (the original German is very strong: *"fingiert"*): this cause is not a thing that can be observed. Yet

No answer can be admitted as epistemologically satisfactory, unless the reason given is an *observable fact of experience*. The law of causality has not the significance of a statement as to the world of experience, except when *observable facts* ultimately appear as causes and effects.

That is why the fiction of an absolute space is as unsatisfactory as the fiction of a universe deprived of large-scale mechanical effects. The above epistemological requirement of observability, which is concomitant with the strict verifiability of the shapes of the bodies as dependent on their motion, can be satisfied only when the system S_1S_2 is reintroduced in the real universe. The system S_1S_2 offers by itself no explanation, and Fig. 2.2 could equally be 'observable' in an otherwise amorphous universe.

Einstein is thus led to see the answer in Mach's theory. The cause must lie *outside* the system S_1S_2 and the spaces of reference:

We have to take it that the general laws of motion, which in particular determine the shapes of S_1 and S_2, must be such that the mechanical behaviour of S_1 and S_2 is partly conditioned, in quite essential respects, by distant masses which we have not included in the system under consideration. These distant masses and their motions relative to S_1 and S_2 must then be regarded as the seat of the causes (which must be susceptible to observation) of the different behaviour of our two bodies S_1 and S_2. They take over the role of the factitious cause R_1.

This answer points to the difficulty in the premises that have been adopted. Two assumptions were involved:

1. There is no interaction between the two bodies under consideration.
2. These two bodies have no interaction with other masses of the universe.

Premise 1. can be called *local*, and 2., *global*. These are typically Newtonian assumptions, because inertia is disconnected from gravitational influence. Now, the thought experiment reveals a situation which is inexplicable in terms of this set. In order to remove the difficulty, Einstein maintains 1. and drops 2. The reason for this is the epistemological law of causality: the system S_1S_2 is the only thing in the experiment which need be observed directly, whereas distant matter need not be observed. In this sense, it is assumed that a wrong conjecture on 1. was absolutely impossible, whereas a mistake on 2. was unwittingly alluring. As a matter of fact, both 1. and 2. taken together do not make *necessary* what is *observable*—the differing behaviour within the system S_1S_2 *and* distant matter itself. Had Einstein dropped premise 1. and kept 2., the existence of a materialized space (distant matter) would not follow from an observed fact which is supposed to be verifiable strictly *within* a local system. He would have come back to the case he was exploring in 1914, where the effective existence and the action of distant matter were not combined.

Ambiguity is here the very strength of Einstein's reasoning. Actually, with these distant masses, the whole problem is to start over again on a new basis. If the system S_1S_2 is suddenly included in the action of some distant masses, the question arises whether the differing behaviour of S_1 and S_2 remains, in the same sense as previously, a phenomenon susceptible to empirical verification. This is extremely doubtful. In order to conceive of the system S_1S_2 as a system whose motion is linked to forces, distant matter must already play the role of, at least, a geometrical reference. Without this, the rotation of S_2 could not even be ascertained: it would always be possible to say that S_2 is an ellipsoid in absolute rest, indifferently with respect to space or any distant matter. But on this basis, Einstein could not have really modified the Newtonian way of coming to terms with the problem of rotation.

Newton himself already settles accounts with the problem of distant matter as the ultimate material frame of reference. This forms the basis of his famous thought experiment where two weights, tied together by a rope, inhabit an otherwise empty universe (1729, pp. 11-12. See Sklar 1974, pp. 183-4). If the two weights rotate about their common centre, this will be indicated by a tension in the rope, i.e. by forces whose origin is strictly local. Now,

if in that space some remote bodies were placed that kept always a given position to one another, as the fixed stars do in our regions, we could not indeed determine from the relative translation of the globes among those bodies, whether the motion did belong to the globes or to the bodies. But if we observed the cord, and found that its tension was that very tension which the motions of the globes required, we might conclude the motion to be in the globes, and the bodies to be at rest; and then, lastly, from the translation of the globes among the bodies, we should find the determination of their motions.

In this experiment, Newton first considers two spheres in a simply possible world. Next, he examines the case of these two spheres suddenly reinserted in a world quite similar to ours. It is a fact of observation that the *observed* velocities of the planets with respect to the stars and the *computed* velocities for stable orbits around the sun are equal. Newton concludes therefrom that the stars are 'truly' at rest, the planets 'truly' in motion. In fact, distant masses come to play a purely geometrical role, they simply confirm the result of local measurements predicted by the theory; the theory is constructed so as to predict them. For instance, were observation and computation not to coincide, it would always be possible to 'correct' the discrepancy by ascribing some rotating motion to the stars. This motion would have observable consequences, like a variation in the relative positions of the stars.

Of course, there is also something more in the experiment. The peculiarity of Newton's principle of relativity is that uniform translation is an absolute motion which cannot be observed; absolute motions that can be observed are accelerations and rotations. Here, the thought experiment is designed to render translation 'observable', or at least measurable. Were there no physical link between the two globes (such as the rope, subject to measurable tensions), the fact of motion could never be established in an otherwise empty universe, nor could it be ascribed to any one system (the two globes or the stars) in a starry universe. In the example treated by Einstein, there is no such thing as a rope or any physical link whatsoever between S_1 and S_2. That is why their different behaviours are verifiable, provided that measurement is subject to an additional, quite specific condition. For instance, an observer located on the line joining the two spheres uses a light ray in order to survey them. Measurements are supposed to be performed *before* distant matter is taken into consideration, that is, subject to gravitational effects—it being assumed that the ellipsoid is *already* under such influence. Similarly, if the observer moves from one sphere to the other in order to survey the surfaces directly, the assumption is that the measuring instruments have not been irreversibly altered during transportation. This amounts to nothing other than presupposing the theory of general relativity. The system $S_1 S_2$ is *first* disconnected from the real world so as to prevent measuring instruments from being embedded in, or disturbed by, the real world *even after* the reinsertion of the system into that world.

Strictly speaking, Mach's project is an extension of kinematic relativity to the domain of dynamics. Thus, "if we take our stand on the basis of facts, we shall find we have knowledge only of *relative* spaces and motions" (Mach 1883, pp. 281-4). Certainly, "when we say that a body K alters its direction and velocity solely through the influence of another body K', we have asserted a conception that it is impossible to come to unless other bodies A, B, C . . . are present with reference to which the motion of the body K has been estimated". The fundamental principles of mechanics can be understood in such a way that centrifugal forces occur even with relative rotations. Thus,

the motions of the universe are the same whether we adopt the Ptolemaic or the Copernican mode of view. Both views are, indeed, equally *correct*; only the latter is more simple and more *practical*. The universe is not *twice* given, with an earth at rest and an earth in motion; but only *once*, with its *relative* motions, alone determinable. It is, accordingly, not permitted to say how things would be if the earth did not rotate.

How are we to account for the observed centrifugal force on the earth in the Ptolemaic viewpoint where the Earth is at rest? Mach's answer is that the cause is similar in both the Copernican and the Ptolemaic systems, because the earth is always in relative motion with respect to the stars. Supposition of a non-rotating Earth is conducive to considering the physical state of the earth in relation to something other than the directly observable system of the Earth and the starry heavens. From this condition of *judgment*, Mach concludes that certain *intrinsic* effects of nature are a necessity. In fact, Einstein's experiment is quite a reversal of this procedure. The two masses are not only initially free with respect to distant matter from the observational point of view, but also with respect to one another from the interactive point of view, which is quite different from the bodies, K and K', in Mach's reasoning, where they are linked mechanically. Also, it seems that the case studied by Einstein is the Ptolemaic and the Copernican viewpoints as *two different* viewpoints. The former is the S_2 viewpoint (with his space of reference, R_2) on S_1, the latter is the same viewpoint with space of reference, R_1. In other words, Einstein's epistemological requirement is a sort of conjuring trick, intended to implement a true *tour de force*: distant matter will be absolutely unavoidable if it must be posited even if every means is lacking to form a judgment on the behaviour of bodies. We are a far cry from the early attempt to derive the general-relativistic way of thinking from naïve and uncommitted observations, but at last a new conception of observability emerges from Einstein's thought experiment: the cause of any given physical effect can always be attributed to some distant matter which need not be observed but is *in principle* observable. Distant masses, as it were, *make possible* the given phenomenon. That is why they must be involved in our *explication*, even though it matters little whether they be actually observed or not.

Einstein had to clear many obstacles before reaching such a conclusion. Newton's theory is clearly his target, but Mach's conceptions are also developed in a new direction. Newton had stated quite clearly that experience can ascertain the state of absolute rest for observable distant matter (the fixed stars), provided that inertia (the *vis insita*) can be conceived as belonging to the internal properties of a body, irrespective of its relations to the external world. Now, Mach has provided the means enabling us to define the inertial state of a given body or system of bodies in relation to the whole universe. Instead of speaking of the direction and velocity of a mass remaining constant, it must be possible to say that the mean acceleration of this mass with respect to other neighbouring masses is zero (1883, pp. 286–7). In fact, as many masses as we please may be introduced in order to maintain the empirical equivalence. The zero mean acceleration can be defined with respect to the *whole* of space surrounding the given mass, provided this space consists of uniformly distributed masses up to infinity. Also, "if two masses. . .exert on each other a force which is dependent on their distance. . .the acceleration of the centre of gravity of the two masses or the mean acceleration of the mass-system with respect to the masses of the universe (by the principle of reaction) remains = 0". This leads Mach to conclude: "Even in the simplest case, in which apparently we deal with the mutual action of only *two* masses, the neglecting of the rest of the world is *impossible*". Thus, there is no longer a difference of reality between inertia and a non-free motion. When the universe is idealized as a uniform, infinite distribution of masses, free motion becomes a particular case of non-free motion. If the physicist wants to adopt the Newtonian form of the laws of nature, he has to choose the heliocentric system because the sun, and not the earth, is unaccelerated with respect to the stars. The relation of the geocentric system to the idealized universe is quite different *only* because it yields far more complex laws.

Einstein's originality lies in the discovery of what is meant by 'observable' when the logical connection between the simple and the complex laws is actually worked out. In his thought experiment, we do not know *which* of the two systems S_1 or S_2 is actually at rest with respect to distant matter. That is why, "of all imaginable spaces R_1, R_2, etc., in any kind of motion relative to one another, there is none which we may look upon as privileged a priori without reviving the above-mentioned epistemological objection". Einstein comes to the point when he deduces therefrom a fundamental principle: *"The laws of physics must be of such a nature that they apply to systems of reference in any kind of motion.* Along this road we arrive at an extension of the postulate of relativity" (1916, p. 113). This is the so-called principle of general relativity, which gets over the difficulty of two types of laws, simple and complex, without logical relation to one another. The distance traversed so far rests entirely on the very strange, and so to speak 'double' status of distant masses. First, in that part of the thought experiment which precedes establishment of the verifiable fact of observation, distant masses being dissembled bring the new theory into conformity with the principle of the old one, namely, there is no physical experience which would ever enable us to ascribe an absolute velocity to any one of the objects of some system (in the example: the constant angular velocity). In other words, the particular role of the velocity of light or the intrinsic rigidity of measuring rods, as *unconditionals* of the special theory, are left untouched in the new theory. Secondly, the fact that distant masses are observable is taken as evidence that they have gravitational effects. This allows the general principle of relativity to be regarded as a mere *extension* or *enlargement* of the special principle.

(d) Antinomy of Mach's principle and the general relativity principle

The thought experiment is designed to show that the general principle rests upon Mach's principle, and Einstein attempts to provide an epistemological justification for the principle of general relativity in Mach's principle. This serves to unify the different aspects of the new physics, so that the numerical equality between the inert and the gravitational mass now appears to be a *logically* inferred consequence, and not simply a *fact* determining the generalization of inertia. As Einstein says in his Four Lectures at Princeton, it is the whole logical reasoning involving Mach and the principle of general relativity that leads to the view of a "unity of nature of inertia and gravitation" (1921, p. 64).

However, in view of the problematical status of distant matter, a careful examination of the outcome of the thought experiment shows how questionable the justification is. In the field of view from S_1, S_2 rotates with a measurable angular velocity against distant matter. Distant matter causally explains the ellipsoidal shape of S_2, but S_1 is locally Euclidean since it remains a sphere. It can be said that the ellipsoid exists *within* the pseudo-Euclidean continuum. On the other hand, if distant matter is also to invest no privilege in one viewpoint over any other, both viewpoints (from S_1 and S_2) should be equivalent centres. In other words, from S_1 or S_2, an observer should see the same thing (that is, S_2 or S_1) as an ellipsoid. Thus, in accordance with the generalization of inertia, the differential effects should *disappear*. Any difference between the shapes of the two bodies is utterly impossible, unless one of the bodies is declared to be 'really' rotating in a quasi-Leibnizian sense, that is, in a way that is interior to the body as distinct from a process in space.

One might see in this no more than an obvious confusion in Einstein's reasoning between two things, Mach's solution to the problem of relative rotation, and an extension of the restricted principle of relativity to the case of inhomogeneous fields. But it is certainly instructive to follow this confusion where it leads us and see how it creates a problem which, if carefully enunciated, may have valuable consequences. The problem can be stated in the following antinomical form: While Mach's principle amounts to making distant matter the physical support of a pseudo-Euclidean continuum, the principle of general relativity, on the contrary, assumes a sort of ductility in the continuum itself in relation to the chosen reference system. Actually the fact that Einstein perceived the tension in this way is clearly shown in the later mathematical and physical sections of his memoir; he himself called it later an "inconsistency" (1934, p. 84) as we have already seen. Einstein comes up with the so-called principle of general covariance: "*The general laws of nature are to be expressed by equations which hold good for all systems of coordinates, that is, are covariant with respect to any substitutions whatsoever (generally covariant)*" (1916, p. 117).* It is clear that general covariance is meant to at least resolve the tension. It gives rise to Einstein's celebrated image of the "reference-mollusc", a term which he used for the first time in his popular exposition of general relativity, completed in December 1916 (1917a, p. 99). The extension of inertial motion to take in all motions is concomitant with the extension of the Cartesian coordinate system to arbitrary coordinates. The mollusc, however, is not itself arbitrary: its 'ductility' has an *intrinsic* meaning, in the sense that it is defined in relation to any other reference system and within the physical neighbourhood of a given

---

*The antinomy has been well exemplified by O. Klein (1962, particularly pp. 294–5). Yet, in Klein's treatment, the antinomy arises from the cosmological question, whereas I try to see it as conditioning the discovery of the question. Friedman (1983, p. 207) points to sliding from Mach's solution to the extension of the principle of relativity as being a flat confusion, and he claims general covariance to be the decisive confusion since Einstein would identify it with the extension of relativity.

point. The transformation of Newtonian equations into rotating coordinates is not covariant, since the equations would contain quantities—such as angular velocity of rotation—which allow no intrinsic interpretation, i.e. they can only be related to another, privileged, reference system, the Cartesian one. Einstein adds that general covariance "takes away from space and time the last remnant of physical objectivity", that is, space and time not only depend upon the chosen reference system but they have no *existence* apart from the actual configuration of matter in a given neighbourhood.

Thus, the new principle offers a solution to the apparent conflict between Mach's principle and the principle of general relativity to which we have already referred, since an *absolute* difference between S_1 and S_2 is now possible: the geometry of the ellipsoid is not Euclidean, which means that the differing shapes do *not* allow us to privilege either viewpoint. If Einstein is at all confused in his reasoning, it is only in the sense that the quasi-explicit ambiguity of the distant masses is being made to serve a totally new purpose. If a satisfactory concept is to be derived, one which would complete Mach's partial solution to the classical dispute between Leibniz and Clarke, a complete account of the problem would have to be seen as overthrowing mathematical analysis as it had been practised. In particular, the mystery of non-spatial processes will be decisively done away with if we start seeing the intrinsic curvature of space as a dynamic process induced by the existence of other masses in space.

3. De Sitter's early critique of the principle of relativity and the nature of boundary conditions

(a) Introduction

The mental revolution involved was so enormous that it took time for the consequences to be articulated, let alone generally understood. To begin with, there was the reaction of the anti-positivist philosophers of science. A typical reaction was Ernst Cassirer's, in his 1921 essay, *Einstein's Theory of Relativity*. Because of his Kantian orientation, Cassirer seems to have confused the existence and the objectivity of the new form of space–time continuum. This led him to overlook the crucial distinction between two kinds of curvature, one that can indeed be eliminated by an appropriate transformation of coordinates, and one that is an independent property of space–time. The former is possible in case of vanishing curvature tensor but non-vanishing Christoffel symbols (i.e. those terms that include the first derivatives of the $g_{\mu\nu}$), while the latter implies vanishing Christoffel symbols but non-vanishing curvature tensor (second derivatives of the $g_{\mu\nu}$). In fact, Einstein maintained for quite some time a connection between the Christoffel symbols and the strength of gravitational fields, without specifying that this implied a careful choice of coordinate system. This delay in spelling things out certainly caused great confusion, but the amalgamation itself enabled Cassirer to save the Kantian aprioricity of Euclidean space by arguing that the curvature could be transformed away in any case (p. 432). In fact, if we are looking at the early contribution of philosophers, a much clearer understanding of the actual meaning of curvature comes from positivists like Schlick and Reichenbach. Significantly, Einstein described Schlick's first edition of *Space and Time in Contemporary Physics* (1917) as "masterful" (letter of Einstein to Sommerfeld, 1st February 1918, in A. Hermann 1968, p. 47); Schlick, indeed, offered a critical account of the a priori in terms of revisability, and this enabled him to understand the curvature as an intrinsic property of space.

In the first days of relativity, scientists themselves failed to perceive the distinction between the two curvatures. But here an anti-positivist like Arthur S. Eddington seems to have grasped the theory well before the others. Initial reception of the theory in England provided the spectacle of an interesting exchange between James Jeans and Eddington in the columns of *The Observatory*, during January and February 1917. Jeans argued that Einstein's theory should be seen as quite independent of metaphysics because the procedure is essentially matter-of-fact in character: "Einstein's crumpling up of his four-dimensional space may, for the present, be considered to be. . .fictitious". Eddington replied that the metaphysics was quite crucial, precisely because the theory brought into being an entirely new picture of the world. What gets crumpled, he said, is "the ordinary space which we are persuaded to discard" (1917, pp. 58 and 94).

Enter De Sitter. In his paper of March 1917, which we will look at later, he writes that "the curvature of space. . .only serves to satisfy a philosophical need felt by many. . .it has no real physical meaning" (1917, p. 1224). Commenting on this latter argument, John North finds that "De Sitter's scepticism may be excused", because he was only one of the many scientists and philosophers who confused an intrinsic curvature of space of three dimensions with a "fourth dimension into which it can curve" (1965, p. 81). But was De Sitter confused? It is true that no clear distinction between the two types of curvature is to be found anywhere in his writings. Oddly enough, De Sitter persisted within his first view until the end of his life. While he discussed with apparent lucidity the ins and outs of the mathematical formalism involved in the different classes of possible universes (static and non-static), he also thought that the radius of curvature was merely mathematical, without counterpart in reality (1932, p. 122). But he did have the great virtue of never getting the theory of general relativity wrong. His great merit was to move within the difficulties, with a clear awareness that what differentiated him from Einstein was that, as he said to Einstein in a postcard of 18 April

1917, "while you have a definite belief, I am a sceptic". There is world of difference between the sceptical pursuit of truth, and confusion over the real meaning of a scientific theory. By asserting his scepticism, De Sitter revealed both the existence and the nature of a significant problem related to the cosmic-scale extension of the relativity principle. For him, it was primarily this extension which created the need for a careful distinction between the mathematics and the physics of the whole theory.

(b) The relativity of rotation and the principle of relativity

So far Einstein's deduction is based on the epistemological principle of observability. The substitution of distant matter for absolute space engenders much more than the mere interchange of two things because, as Einstein says elsewhere, the aim is now to eliminate all "independent property" (1921, p. 61). Absolute space is to be discarded altogether, since physical science seeks to ascertain verifiable conditions only with the aid of its own concepts. Einstein takes 'observability' as the most vital of these concepts. This implies a subtle distinction between actually observed and observable distant matter. Take, for instance, a single rotating disc, very remote from all other matter. There *is* a non-homogeneous inertial field (centrifugal acceleration) on that disc, and thus also a field of gravitation, because the motion of the disc should be referred to as a motion relative to some distant matter. Distant matter is the condition without which rotation of that disc could not even be established by an *internal* experience. The relative motion being accepted as cause, the same physical phenomena should be describable in two coordinate systems, that where the disc is rotating and that where distant matter is rotating. Furthermore, in an otherwise empty universe, it should not be possible to distinguish dynamically between a coordinate system where a single source is at rest and another system where the source is rotating. Suppose only one material point exists in the universe (both observed and observable): space–time would then be Euclidean. Because of the dynamic impossibility of such a single point, this space–time is deprived of objectivity. Only the Riemannian continuum is objective: it is much more than a purely mathematical entity; between two mutually accelerated points, the common space–time is the Riemannian one.

De Sitter's approach to the question of relative rotation is particularly profound and new. Relative rotation is examined by De Sitter with respect to the Earth. Its rotation is represented in two different ways, which are certainly an analogue of Einstein's two fluid bodies.

First a preliminary remark. The very delicate question of the rotating disc is a problem which has not as yet received a fully rigorous solution in the current literature (for a historical perspective on the problem, see J. Stachel 1980). *Within* the disc, the solution is rigorous: the three-dimensional reference system is physically limited, so that the angular velocity simply defines a field of gravitation. The problem is to know with respect to what the field is defined. The difficulty is that it is within the very same Euclidean space-time that the two reference systems are considered, one attached to the spherical body at rest and the other attached to the same body in motion. But these two systems should yield two different physical spaces. That is the reason why the resulting relations have a purely local meaning, it being impossible to integrate them in one space-time which would be that of the body under consideration. Instrinsic geometry of the rotating disc should be non-Euclidean; only such a geometry would allow the connection between the different, locally Euclidean space-times to be realized. (A rigorous solution is sketched by M.A. Tonnelat 1964, pp. 68–81.) De Sitter's analysis, because it is translated into Minkowski space-time, offers an approximate solution only. This, in fact, is only yet another facet of the fundamental difficulty of how to distinguish intrinsic from non-intrinsic curvature.

I refer here to the very first paper by De Sitter on the question, which was communicated on 30 September 1916. The second paper published in *Monthly Notices*, December, offers a somewhat refined argument, complemented by new insights into the fundamental issues. It is quite probable that these insights were worked out in the light of Eddington's early comments. (The relevant letter of Eddington, which I have found in the De Sitter papers at Leiden, is dated 13 October 1916.)

De Sitter takes the coordinate system $x_1 = r, x_2 = \theta, x_3 = z, x_4 = ct$, where z is the Earth's axis of rotation, r and θ are polar coordinates in the plane perpendicular to that axis. What are the $g_{\mu\nu}$ prevailing on the Earth's surface? Were the Earth not to rotate, the $g_{\mu\nu}$ would take the values

$$\begin{matrix} -1 & 0 & 0 & 0 \\ 0 & -r^2 & 0 & 0 \\ 0 & 0 & -1 & 0 \\ 0 & 0 & 0 & +1. \end{matrix} \qquad (2.1)$$

This is nothing more than the Minkowski ds^2 in terms of polar coordinates:

$$ds^2 = c^2dt^2 - dr^2 - r^2d\theta^2 - dz^2. \qquad (2.2)$$

Transforming to rotating axes, the components $g_{\mu\nu}$ for the new system $\theta' = \theta - \omega t$ (where ω is the angular velocity of rotation) are

$$\begin{matrix} -1 & 0 & 0 & 0 \\ 0 & -r^2 & 0 & -r^2\omega \\ 0 & 0 & -1 & 0 \\ 0 & -r^2\omega & 0 & +1-r^2\omega^2. \end{matrix} \qquad (2.3)$$

The interval is now:

$$ds^2 = -dr^2 - r^2d\theta^2 - dz^2 - 2\omega r^2 d\theta dt + (1 - r^2\omega^2)c^2dt^2. \qquad (2.4)$$

In this expression, if $\omega = 0$, the interval is reduced to (2.2), with space sections perpendicular to the time axis. If $\omega \neq 0$, the orthogonality is no longer satisfied and terms appear (such as g_{24}) where space and time are interwoven. The aim of the general theory of relativity is to build a world-view where the two situations are equivalent.

But, De Sitter writes,

"It is found that the set (2.1) does not explain the observed phenomena at the surface of the actual earth correctly, and (2.3) does, if we take the appropriate value for ω. This value of ω we call the velocity of rotation of the earth. Then relatively to the axes (2.3) the earth has no rotation, and we should expect the values (2.1) of $g_{\mu\nu}$." (1916b, p. 528).

Thus there is a difference between axes 'at rest' and 'rotating' axes, in the sense that the components of the metrical tensor are not identical. How would it be possible to avoid non-observable absolute space, granted that this difference be insuperable? De Sitter explains that Einstein's solution is: "The g'_{24} and the second term of g'_{44} in (2.3) therefore do not belong to the field of the earth itself, and must be produced by distant masses". Preserving absolute rotation leads to differentiation of the spaces of reference, (2.1) and (2.3). For De Sitter, the introduction of distant masses is as contrary to the spirit of general relativity as would be the withdrawal of absolute space in Newtonian mechanics. These masses, exactly like the absolute space in Newtonian theory, are *independent* of the reference system (p. 527).

De Sitter wants to develop an interpretation more appropriate to the spirit of general relativity. As an example, he takes the component g_{24}. The solution of the differential equation which determines that quantity is

$$g_{24} = kr^2 \qquad (2.5)$$

where k is an arbitrary constant of integration. The Einsteinian theory requires g_{24} to be of such a form, but it does not prescribe the value of the constant: "the *differential* equation is the fundamental one, and the choice of the constants of integration remains free". Accordingly, the Newtonian theory makes a prescription on the values of the constants, and that is its absolute character. A consistent theory of relativity leaves the constants free; those constants

must, of course, be so determined that the solution represents the observed relative motions of material bodies and light-rays as described in the adopted system of coordinates. The particular solution which does so in one system must, if this system is transformed into another, by the same transformation be reduced to the particular solution which fits the observed phenomena in the new system (p. 529).

What De Sitter points out is that the requirement of relativity is purely formal: the differential equations which govern all phenomena of rotation keep the same form in every coordinate system. The components of the metrical tensor have values which may vary from one system to another, without jeopardizing dynamic relativity. Thus, general relativity "implies that the constants of integration are also subjected to the transformations, and are therefore as a rule different in different systems of coordinates" (1916c, p. 179). Resorting to distant masses originates from a confusion between the equations and their solutions: "The flaw in the argument used above was that (2.1) was considered to be *the* solution, instead of (2.5)" (1916b, p. 529).

In practice, and broadly speaking, De Sitter describes the situation as follows (1916b, pp. 529–30):

Suppose that we have originally taken a system of coordinates relatively to which the earth has a rotation ω_1. This ω_1 is, of course, entirely arbitrary and, as our coordinate axes cannot be observed, it must in a true theory of relativity disappear from the final formulae. Now we have $g_{24} = kr$, and to determine k, we transform to axes relatively to which the earth has a rotation by $\theta' = \theta - \omega_1 t$. Then in the new system

$$g'_{24} = (k - \omega_1)r^2. \tag{2.6}$$

Observation shows that the correct value of g'_{24} in this system is $-\omega r^2$, therefore

$$k = \omega_1 - \omega. \tag{2.7}$$

The quantity ω_1 must disappear from the final formulae, that is, from the solutions, inasmuch as it is as arbitrary as the initial choice of coordinates. The constant k must then be determined in Eqn (2.5), that is to say, in that coordinate system where the Earth has a rotation ω_1. With a transformation towards axes where the Earth does not rotate, (2.5) becomes (2.6). Observation of forces indicates the value $-\omega r^2$ for g'_{24} in that new system. From this, we deduce that $g'_{24} = (k - \omega_1)r^2 - \omega r^2$ and, as a result, the value (2.7). The idea according to which the 'true' value of k is zero—and, thus, that there exists a privileged system where k takes that value—is grounded in the following fact: relative to the last coordinate system, where the Earth rotates with ω_1,

an average star has a rotation $\omega_1 - \omega = \omega_2$. If we transform to axes relative to which this star has no rotation, we find

$$g''_{24} = (k - \omega_2)r^2 \tag{2.8}$$

Observations of stars require very approximately the value (2.1), i.e. $g''_{24} = 0$. It is assumed that zero is the exact value, and consequently

$$k = \omega_2 = \omega_1 - \omega \tag{2.9}$$

i.e. we find the same value as above in (2.7) (1916b, pp. 530–1).

This time, a kinematic observation provides the value (2.9), so that we find again the value (2.7). In short, the following values are obtained:

(1) In the coordinate system where the Earth rotates

$$g_{24} = (\omega_1 - \omega)r^2;$$

(2) In the system where the Earth does not rotate

$$g'_{24} = (\omega_1 - \omega - \omega_1)r^2 = -\omega r^2;$$

(3) In the system where the stars do not rotate

$$g''_{24} = (\omega_1 - \omega - \omega_2)r^2 = 0.$$

The general solution is $g_{24} = kr^2$, where k varies from one coordinate system to the other. It is only in the last coordinate system that $k = 0$. This zero value for k in the system where stars are at rest is a result of the observed coincidence between the value of inertial forces, as measured at the surface of the Earth, and the value of these forces, as computed from the observed rotation of the Earth with respect to the stars. Concerning the coincidence between (2.7) and (2.9), De Sitter explains that "this is certainly a remarkable fact, and a confirmation of Einstein's theory". The belief in absolute space, together with the assumption that the system of fixed stars has no real rotation, implies the 'true' value of ω_2 to be zero, so that (2.9) gives $k = 0$. But that $k = 0$ is a property of space (or of the "ether", as De Sitter surmises) is just the result of this conviction. If we believe that there is no absolute space, De Sitter says, "we must regard the differential equations as the fundamental ones, and be prepared to have different constants of integration in different systems of reference". The substituion of distant masses for an absolute space comes from the belief that any value of k different from zero in an arbitrary coordinate system should be 'explained' by something which is declared to be 'observable'.

Einstein's differential equations are universally invariant, but this is not true for their solutions, whose constants of integration are different in different coordinate systems. It would be alluring to believe that the system where k takes the value zero bears a privilege analogous to that of inertial systems, since the metric takes a particularly simple diagonal form. The significance of this first contribution by De Sitter lies in its ability to show that no such thing can happen. The privilege of inertial systems is based upon the very simple form of the differential equations in these systems. In the case under consideration, the field equations have the same form in the coordinate system associated with the stars and in every other system as well; only the solution is far more simple in the former. This does not confer a privileged status to the system of stars.

De Sitter thus maintains that rotation is relative in Einstein's theory, and even as relative as a linear translation. Both rotation and translation are susceptible to being transformed away. None the less, a difference persists:

If a linear translation is transformed away (by a Lorentz transformation), it is utterly gone; no trace of it remains. Not so in the case of rotation. The transformation which does away with a rotation, at the same time alters the equation of relative motion in a definite manner. This shows that rotation is not a purely kinematical fact, but an essential physical reality. Its amount ω is a physical constant, proper to the earth, like its mass (1916c, p. 180).

There is a purely kinematical aspect of rotation, which makes itself felt in the variability of constants of integration in the solutions of generally covariant equations. But there is also a more "essential" aspect, as De Sitter says, which commands a determinate change in the equations themselves when passing from one coordinate system to the other. This essential aspect is independent of all theories (1916b, p. 532). In the chosen example, the value of ω is an objective fact, the only quantity directly observable in the phenomenon of relative Earth/stars rotation, but each theory seeks to represent it in its own manner. Newton 'explains' the fact by eliminating ω_1 with k: there is a constraint on the very *form* of the equations, that is, on the allowable type of coordinates. Einstein seeks to 'adjust' ω_1 to k, so that rotation is always relative to observable masses: this is a sort of constraint on the very *content* of the universe. De Sitter, on the other hand, clarifies what he claims to be the true originality of general relativity: it enables us to get rid of any explanation whatsoever:

Newton 'explains' it by his law of inertia and the absolute space. For Einstein, who makes no difference between inertia and gravitation, and knows no absolute space, the accelerations which the classical mechanics ascribes to centrifugal forces are of exactly the same nature and require no more and no less explanation, than those which in classical mechanics are due to gravitational attraction (1916b, p. 532).

It is precisely at this point that De Sitter indicates a major flaw in Einstein's own presentation of general relativity. Formulated in the terms of the 1916 memoir, it is clear that the principle of general relativity and the principle of general covariance amalgamate two different things: the physical reference systems, and the coordinate systems. In fact, this is a flaw which relativists did not perceive until much later. But in accordance with his conclusion, De Sitter's demand for a higher consistency already implies a distinction to be drawn (in any physical theory) between arbitrary and physical coordinates. In many points of his discussion, De Sitter urges the superiority of Einstein's theory to be seen as demonstrating "the irreality of the coordinates"; what Einstein has perceived is "the irrelevance of the representation by coordinates", to the effect that it penetrates "to the deeper realities which lay hidden behind it" (1916c, p. 178 and 183). Arbitrary coordinates are the purely mathematical requirement of the principle of general covariance; physical coordinates are specific among arbitrary coordinates, imposed by certain physical quantities inherent in nature.

De Sitter's criticism strikes at the root of Mach's interpretation. The deficiency of that interpretation lies in tacit assuming the Minkowski ds^2 to be the true solution, and divergence from it (in an arbitrary coordinate system) to be explained by the distribution and motion of masses in the universe. In particular, the motion of masses in the Einstein–Mach universe does not determine the totality of metrical components, but only the *non*-diagonal terms and the second term $-r^2\omega^2$ in g_{44}, that is, those terms in which the relative Earth/stars velocity occurs. In order to prove that the distribution and motion of masses determine the metrical components, at least partially, one should be able, in a given coordinate system, to introduce changes in the distribution and motion of these masses. This is, of course, utterly impossible. That is why a true causality induced by distant matter seems to be definitely undermined by De Sitter. The value of the constant k in a given coordinate system could, at most, be computed from the observed motion of the stars in that system, independently of the measurement of local dynamic effects. But how does De Sitter discover the value of k in each of the three systems—those where the Earth rotates, where it does not rotate, and where the stars do not rotate? He begins with a coordinate system in which the Earth has an angular velocity ω_1, and then he makes a coordinate transformation in which the Earth is at rest. In the latter system, observations show the correct value of g'_{24}, i.e. $-\omega r^2$; these are dynamic observations, and that is why De Sitter does not mention here observation of stars. Next, he introduces the motion of stars, from which the value ω, measured in the system where the Earth has no rotation, is interpreted in terms of a relative Earth/stars motion. The single kinematic observation of stars, independent of all dynamic measurement, does not immediately provide us with the value of k in a certain system. Only after having ascertained the coincidence of the value of $k = \omega$ (obtained dynamically in the system where the Earth is at rest) with the value of rotation of stars in that system, is it possible to find out the value of k in any other system, given the motion of stars in that other system: that latter value is identical to their angular velocity. In other words, the value of k is *not* deducible only from the field equations and observation of the motion of stars.

(c) On the origin of inertia

De Sitter's conclusion is very impressive, since it reveals a contradiction in Mach's objection to Newton's interpretation of the bucket experiment. Mach thought that we did not have to accept Newton's interpretation because there was no such thing as direct 'experience' of the whole universe. What De Sitter's speculations tend to reveal is that a consistent Machian theory involves the same kind of impossible relation to the whole universe.

What, then, of the mathematical aspect of Einstein's theory, which De Sitter sees as its true originality? Is it really free of all this total vision business? From the letter which he sent to De Sitter on 13 October 1916, while waiting for a reprint of De Sitter's paper on relative rotation, it is clear that Eddington already perceived how Einstein's wish to abolish absolute rotation brought with it a batch of new conceptual difficulties. For Eddington the important point was what this meant for the problem of definite boundary conditions governing the differential equations of gravitation. He surmised that "when you choose axes which are rotating relatively to Galilean axes, you get a gravitational field which is not due to observable matter; but is of the nature of a complementarity function due to boundary conditions—sources or sinks—at infinity...That seems to me to contradict the fundamental postulate that observable phenomena are entirely conditioned by other observable phenomena". These remarks contain the seeds of De Sitter's more sophisticated strategy, in the December paper.

The fact that the constants of integration may vary from one coordinate system to another casts quite a new light on the consistency of general relativity. This variability in values of constants allows an adjustment in the equations of motion so that a quantity like ω is always the same for any observer. In particular, no a priori identity governs these values at *infinity*. As De Sitter says: "The condition that the gravitational field shall be zero at infinity forms part of the conception of an absolute space, and in a theory of relativity it has no foundation" (1916b, p. 531). No observation can ever teach us anything about the infinite—and this is certainly valid of any possible theory. But the difference between the Newtonian theory and a consistent theory of relativity is that, even though the infinite is evidently an 'ideality', it is no longer a priori identical for all observers in the latter. This means that the problem of boundary conditions in general relativity cannot be dealt with as though it were analogous to its 'classical' treatment.

Yet, the tracing of such a parallel is an attractive approach to the problem of boundary conditions, particularly in view of Einstein's commitment to Mach's ideas. When, in Poisson's equation, the potential function at infinity for any finite system is set at zero, it can be asserted that this is done in order for the *whole* of the numerical value of the potential at any point to be derived solely from *material sources*. In the theory of general relativity, such ideal boundary conditions should give place to the structural conditions of space–time. Following Einstein's own methodology you could as easily start with the Newtonian approximation. Throughout those portions of space–time which are sufficiently large, the pseudo-Euclidean continuum holds sway. Thus, at infinity, we would expect the following values for the potentials:

$$\begin{matrix} -1 & 0 & 0 & 0 \\ 0 & -1 & 0 & 0 \\ 0 & 0 & -1 & 0 \\ 0 & 0 & 0 & +1. \end{matrix} \qquad (2.10)$$

Of course, all 'real' potentials differ from the values at infinity. In keeping with classical mechanics, any divergence from these values in the neighbourhood of matter would be expressed by modifying the value of g_{44}. Indeed, if general relativity could be said to describe local distortions of the flat space–time, it remains possible to speak of a 'natural' separation of the curvature of space from that of time, i.e. 'natural' coordinates are just as meaningful as they are in the Minkowskian space–time. Because the curvature of space is extremely small by comparison with the curvature of time, only the time potential can be responsible for the occurrence of a field of force in flat space–time. Instead of the value 1, the g_{44} at finite distance would be

$$g_{44} = 1 + \gamma. \qquad (2.11)$$

As with Poisson's equation, all $g_{\mu\nu}$ are determined by differential equations which represent matter (the material tensor). In this sense, the set (2.10) seems to meet the requirements of Mach's principle. The thought experiment involving the two rotating fluids had already pointed to the conclusion that distant matter was the physical support of the pseudo-Euclidean continuum.

Once again, De Sitter raises the question of what would constitute a genuine relativistic consistency:

Thus matter here also appears as the source of the $g_{\mu\nu}$, i.e. of inertia. But can we say that the *whole* of the $g_{\mu\nu}$ is derived from the sources? The differential equations determine the $g_{\mu\nu}$ apart from constants of integration, or rather arbitrary functions, or boundary conditions, which can be mathematically defined by stating the values of $g_{\mu\nu}$ at infinity. Evidently we could only say that the whole of the $g_{\mu\nu}$ is of material origin if these values at infinity were *the same for all systems of coordinates*" (1916c, p. 181).

If the values of $g_{\mu\nu}$ at infinity were the same for all coordinate systems, that is, if they were both *arbitrary* and *generally covariant* constants of integration, then and only then, could any observer distinguish local values from values at infinity in one and the same manner. But it is quite clear that the set (2.10) is not generally covariant: the values in question work only to satisfy the principle of special relativity.

Because the search for a physical interpretation of boundary conditions is now seen as vital to any viability of Mach's principle, Einstein demonstrated a certain obstinacy and made one last attempt to find boundary conditions which are generally covariant. De Sitter relates a conversation he had with

Einstein, shortly after his own first paper on the relativity of rotation became known to Einstein (but before its publication). This is a conversation which took place on 29 September, and which appeared as an addendum to the paper on the relativity of rotation (1916b, p. 531. A short commentary on these debates can be found in P.J.E. Peebles 1980b.) Einstein had suggested that the $g_{\mu\nu}$ converge at infinity towards the following values:

$$\begin{matrix} 0 & 0 & 0 & \infty \\ 0 & 0 & 0 & \infty \\ 0 & 0 & 0 & \infty \\ \infty & \infty & \infty & \infty^2. \end{matrix} \quad (2.12)$$

This amounts to a complete separation of space and time. At infinity, it is still possible to introduce arbitrary functions t' of t, but no longer of x_1, x_2, x_3. At a finite distance, relativity is total—all four dimensions are affected by it—but a kind of absolute time is again operative at infinity. When he attempted to explain these values in the September addendum, De Sitter was still talking about "absolute space" at infinity too: the separation of space and time "undoubtedly has some of the characteristics of the old absolute space and absolute time. The hypothesis can thus be said to make space and time absolute at infinity, although arbitrary transformations of three-dimensional space are still allowed". But from a letter he wrote to Einstein, on the 1st November 1916, it is clear that De Sitter had some difficulty with this interpretation. He asks Einstein if the meaning of the set (2.12) can be accurately put like this: the hypothesis not only asserts that the g_{ij} degenerate according to the given values for infinite space variables, but also for an infinite time variable. De Sitter writes: "Is this correct, or do the g_{ij} remain Galilean—or approximately Galilean—when x_4 and x_1, x_2, x_3 are finite?" Significantly enough, as we shall see later, Einstein begged the question in the first instance because he was so shocked by the use of the word "finite" in this context. Trying to make sense of (2.12) for himself, De Sitter is much clearer in his December paper: "At infinity we would thus have an absolute time, but no absolute space. Of course, so far as our observations reach (in space and time), there would still be complete four-dimensional relativity, but at very large distances from all matter the $g_{\mu\nu}$ would gradually converge towards the degenerated values (2.12)". With these values, the discrepancies which make the relativistic theory distinguishable from the Newtonian are no longer allowable at infinity. Does this mean that physics must necessarily be 'classical' at infinity? The metrical components which govern the computation of time become infinite at infinity: the larger the distance from a point of mass taken as a centre, the longer the time interval. To use a metaphor, space becomes 'engulfed' in time. Because the relative measurements of time at infinity are infinite, time is like the 'bottle-neck' of space. At infinity, there is only time, an instant which is eternal. What Einstein has written is as novel as it is bewildering, since the boundary conditions (2.12) constitute a *nothing*—not in the fictive manner of the mathematical device, as in classical physics, but a nothing which is physically representable as being somehow different from the *empty* Minkowski space-time.

In Einstein's view, the set (2.12) was the expression of an ultimate version of Mach's principle. In his cosmological memoir, he recalls this set as being a most decisive step in the "rough and winding road" towards cosmology (1917b, pp. 179-183. A more literal translation of "indirekten und holperigen Wege" would be "indirect and bumpy".). The first thing is the mathematical formulation of Mach's requirement that a test body, sufficiently distant from all other masses in the universe, has zero inertia. Starting from the elementary space-time interval $ds^2 = g_{\mu\nu} dx_\mu dx_\nu$, spatial isotropy, that is the invariability of the solution of the field equations under an orthogonal transformation of coordinates x_1, x_2, x_3, allows the interval to be written:

$$ds^2 = -A(dx_1^2 + dx_2^2 + dx_3^2) + B(dx_4^2). \quad (2.13)$$

Moreover, the condition for local Euclideanity can be written as $\sqrt{-g} = 1$.*
This leads to

$$\sqrt{-g} = 1 = \sqrt{A^3 B}. \quad (2.14)$$

On the other hand, the weakness of stellar velocities with respect to the velocity of light leads to

$$ds^2 \approx g_{44} dx_4, \quad (2.15)$$

since g_{44} contains the time component which (because it is multiplied by c^2) becomes preponderant. These restrictive conditions have the following consequences on the form of the components of the energy-momentum tensor. The components of momentum are given by the three first components of the covariant tensor multiplied by $\sqrt{-g}$, i.e. $m\sqrt{-g} \cdot g_{\mu\nu} dx_\mu/ds$; they can now be written

---

*See 1916, pp. 129-30. Take $d\tau$ to be a four-dimensional element of volume. Calculation shows that $\sqrt{g} \cdot d\tau = \sqrt{g'} \cdot d\tau'$, where g is the determinant of the $g_{\mu\nu}$. In other words, this is an invariant for all possible transformations. If the $g_{\mu\nu}$ at any point can always be reduced to the Minkowskian values ($ds^2 = -dx_1^2 - dx_2^2 - dx_3^2 + dx_4^2$), g is always negative if we limit ourselves to those transformations for which $d\tau$ remains real and positive. The quantity $-g$ being always finite and positive, it is quite natural to determine *a posteriori* the coordinates so that $-g$ is always equal to 1. As De Sitter explains in 1916a, p. 706, a great simplification then appears in the field equations. The equations tantamount to Poisson's equation are

$$R_{\mu\nu} = \partial^2/\partial x_\mu \partial x_\nu \log \sqrt{-g} - \sum_\alpha \left\{ {\mu\nu \atop \alpha} \right\} \partial \log \sqrt{-g}/\partial x_\alpha.$$

With the simplification $\sqrt{-g} = 1$, the field equations are expressed under their well-known form: $R_{\mu\nu} = -k(T_{\mu\nu} - \frac{1}{2} g_{\mu\nu} T)$.

$$m\frac{A\mathrm{d}x_1}{\sqrt{B}\,\mathrm{d}x_4},\ m\frac{A\mathrm{d}x_2}{\sqrt{B}\,\mathrm{d}x_4},\ m\frac{A\mathrm{d}x_3}{\sqrt{B}\,\mathrm{d}x_4}. \quad (2.16)$$

The component of energy, which is the fourth of the covariant tensor, is (in the static case), $m\sqrt{B}$. These are thus the components of the energy–momentum tensor in their geometrical expression (A and B are metrical elements). In order to have a full physical meaning, they are not to be contradicted by the conditions pertaining to the material tensor. Mach's principle acts at that level. The quantity $A/\sqrt{B}$, in the expression of the momentum, plays the role of the inertia coefficient, while $mA/\sqrt{B}$ represents mass at rest in the static metric. Einstein deduces:

> As m is a constant peculiar to the point of mass, independently of its position, this expression, if we retain the condition $\sqrt{-g} = 1$ at spatial infinity, can vanish only when A diminishes to zero, while B increases to infinity. It seems, therefore, that such a degeneration of the coefficients $g_{\mu\nu}$ is required by the postulate of relativity of all inertia.

This justifies the set of values (2.12), which Einstein had suggested in his conversation with De Sitter. There is a physical reason that accounts for the invariance of (2.13) throughout space (in particular, at infinity): if A tends to zero for a test body very remote from all other matter, B must simultaneously tend to infinity; only this allows the $\mathrm{d}s^2$ to remain invariant.

If the hypothesis (2.12) is to be adopted, De Sitter says, "...then in any system of coordinates the $g_{\mu\nu}$ would have exactly determined values, there being no constants of integration left by which they can be made to fit the observed phenomena" (1916c, p. 182). This is De Sitter's way of translating Einstein's ultimate version of the boundary conditions, a version which neatly circumvents his own earlier objections by pointing out that the constants of integration comprise the only element of mathematical contrivance, not the particular *values* which are ascribed to them so as to meet with the given phenomena. In every physical law, expressible in terms of functions of coordinates, the influence of the total material universe should be understood as fixing the values of the constants of integration, *prior* to any measurement. The fundamental pre-condition for arriving at this view would be the generally covariant form of the laws.

Now, observation shows that the values of $g_{\mu\nu}$ are not very different from (2.10) at the remotest distances. This forms the basis of De Sitter's next counter-argument which, as we will see, is not at all restricted to the rather short distances then explored by astronomers. On the contrary, it is an argument of remarkable philosophical pertinence. On the subject of observed values, De Sitter comments that "on Einstein's hypothesis these are special values which, since they differ from (2.11), must be produced by some material bodies. Consequently there must exist, at still larger distances, certain unknown masses which are the source of values (2.10), i.e. of all inertia''. If the infinite is a 'nothing' that differs from an empty, Minkowskian space-time and if, furthermore, this hypothesis "has arisen from the wish to explain not only a small portion of the $g_{\mu\nu}$ (i.e. of inertia) by the influence of material bodies, but to ascribe *the whole of the $g_{\mu\nu}$* (or rather the whole of the difference of the actual $g_{\mu\nu}$ from the standard values (2.12) to this influence'' (1916b, p. 531), it then follows that the $g_{\mu\nu}$ pertaining to the Minkowski metric must be produced by material sources. The values at infinity are postulated in order that there be distant masses, but they only succeed in moving these masses beyond the field of *any* effective observation. Indeed, all measured values must be different from (2.12), since the infinite is not measurable. However large our picture of the measured universe may be, distant masses will always fulfil the same task, that is, of explaining the values (2.12) (or any other local values that could well be measured in the future) for the $g_{\mu\nu}$ at a finite distance. Rather than real objects, these masses are therefore nothing more than objects which remain always beyond the field of effective observation. And yet, in principle, because the infinite is declared to be 'physically' nothing, the whole hypothesis "implies the finiteness of the physical world, it assigns to it a priori a limit, however large, beyond which there is *nothing* but the field of the $g_{\mu\nu}$ which at infinity degenerate into the values (2.12)'' (1916b, p. 532). The limit is that which is formed by the hypothetical masses. But, even though it is fixed a priori, the limit may stay as large as we like, it may have a dimension that is co-extensive with any conceivable experience. The theory admits of no secure method for fixing a limit.

In general terms, by seeking the true consistency of Einstein's theory, De Sitter pointed to the consequences of trying to accept and maintain both premises in the thought experiment of the two rotating fluid bodies. Rather than dispensing with one of the premises on the grounds of the problematical status of distant matter, as Einstein did, De Sitter reveals that the proper way of contemplating the possible existence of distant matter is to identify it with some matter *beyond* the system of directly visible stars. In his correspondence with Einstein, De Sitter calls these masses "*übernaturliche*", supernatural. This terminology tends to reveal two different things simultaneously. On the one hand, the flaw of the supposed solution to the antinomy between Mach's principle and the principle of general relativity is that it seems doomed to remain purely local. On the other hand, the available theory cannot possibly offer a global solution. The hiatus is a blow to Einstein's criterion of observability. In his October letter to De Sitter, Eddington had already emphasized that our ignorance of the infinite should not bother us. But he also pointed out that the Machian conjectures put forward by Einstein as to the problem of relative rotation led inevitably to just that question: "Where is infinity according to the new conception?" As he went on to say, this question is unanswerable, but the very fact that it rears its head as an explicit part of the new theory is a proof that "we must bring in something outside observation".

Because he is concerned with the whole problem of the origin of inertia, and precisely because he makes his criticism a matter of principle, De Sitter is inclined to an epistemological position similar to that which governed the dynamic equivalence of coordinate systems in the problem of rotation. There the belief in a true value for k was conducive to a privileged system in which k took that value; consequently an explanation of the other values of k in other systems was called for. But the dynamic equivalence between coordinate systems deprived the divergence from a strict diagonal metric of any causal explanation. The same is true of a purely material origin of inertia: "practically it makes no difference whether we explain a thing by an uncontrollable hypothesis invented for the purpose, or not explain it at all".

Returning to the whole question of what constitutes the real difference between a rotation and a translation, De Sitter touches on an ultimate significance:

The case is somewhat analogous with the gravitational effect on planetary motion of a rotation of the sun. Whether the solar system has a translation or not can only be ascertained by observations of bodies *outside* the system [my emphasis]. Therefore such a translation can have no effect on the relative motions within the system. But whether the sun rotates or not can be ascertained by observations within the solar system, and accordingly the principle of relativity does admit an effect of the rotation of the sun on the relative motions of the planets" (1916c, pp. 180–1).

A thoroughly new meaning is attached to the epistemological criterion of observability: everything that depends exclusively on data exterior to a system can be of no influence on that system. This is decisive, but (as Einstein's quick reaction showed) it was by the same token just as problematical. Indeed, the question remains open: what is the extent of a system viewed as *interior*, what differentiates an interior from an exterior? The arbitrariness of the boundary is quite reminiscent of the way Newton isolated the solar system from the rest of the universe, basing himself on the assumed immovability of the centre of the solar system (see Chapter 1). In this sense, De Sitter's criticism reflects the classical habit of implicitly dissociating the concrete problem of the total structure of the universe from the foundation principles of physical science. But the implications went beyond anything envisaged by De Sitter. It seems that, however unwittingly, De Sitter sowed the seeds that were to lead to his theory being transcended by an even more radical approach. In actual fact, De Sitter himself noted that nothing which is effectively observable allows for the immediate deduction of the value of k, *except* the distribution and the motion of masses in the *total* universe. And when he spoke of the importance of Einstein's hypothesis in terms of the demonstration of the *possibility* of an entirely material origin for inertia, he underlined in a footnote that by possibility he means *logical* possibility. What remained to be investigated, he added, was the idea of "an actual distribution of ordinary physical matter having the desired effect".

Einstein, with his new, properly cosmological, conjectures, was to put his finger on this relevant exception which relates to the real, as distinct from the merely logical, possibility of relativity. But at this stage, he was certainly very far from even foreshadowing such a breakthrough. In a letter dated 31 October 1916 to his faithful friend Michele Besso, he explains his feelings about the actual meaning of the principle of equivalence. He does not like the generalization of relativity to be conceived as starting from the behaviour of physically different rods or clocks. He says that "this way of conceiving is quite unpleasant, in the sense that one should start from *the universe regarded as a whole*. Here [i.e. with the hypothesis of equivalence] it is much more convenient to begin with a *part*, without specifying the boundary conditions" (A. Einstein and M. Besso 1972, p. 86). The specification of boundary conditions is seen by Einstein as implying an overview of the whole universe. His unsuccessful attempts to come to terms with boundary conditions seems to have only reinforced in him the old view that the whole universe cannot be a physical problem.

4. The invention of cosmology

(a) General relativity before cosmology

The path followed so far may be puzzling. The normal presentation of cosmological arguments and theories tends to begin with the basic facts of the universe, and then goes on to rehearse the fundamental assumptions needed to give a theoretical account. Discussion of boundary conditions would be dealt with in an entirely different framework, emphasizing the apparent contradictions of pre-relativistic cosmology. And any new discussion of boundary conditions would be picked up only in some later, critical re-examination of the initial assumptions; as is the case with mass distributions which are less symmetrical than, say, Mach's idealized universe (or than Einstein's own model, which will preoccupy us for the rest of this chapter).

The whole issue centres on the occurrence of the infinite as a *problem*. Let us try to perceive its significance by way of an overall view of what has been argued so far.

The background, fundamentally and massively, is always Newton's physics. Newton assumes absolute space, from which it follows that there is a distinction between relative linear constant velocities and absolute ones, between relative linear accelerations and absolute ones, between relative and absolute rotation. The only potent arguments (that is, arguments which are physically verifiable) introduced by Newton deal with the last; arguments of pure intelligibility are offered for the first two kinds of distinction. Newton's arguments, therefore, are not enough to validate the idea of absolute space, but they do suffice to refute the more radical versions of relationism that Leibniz had tried to impose. Now, as we have seen in Chapter 1, the assumption that there are privileged velocities comes into conflict with the electromagnetic theory of the nineteenth century, and this in turn gets patched up by the special theory of relativity. However, the special theory provides no alternative account of why some kinds of acceleration are privileged over others.

This creates a new problem, quite separate from that solved by the special theory. How is it possible to modify a Newtonian universe so that absolute rotation disappears? Over large portions of the universe, it would seem reasonable to begin the analysis with the modifications imposed upon a Newtonian conception of space and time. What both Einstein and De Sitter discover in the course of their respective discussions is that, because the relationship between coordinates and measurements becomes severed in general relativity, it suddenly becomes unclear what is meant by infinity.

De Sitter maintains that a return to Mach's original insights is no help in a critical account of Newton's theory. He finds unjustifiable Mach's attempt to 'physicalize' Newton's absolute space in terms of inertia created by ponderable masses, though such an attempt was revived by Einstein after he himself had shown the failure of a different physicalization by means of ether. It is almost as if De Sitter was here spelling out for the first time the implications of what Newton himself had already hinted at, namely, the essentially 'mathematical' character of space as an absolute entity. Of course, in Newton's time, the radical consequences of this enunciation could not be realized, since the actual background of the conception of space was more metaphysical than physical. Nowhere does De Sitter come to a clearer formulation of his basic convictions than in a 4 November 1920 letter to Einstein:

As far back as when I was a student (that is, round about 1894) I was always hitting up against the idea of a mechanical explanation, that is, you were first of all supposed to explain matter in terms of ether or electricity. And then you were supposed to explain ether in terms of matter! This always seemed nonsense to me.

And writing to Einstein on 4 November 1916, De Sitter was already comparing the unobservability of distant masses with that of the old ether wind.

(b) The implications for astronomy of the relativity of time

In the December paper, De Sitter takes the first steps towards something more than mere logical possibility in the account of the origin of inertia: that is the connection between the new theory and the actual distribution of matter. He is concerned with the astronomical problem of how the theory of relativity might yield some insight into the computation of the total mass of the system of stars. In the process, he reanimates the problem which had haunted astronomers ever since the rise of spectroscopy in the nineteenth century, and tries to show how relativity may bring a new perspective to bear on it.

As early as 1911, De Sitter showed interest in the principle of relativity insofar as it interacted with the problems of practical astronomy. He sought to evaluate its claims to universal application by way of empirical tests. He was one of the first to investigate certain discordances between astronomical observation and Newton's theory of gravitation—the motion of the perihelion of Mercury as well as irregularities in the motion of the Moon—which led him to suggest that some modification of the law would be desirable. In particular he stated the possibility of gravitation being absorbed by some intervening matter as far as the Moon's motion is concerned. But no definite conclusion was reached at this stage. Interestingly enough, De Sitter noted that the theoretical requirement governing the relativity of time would invalidate all possible counterparts in reality of any definition (for practical purposes) of astronomical time. We have just seen how the discussion on the status of time became the key to interpretation of relative rotation. Similarly, in 1917, denial of a distinctive character for the time coordinate was to pave the way for the notorious empty solution of Einstein's cosmological equations. (More of this in Chapter 3.)

This early contribution by De Sitter is also interesting because it provides clues to his own philosophical influences. The purely conventional aspect of the measurement of time had already been worked out by Poincaré by the end of the nineteenth century (H. Poincaré 1898). Poincaré had the idea of extending his early concepts of conventionality in the foundations of geometry to the analysis of time measurement, as well as to what he called the principles of physics (i.e. the law of inertia, the conservation principles, the principle of relativity, etc.) (H. Poincaré 1902, Part III, Ch. 6 to Ch. 8 and Part IV, Ch. 10. See also 1905b, Parts II and III). In fact, as De Sitter was himself to say, "the starting-point of my investigations has been the papers by Poincaré and Minkowski" (De Sitter 1911, p. 389. He is referring to Poincaré 1905a and Minkowski 1908). The name of Einstein is not to be seen here. The closest thing to a reference is an allusion to an important paper by Lorentz where some aspects of Einstein's special theory are discussed (H. Lorentz 1910). Poincaré celebrated article, "Sur la Dynamique de l'Electron" (1905a), has been extensively discussed over the past years by historians and philosophers of science (see the exhaustive analysis by A.I. Miller 1975) in terms of the dispute over whether Poincaré or Einstein (or even Lorentz), had first formulated the basic principles of the so-called special theory of relativity. Clearly, De Sitter's reference to Poincaré rather than Einstein is disingenuous. Poincaré's article had the advantage of providing an account in which the principle of relativity was applied to the very topics that were of most interest to De Sitter, namely, celestial mechanics and the theory of gravitation. Even more importantly, this reference indicates how De Sitter may well have been concerned with the crucial idea of an *observational* equivalence between two such theories as those of Poincaré and Einstein, rather than plumping for a possible non-equivalence on *conceptual* grounds alone.

This impression seems to be confirmed in another remark by De Sitter in his paper, as well as from his work over the next two years. For the physical meaning of the principle of relativity, De Sitter refers to papers by H.C. Plummer 1910 and E.T. Whittaker 1910, who both "make free use of the word 'ether'". De Sitter goes on, however: "it may be well to point out that the principle of relativity is essentially independent of the concept of an ether...Astronomers have nothing to do with the ether...All Mr. Plummer's results remain true, and retain their full value, if the 'ether' is eliminated from his terminology. And also in Mr. Whittaker's note the word 'ether' is not essential, except, of course, from an historical point of view" (De Sitter 1911, p. 389). This is the case of two conceptually different theories, without the difference having an observational incidence. In a later work, De Sitter goes on to offer an observational test for discriminating between two other, conceptually non-equivalent theories.

In 1912, H.A. Lorentz resigned from his chair of theoretical physics at the University of Leiden and took up a new research position at Haarlem. His successor was Paul Ehrenfest, a young Viennese physicist who had lived in Russia since 1907. In his inaugural lecture on 4 December 1912, Ehrenfest addressed himself to the crisis of the ether hypothesis. The heart of his argument was Einstein's theory, because it was this theory which simply abolished the ether by postulating that light was an independent entity travelling through empty space. Ehrenfest also reviewed an alternative attempt to overcome the riddle of the ether by simply doing away with it. In 1908, Walter Ritz proposed a theory which differed from Einstein's in assuming that the velocity of light depended upon the velocity of its source, thus resurrecting the possibility of an emission theory of light. Ehrenfest closed his lecture by showing how much work needed to be done before the ether problem could be solved. (See P. Ehrenfest 1913 in M.J. Klein 1959, particularly pp. 319–23.) At Ehrenfest's prompting, De Sitter planned a crucial experiment to distinguish between the assumptions of Einstein and Ritz. In 1913, he published the result of his observations. Light emitted from eclipsing binary stars showed no peculiarities. This was a conclusive proof of constancy in the velocity of light. De Sitter's result has been taken ever since as a principal piece of evidence against emission theories. (De Sitter 1913a,b,c. R. Tolman had suggested this type of experimental test in 1910. See also Einstein's remarks in his popular exposition of 1916, p. 17. On Ritz, see A.I. Miller 1981, pp. 280-1.)

Poincaré's influence on De Sitter also seems evident in the very way De Sitter took exception to Einstein's argument about relative rotation. For De Sitter does echo the position which had led Poincaré to reject any sense of physical cosmology in terms of our mathematical physics. In his 1912 conference entitled 'Space and Time', Poincaré defended an operational or instrumentalist view of space and time: these have no properties of their own in as much as they are relative to our measuring instruments. Poincaré then raised the question of why physical relativity is restricted when a system is placed in relation with *rotating* axes. What we take note directly of are the finite (integral) equations of motion, in the sense that these equations immediately translate observable phenomena. Of course, they include constants of integration, and it is from this that we deduce differential equations. Only the latter do not change when passing from an inertial to a rotating reference system; by contrast, the finite equations, as Poincaré aptly pointed out, are *not* universally invariant as their constants do change from one system to the other. Therefore, Poincaré added, the task of understanding what exactly in Nature itself allows to proceed from the finite to the differential equations is as difficult as the solution to the cosmological problem. In Poincaré's terms: "there was only one copy of the universe ever printed, and that is a prime difficulty" (1912, p. 19). For we should be able to know several integrals of the same phenomenon which differ only by the values of the constants of integration; and this implies that, in addition to those differential equations which are realized in nature, we should know a whole range of those which are merely possible. Poincaré's suggestion for a way out of this situation consisted in abandoning the ideal of absolute rigor (p. 20): our physics being local only, it is sufficient to assume that, for instance, a force like gravity vanishes at large distances from the Earth. That is, the way out is to break up the world (which by itself is a whole of interdependent parts) into smaller worlds compatible with what we do in our laboratory. The powers of mathematical physics give out when it comes to embrace such a system as the whole universe: our mathematics, with its methods of approximation, would divorce from physics if the universe were not said to exist in more than one copy.

There is no explicit evidence of De Sitter's familiarity with this part of Poincaré's work, but the affinity is a very striking one. As an astronomer interested in the calculation of global effects, however, De Sitter would not leave the matter there. In his second paper for *Monthly Notices* of 1916, he endeavours to provide some hints to the total mass of stars, by way of relativistic predictions of the gravitational field produced by the stars. He does this in the wake of the solution to the gravitational field around the sun, which Schwarzschild had given early in the year (1916c, pp. 175-7). He supposes the distribution around the Sun to be of spherical symmetry. Even though this distribution is certainly very remote from what the galactic system really is, the order of magnitude of the resulting numerical relations, De Sitter surmises, should be quite acceptable.

Imagine a hollow spherical shell. Calculation of the gravitational effects produced by the shell runs as follows. The approximation involved is that these effects generate a 'distortion' of the pre-existing Minkowski space-time, i.e.,

$$ds^2 = (1+\gamma)c^2dt^2 - (1-\gamma)d\sigma^2, \qquad (2.17)$$

where $d\sigma$ is the spatial line element and γ represents the contribution of matter. The $g_{\mu\nu}$ become, in accordance with an earlier analysis in 1916a, pp. 711–17:

$$\begin{matrix} -1+\gamma & 0 & 0 & 0 \\ 0 & -1+\gamma & 0 & 0 \\ 0 & 0 & -1+\gamma & 0 \\ 0 & 0 & 0 & +1+\gamma \end{matrix} \qquad (2.18)$$

This expression can be written

$$ds^2 = -d\sigma^2 + c^2dt^2 + \gamma(d\sigma^2 + c^2dt^2). \qquad (2.19)$$

De Sitter develops a whole series of calculations which yield the expected conclusion: the theory predicts a relativistic analogue to the well-known theorem of Newton's mechanics, according to which a homogeneous, spherical layer of matter exerts no gravitational effect in its inside. Within the shell of stars, $\gamma = \gamma_0 = cst$. As a result, it is not possible to measure spatial effects inside, since the space is flat. Next, De Sitter performs the coordinate transformation which gives the *true* inertial coordinates, and it is these which enable us to find the Minkowski ds^2 from (2.19):

$$x'_i = x_i\sqrt{1-\gamma_0}; \; t' = t\sqrt{1+\gamma_0}. \qquad (2.20)$$

The consequence is, as De Sitter says, that "the units of space and time will be changed, but this cannot be verified by observations of the motions of bodies inside the shell relatively to each other".

The situation is quite different when we come to consider stars located on the surface of the shell or even beyond it. Here, the difference in the unit of time can be verified. Consider a fixed point in the three-dimensional space, so that $d\sigma = 0$ and thus $dt/ds = 1/c\sqrt{1+\gamma}$. It is clear that the measurement of time depends on the gravitational potential and is different at different places in the gravitational field. Consequently, the frequency of a periodic phenomenon, which is constant when expressed in the natural measure or proper time ds, is variable when expressed in coordinate time t. In his first paper (1916a, pp. 719-20), De Sitter had already established that the frequency of light vibrations emitted by a star is proportional to the quantity $\sqrt{1+\gamma}$. By virtue of the transformation of x_i into x'_i and of t into t', the coordinates in proper time τ are $d\tau = \sqrt{1+\gamma}\, dt$. The frequency γ_0 measured at the point of observation is thus in the ratio $v_0 \sqrt{1+\gamma}$ to the proper frequency. Because the quantity γ is no longer a constant for the matter within the layer under consideration, the light coming from a source belonging to this layer is affected by a change in the time units. The ratio between frequencies will be of the order of $\sqrt{1+\gamma_1}/\sqrt{1+\gamma_0}$, where γ_1 represents the value of γ at the source. The frequencies of such a source should exhibit a shift towards the *violet* compared with those which originate from a source located somewhere between the centre and the layer.

General relativity had already announced the theoretical prediction of a *red*shift of the spectral lines produced by the stars. This shift is quite independent of any choice of reference system: it is of *gravitational* origin, directly connected to the mass of the star. In 1916, it was still largely theoretical—the empirical evidence came somewhat later. It had taken the solar physicists a few years before they could disentangle the small predicted redshift from many other anomalies observed in the solar spectrum. (Einstein 1911. For the historical perspective on measurement of the redshift, see Earman and Glymour 1980.) A displacement of spectral lines can also be conceived in *kinematic* terms, as resulting from the relative motion of two sources: this is the well-known Doppler effect. Thus, De Sitter explains, displacement towards the red end of the spectrum for a star like the sun, as an effect due to its intrinsic field of gravitation, can easily be confused with a receding motion from the Earth. As for the violetshift predicted by De Sitter, no measurement offers anything like a systematic displacement. On the contrary, the stars generally exhibit a small displacement towards the red. De Sitter's assumption is that the internal, gravitational effect towards the red produced by each star cancels the purely relational effect towards the violet. In fact, it is precisely the slight excess of the measured redshift over the estimated violetshift that enables De Sitter to deduce an upper limit for the ratio of frequencies between γ_1 and γ_0. This, in turn, provides a rough estimation of the actual quantity of matter contained within the observable part of the universe.

(c) Einstein's very last doubts

In the note added 29 September after a conversation with Einstein, De Sitter mentions in his 1916 paper on relative rotation the set (2.12) as implying the finiteness of "the physical world", with the degenerated field of $g_{\mu\nu}$ making "space and time absolute at infinity". Accordingly, "if we wish to have complete four-dimensional relativity for the actual world, this world must of necessity be finite". A shift in the use of words is quite noticeable. De Sitter's account of Einstein's hypothesis (2.12) is in terms of the *physical* world, while his reference to four-dimensional relativity is to the *actual* world, thus stressing the fulfilment of relativity as accommodating both physics and geometry.

What is the meaning of 'finite' in this context? In fact, Einstein wrote as early as 14 May 1916 to Besso, well before he discussed the matter with De Sitter, that he now tries, in the quiet period following the establishment of the field equations, "to determine the boundary conditions at infinity; it is interesting to ask how a *finite* universe can exist, that is, a universe whose finite extent has been fixed by nature and in which all inertia is truly relative" (Einstein and Besso 1972, p. 70). As he himself recalls in his Four Lectures of 1921: "The question as to whether the universe as a whole is non-Euclidean was much discussed from the geometrical point of view before the development of the theory of relativity" (Einstein 1921, p. 109). Einstein thus explicitly acknowledges his debt to the mathematicians of the second half of the nineteenth century, without mentioning their names. He goes on to say that "with the theory of relativity, this problem has entered upon a new stage. . ." True, Riemann himself had best depicted the nineteenth-century divorce of both field physics and differential geometry from large-scale considerations, when he proclaimed that "the questions concerning the incommensurably large are, for the explanation of nature, useless questions. . . Knowledge of the causal connection of phenomena is based essentially upon the precision with which we follow them down into the infinitely small" (1854, pp. 423–4).

Throughout 1916, application of the geometrical idea of global finiteness to physics remains a matter of mere intuition. Because Einstein mixes the discussion of boundary conditions at infinity with the idea of a finite universe, it is not really clear that the meaning of 'finite' is fixed. The actual path of ideas is, however, decisive since it will determine the later formalization.

In March 1917, De Sitter, in a revealing note, acknowledged that the idea of conceiving a spherical four-dimensional world had been suggested by P. Ehrenfest in a conversation in Leiden a few months earlier (W. De Sitter, 1917, p. 1220). It would appear, therefore, that Einstein was very probably aware of the idea. At any rate, it is obvious as far back as these early discussions that the word 'finite' could carry a decidedly non-Euclidean meaning. In fact, in a letter to De Sitter dated 18 April 1917, Ehrenfest confessed that the whole idea of a curved four-dimensional universe had crossed his mind as early as the summer 1912, when he was still in Russia. He subsequently abandoned the problem rather quickly, however, because he could see the overwhelming difficulty this kind of curvature would present. It is true that Ehrenfest did not publish a single line on the subject, neither at that time nor later when the difficulty had apparently been overcome, partly it seems because he thought the very concept of time would become quite unintelligible. Yet he still believed the curvature of time together with that of space to be the straightforward consequence of what he called the Lorentz–Einstein–Minkowski relativity. It was the only possible answer to a problem which Ehrenfest put like this: "The Newton–Galileo relativity (with uniform translation) will be completely annihilated, if we pass from Euclidean to non-Euclidean space (for instance, the *spherical* space). Now I asked myself: what will become of the Lorentz–Einstein–Minkowski relativity if we transform to curved space? I saw then, that the whole four-dimensional world had to be curved" (Ehrenfest to De Sitter, Leiden Observatory).

That Einstein had a strong *theoretical* motivation for assigning a special role to time is shown by his reaction in a letter to De Sitter, dated 4 November 1916. This letter presents a partly corrective response to De Sitter's views in the note of 29 September. A few days before Einstein's letter arrived, on the 1st of November, De Sitter had worked out a lot of interesting further details on how he understood the proposed boundary conditions (2.12). If the $g_{\mu\nu}$ degenerate to these values when both the space *and the time* variables become infinite, then "the hypothesis would make the world finite not only in space but also in time". De Sitter adds that nothing can be known from the infinite past or from the infinite future. Does this make the hypothesis comply with the requirement of observability? No, because the limits "always remain hypothetical, and can never be observed". De Sitter calls these limits the "shell", in reference to his own calculation of the gravitational field produced by a hollow sphere of stars: "the shell is *not* a physical reality", since we will never have the right to assume that the knowledge of larger portions of the universe (of what lies outside the Milky Way) will confirm the already computed relations within the known part.

In fact, De Sitter is ready to accept the limitedness of the universe as a general principle (*"die prinzipielle Beschränkheit"*), but he cannot see how this stops the endless chain of questions: *Where* are the distant masses? *What* is their constitution? *How* do they influence the inertia here? He ends up suggesting that a true explanation of inertia might be found in the infinitely *small* rather than in the infinitely *large*, but shrugs it off: "I am not a physicist, and this is probably just hallucination" (De Sitter's letter to Einstein, 1 November 1916). De Sitter is perhaps echoing what *should* have happened to field physics after the revolution brought about by differential geometry. Weyl has described this in terms of a theoretical requirement: "The transition from Euclidean geometry to that of Riemann is founded in principle on the same idea as that which led from physics based on action at a distance to physics based on infinitely near action" (1918, p. 91). Obviously, Einstein's proposed explanation of the origin of inertia falls short of this analogy, and De Sitter even seems to insinuate that Mach's principle is no better than action at a distance.

Einstein's reply of 4 November shows how harassing the whole question has become. Towards the end of the letter, he almost apologizes for his

Machian commitment, saying it is no more than one of his innocent fads, and adding that in no way "do I demand that you share my curiosity". The question of boundary conditions, Einstein now says, is a question of taste which seemingly cannot receive a scientific meaning in any case. Yet, he insists he would never have thought of the world as being of finite *temporal* extent. Nor is there finite extent with regard to *space*. Einstein generalizes De Sitter's conception of the "shell". Take a purely spatial shell (without matter); in four dimensions, this becomes a "tube" ("*Schlauch*") since time is *not* curved. The shell can be taken to be so large as to include most of the inertia in the world; at any rate, the inertia outside it will be as small as we want. *Inside* the shell, it can now be said that only the existing masses will determine the inertia. Einstein is well aware that, if his argument is to have any validity with regard to observable facts, it must be assumed that the observed part of the world is *extremely small* relative to its actual size. Although this is all he says, we can easily infer that what he has in mind here is, actually, one of the strongest arguments that has ever been put forward in favour of Mach's principle. The isotropy of inertia is observed to a very high degree of accuracy in our part of the universe. But the matter in our *immediate* vicinity (planets, sun, stars) is patently not isotropic. For this reason, if inertia is produced by matter, an overwhelming part of the effect must come from *distant* matter which must, of necessity, be isotropically distributed. As for the constitution of these masses, Einstein is not worried because he readily assumes that all the sources are stars. The shell, as he says, has nothing distinctive ("*besondere*") about it. We find in these statements the first hints of what was later conceptualized as *the cosmological principle*: either one begins with the assumption that the basic constituents of the universe are the same, in which case there is little reason to deny a uniform distribution, or one begins with the uniform distribution and finds the constitutive uniformity quite plausible.

Einstein goes on to distinguish practical and theoretical implications of the problem of boundary conditions. On practical grounds, he is reluctant to give any validity to the problem of the 'whole'. In any designated part of the world, only the totality of masses which happen to exist in it may be said to determine the inertia of a given body. In the delineated section, the $g_{\mu\nu}$ are determined by these masses and by the $g_{\mu\nu}$ at the periphery. The inertia of the masses and the boundary conditions thus depend either on the system which happens to be selected, or more simply on the part of the world that we can describe. Einstein, however, clearly catches a glimpse of the only way meaning can be bestowed on the theoretical aspect of the problem of the totality, when he shows that it is tantamount to the following question: can I possibly think of the world in terms that would make the inertia depend on the masses and not on the boundary conditions? None the less, because the observed part of the universe is so small, Einstein is led to weaken his own desire to articulate a full relativity of inertia. He explains that this smallness was the *psy-*

chological motivation that dictated the drive to secure generally covariant laws.* General covariance having been gained, "there is no longer any ground to attach such a big weight to the full relativity of inertia". Behind his apparent care, one can see that Einstein is now putting into practice the remark he made to Besso in October, in which he said the distinguishing character of his local physics was its independence from the specification of boundary conditions. Indeed he now raises the question of whether the same kind of procedure could be applied to the entire universe. As he says to De Sitter, "You must not scold me if I am still quite interested in that kind of question".

In the way he opposes the practical and the theoretical sides of general relativity, it is all too clear that Einstein has some inkling of the lack of consistency between Mach's principle and the purely mathematical requirements of general covariance. Or, rather, he now realizes that the formal universality of general covariance is not sufficient to overcome completely the antinomy between the local and the global (what is here called the practical and the theoretical).

Of course, it is hard for us today to follow Einstein and De Sitter at every step of their discussions. They do not seem to be troubled by mixing up boundary conditions at 'infinity' (when they are describing the whole universe) with other statements over the truly 'finite' character of the universe. It may well be that the very word 'finite' still retained tenuous links with the classical conception of the universe. There is also no explicit distinction made between coordinates and structure. The confusion, however, has the salutary effect of driving the two protagonists to a radical question: What does it mean to speak of the whole universe as a reality for physics? Classically, and in all probability ever since the time of Kant's antinomies, the dispute over the meaning or non-meaning of such an expression as 'the whole universe' was deemed to have a speculative significance of its own, irrespective of what natural science could teach us.

(d) Half-way to the solution: the true reasons for the impossibility of Newtonian cosmology

In December 1916, Einstein completed his popular exposition on the theory of relativity, which became his most widely known book. The third edition (1918c) included three new chapters, all concerned with "Considerations on the Universe as a Whole". Einstein wanted to show that, in addition to its

---

*A year later, E. Kretschmann argued that general covariance has no physical meaning by itself: any equation can be expressed under a generally covariant form, provided that suitable physical quantities are added. (See Kretschmann 1917.) In his reply to Kretschmann, Einstein now stated that general covariance had a powerful *heuristic* role in establishing the theory of general relativity: see Einstein 1918b.

very definition of the principle of relativity, the Newtonian theory is affected by yet another kind of fundamental difficulty related to celestial mechanics. Einstein definitely assents here to Ehrenfest's proposal that finiteness of the universe is to be understood in terms of non-Euclidean geometry. So, being from now on converted to a picture of the large-scale properties of space which owes nothing to an extension of Newton's theory, it is not surprising that Einstein would pay much attention to the problems of Newtonian cosmology in order to justify his own radical departure from it. In the elementary exposition, the arguments pinpointed the ideas that the cosmological memoir of February 1917 had developed more technically. I shall comment on these arguments by following the increasing importance imparted to them by Einstein himself.

In fact, these problems of Newtonian cosmology are really *his* problems. Throughout the nineteenth century, the failure of Newton's physics to meet the requirements of so all-encompassing a problem as the whole universe was perceived in terms of the inadequate fit between a homogeneous, uniform model, and the physical laws, whether optical or mechanical. Thus, Olbers' speculations on the sky being ablaze with infinite light are dominated by the idea of a pre-given, infinite homogeneity of stars. The same is true for Neumann or Seeliger who, by the end of the nineteenth century, showed that a mean density of matter can be constant to infinity, if the law of gravitation is modified so as to allow equilibrium at great distances. Here too, the pre-given conception of a uniform, infinite distribution of matter is dominant.

One of the profound innovations in Einstein's new conjectures is the critique of just this traditional attitude. Einstein clears his mind of what had become a common prejudice. He does not set out to vindicate a particular model but simply explores the actual consequences implicit in the large-scale predictions of the theory. There had been various attempts to do this sort of thing before, like Charlier's early speculations on a hierarchical model, but always with the aim to justifying an infinite universe (see Chapter 1). The absence of a real *theory* of the island-like universe is very striking, given the importance of this figure for the astronomers. By the turn of the century, the old eighteenth-century image of an island universe had undergone its major revival, in response to the problems raised by the new techniques of distance measurements and the growing controversy over the existence of extra-galactic nebulae. But it seems unlikely that this mutating image of contemporary astronomy should have acted as a source for Einstein's radically original attitude. The belief in extra-galactic nebulae merely changes the *units* of the universe, and Einstein's units are still the stars. It precisely falls to Einstein to bring the two aspects together, i.e. (to use the words of his 4 November 1916 letter to De Sitter) the *theoretical* and the *practical* are first of all connected in the framework of Newtonian theory. In the first place, Newton's law predicts the existence of a centre where the density of stars is maximum; this density decreases as the distance from the centre increases (Fig. 2.3). The universe is thus an island of stars immersed "in the infinite ocean of space" (1918c, p. 106). The proof advocated by Einstein is borrowed from an application of Gauss's theorem. According to this principle, any mass is a point towards which lines of force converge. The number of lines of force coming from the infinite and reaching a mass m is proportional to m. If the density distribution of matter in the universe is a constant ρ, a sphere of volume V has a mass ρV, so that the number of lines of force passing through the sphere is proportional to ρV. But the surface of the sphere is proportional to R^2 and its volume to R^3. As a result, the number of lines of force which pass through each surface unit of the sphere is proportional to ρR. The intensity of the gravitational field on the surface thus increases indefinitely with the radius R of the sphere.

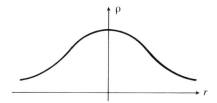

FIG. 2.3. The global Newtonian distribution of star density (ρ) from a centre O according to Einstein.

As he wrote it to Besso (A. Einstein and M. Besso 1972, p. 99), Einstein intended his argument to be a proof that the *symmetry* of a homogeneous, infinite distribution of matter is not sufficient a condition to do away with a cumbersome infinite field at infinity. (Newton had argued almost similarly in his reply to Bentley: see Chapter 1, Section 2.) In the first part of his cosmological memoir, Einstein went further and put forward three separate arguments against the inverse possibility of an island universe (1917b, pp. 177–9).

The first argument deals with a spherically symmetrical gravitational field around some point of mass. The gravitational force of a homogeneous sphere of matter acts as if the whole force were concentrated at its centre. The difference in potential between any point exterior to the sphere and this centre tends to be independent of the distance, i.e. it tends to be constant, only when the density of successive layers decreases more rapidly than $1/r^2$. What we arrive at, then, is a distance within which the exterior point is no longer exterior: it is indistinguishable from the original sphere. In this sense, Einstein claims, one can speak of the universe as a physically self-consistent system. Mathematically, this is expressed by the condition that the potential (which varies as $1/r$) shall tend to a fixed value at infinity. In other words, the Newtonian theory predicts a non-uniform distribution of matter. Once a point can be found, such that the gravitational field around it is isotropic, then that point becomes unique. The universe is therefore affected by a disjunction between space and matter: all points inasmuch as they are centres are equivalent with respect to space, but not with respect to the distribution of matter. Now, it could be argued that this disjunction is not really such a problem, at least in relation to boundary conditions. A constant, zero potential at infinity is a convenient device used in ordinary physics, which enables physicists to ignore the incalculable influence of very remote regions on a given local system. So far as a local system is considered, as we know, the assumption is that any 'rest' of the universe is 'nothing'. Now, the image of an island universe seems to be a natural corollary of such a view: if, for all practical purposes, there is absolutely no matter to worry about in the infinite, then the universe is directly comparable to the largest 'local' system that may be imagined.

The conception of an island universe, Einstein says, "is not very satisfactory" (1918c, p. 107), since it runs against our *esthetical* demands of no privileged centre in the universe. It is precisely in order to avoid this "distasteful conception" (p. 107) that the traditional attitude could have prevailed. And in the two succeeding arguments of his cosmological memoir, Einstein manages to highlight the difficulties inherent in that apparently comfortable collusion between the methods of mathematical physics and the concrete figure of the universe. The snag comes in the form of applying another kind of physical law. With relative velocities of stars so small compared with that of light, how is it that the stellar universe manages to keep itself going instead of scattering itself into infinite space? Radiation emitted by the peripheral stars leaves the Newtonian system, becomes "ineffective and lost in the infinite" and Einstein fears that even entire heavenly bodies might end up the same way. A finite kinetic energy is certainly sufficient to overcome Newtonian forces of attraction. According to Einstein the reality of that process can be established by the logic of statistical mechanics: "this case must occur from time to time, as long as the total energy of the stellar system—transferred to one single star—is great enough to send that star on its journey to infinity, whence it can never return". If the density decreases at a sufficient rate with the increasing distance from the centre, then the difference of potential between the centre and the infinite is, by construction, finite. Now, statistical mechanics shows that the relation between stellar populations at any two points A and B is $p(A)/p(B) = e^{[E(B) - E(A)]/kT}$ where E is kinetic energy, k is Boltzmann's constant and T is the temperature. If the potential at A remains finite while A tends to infinity, and if, furthermore, it is supposed that kinetic energy is everywhere finite, it follows that the population at infinity is not zero. Stars are apt to leave the supposedly finite system, quite apart from the radiations emitted by those bodies. The contradiction Einstein has in mind works like this: a constant value for the potential at infinity implies, according to physical arguments, a finite universe; but, again on physical grounds, this universe cannot remain finite in any strict sense. So the largest system is never large enough.

At this point, and quite independently of the use of statistical arguments, Einstein's conclusion seems to dramatize, with a striking singularity, the problem that lies at the heart of any cosmological theory. Let us take the homogenous and infinite model first. Starting from a certain point chosen as the centre, we draw concentric shells of matter. Any particle that is located at a certain distance from the centre will be subject to contradictory influences. Any matter outside the shell, whose radius is the centre-to-particle distance, will not act upon the particle, only the matter inside it can act upon it, as though all matter were concentrated at the centre (see Fig. 2.4). Now, if we draw the concentric shells around the particle taken as centre, no matter should act upon it since the state it is in may be thought of as analogous to an inertial state. This puzzle led J.L. Synge (1937, pp. 94–5), following on from an early suggestion of J.C. Maxwell's (1887, p. 85), to state the following paradox: there are postulated fields of gravitation incapable of detection by any observer, since it is always possible for a given observer in 'free fall' to consider himself inertial, provided he takes it upon himself to describe the world in terms of a gravitational field which is everywhere null, instead of the non-null field which 'really' exists. All we can say of any particle in free fall which is taken as an inertial centre, is that matter which acts or fails to act upon it is not 'physically' the same in both situations—it is only 'ideally' the same, which amounts to transferring the ideality of inertia to gravitation so far only a hypothetical 'reality' (Fig. 2.5). And in terms of the cosmological question, this means that the uniform model entails difficulties of *local* scale, it being assumed that the boundary is in any case purely 'ideal'. From this point of view, Einstein's argument is far-reaching enough, since it shows that the non-uniform model, with its 'physical' picture of the boundary, is simply

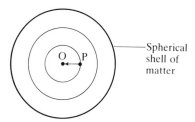

FIG. 2.4. P is attracted by matter situated between O and P, as if all the matter were concentrated in O. The influence of exterior shells is immaterial.

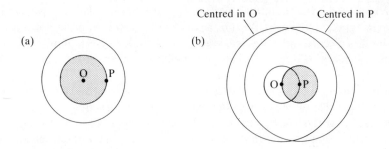

FIG. 2.5. (a) The shaded area represents matter having gravitational influence on P from O's viewpoint. (b) If P is taken as centre, it is free of gravitational action but with respect to other matter.

the inversion of the original problem: the paradoxes are now occurring on a *cosmic* scale, because the physics of the boundary cannot stay in place.

Yet, the validity of Einstein's argument remains questionable. Indeed, Einstein speaks of the infinite as if it were some kind of place within the reach of stars and radiation. Commentators have been struck by the highly dubious cogency of the argument (see J. North 1965, pp. 70–1). As a matter of fact, the idea of some form of 'depopulation' of the star system is not borne out by the kind of statistics he marshalls in his arguments (useful comments are given by W. Pauli 1958, p. 162). A fully consistent application of the very same principles should have led him to predict a statistically equal number of stars *coming back* from the infinite. But the relevant significance of Einstein's argument is its general thrust rather than any disputed validity of its detail. In essence, the idea is that no definite periphery can be assigned to the centre of the island universe, except by equating the infinite with a specific location that remains within reach. While the disjunction between space and matter caused little difficulty in the first argument, it now becomes cumbersome: the unique point as centre of the mass distribution is not related to a definite periphery, which means that the centre itself might lack definition as well, even though the island-like universe still necessitates a centre. Einstein proceeds to annihilate one possible way out of this quandary. One suggestive solution runs like this: "We might try to avoid this peculiar difficulty by assuming a very high value for the limiting potential at infinity". All bodies move from higher to lower potentials. Even though the potential decreases from the centre to the periphery, the boundary potential could be so high as to hinder celestial bodies from crossing it—in short we could posit a definite periphery (Fig. 2.6(a)). Alternatively, we could see the potential as increasing from the centre to the periphery until it reached a particular limiting value and as a result generated a movement towards gravitational collapse (Fig. 2.6(b)). This latter view is never for a moment entertained by Einstein.

He only says to Besso, in December 1916, that "Jehovah has not built the universe on such an extravagant basis" (A. Einstein and M. Besso 1972, p. 99). He pre-empts the entire line of reasoning by invoking *facts* presented to astronomers: the general weakness of stellar velocities imposes an overall distribution of potentials which must not be drastically different from those prevailing in the immediate vicinity of the earth. The logical slide is all the more amazing for being so inconsistent: Einstein, who had begun by demarcating a type of argument which is grounded in the predictions of the theory, finds it necessary, in order to oppose one of the possible predictions, to have recourse to nothing better than an accidental fact. Of course, he does have something in mind which acts, however latently, as a decisive justification—the space/matter disjunction cannot be refined out of existence by some sort of break between local and putatively 'cosmic' conditions. If cosmology is to be a science, it ought to be anchored in the soil of local physics.

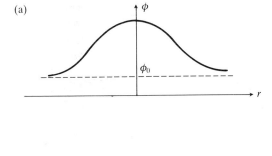

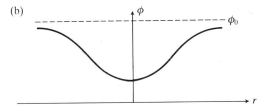

FIG. 2.6. (a) and (b) Two ways of securing a limiting value ϕ_0 for the potential at infinity.

This becomes quite clear in the third and final argument, where theoretical requirements are no longer applied to the behaviour of single bodies *within* the universe. Einstein finds that the application of statistical procedures to the *whole* of the island universe is devastating, because that universe is thereby deprived of existence: "If we apply Boltzmann's law of distribution for gas molecules to the stars, by comparing the stellar system with a gas in thermal equilibrium, we find that the Newtonian stellar system cannot exist at all. For there is a finite ratio of densities corresponding to the finite difference of potential between the centre and spatial infinity. A vanishing of the density at infinity thus implies a vanishing of the density at the centre". This argument takes no account of the *fact* that there are only slight differences of potential in different parts of the universe. Einstein is quite concerned with the very idea of a finite difference between the infinite and any centre. If any difference of potential between r_∞ (r at an infinite distance) and r_0 (the centre) leads to a finite quotient of densities, then

$$\frac{\rho(r_0)}{\rho(r_\infty)} = e^{-\frac{\phi(r_0) - \phi(r_\infty)}{kT}}$$

where ϕ is the potential. If $\rho(r_0)/\rho(r_\infty)$ is a finite number, a zero density at infinity implies a zero density at the centre. It is to be emphasized that a potential is defined by a quotient of the form $1/r$ *and* a constant of integration. A sufficiently large value for the constant would have the effect of invalidating the whole statistical procedure, because the non-vanishing density at the centre would then be equal to the density at infinity.

Einstein either could not see the difficulty or he pretended not to see it. To Besso he writes that "we can also do without statistical considerations". It is only "a more elaborate train of thought that shows the correctness of my application of the statistical procedure to those questions I am interested in" (p. 97). Einstein does not say what he means by "more elaborate", but it is true that the last argument is meant to round off the two preceding arguments. Its effect is to demonstrate that the presence of matter does not enable us to distinguish between the periphery and the centre of the island universe. In fact, the identity of centre and periphery is the distinguishing feature of an *infinite* universe. However alien to infinite homogeneity the predictions of Newtonian physics may be, Einstein wants to be able to provide a demonstration that, in addition, these predictions are no better as physics either. Of course, it is primarily because he keeps in mind his own Machian commitment that Einstein first points out the disjunction between matter and space, and that he then constrains the Newtonian system with its centre so as to deprive it of any definite periphery. There was apparently no other way of bringing together the *form* of Newton's laws with a problem a priori independent of it, the physical interpretation of its boundary conditions. The important result is that, in Einstein's mind, this impossible physics leads to an inevitable return of the more 'natural' conception (independent of all theory) by which we spontaneously assume that, for the universe as a whole, the distribution of matter should be uniform.

(e) Boundary conditions: mathematical or physical?

Having disposed of Newtonian cosmology, Einstein explains that, with the help of the mathematician J. Grommer, he returned to general relativity, and had time and again tried to make sense of static and symmetrical gravitational fields, with the values (2.12) for spatial infinity. The components of the energy–momentum tensor were derived from the $g_{\mu\nu}$ (2.16) and (2.17). The tensor representing ponderable mass is, in the most general case, $T_{\mu\nu} = (dx_\mu/ds)(dx_\nu/ds)$. Substituting $\sqrt{g_{44}}\, dx_4$ for ds, in compliance with the weak stellar velocities, the result is that only T^{44} comes to play a dominant role. This means that the stars form a system which is quite comparable to a perfect gas. The important point, though, is that "it was quite impossible to reconcile this condition with the chosen boundary conditions" (1917b, p. 181). Einstein goes on to explain that, in retrospect, this is not really a surprise. The hypothesis (2.12), that is, A approaching zero while B tends to infinity, might seem quite satisfactory since it predicts an infinite potential energy at infinity (the expression for the potential energy is $m\sqrt{B}$). This prevents bodies from leaving the supposedly island-like universe, the flaw in Newtonian theory. However, there is another prediction which is quite unacceptable: this is the increasing velocities of the stars with distance. Einstein explains that the small velocities of the stars, concomitant with a fairly uniform value for the potentials throughout space, rule out such an enormous discrepancy as predicted by (2.12). In fact, Einstein's insistence on this data of experience, even if it is certainly quite reasonable in view of what he seems to have known of the astronomical researches of his epoch, is much more than a mere matter-of-fact justification. We find the same insistence in the cosmological memoir itself and also later in Einstein's letters to De Sitter—for instance on 14 June 1917, when he repeats that "a sensible conception of the universe before us necessarily requires the approximate constancy of g_{44} in space, owing to the small relative motions of the stars". (In fact, by 1917 Slipher˙ had already measured some of the surprising redshifts in the radiation of nearby galaxies.) Returning several years later to this aspect, in one of the appendices to his popular exposition, he explains revealingly that the problem was not so much factual as theoretical: the static hypothesis "appeared unavoidable to me at the time, since I thought that one would get into bottomless speculations if one departed from it" (Einstein 1954, p. 133).

This is already clear from the examination of values (2.12). They do not seem to make any physical sense at all. Only if the stellar velocities were to increase with distance would these $g_{\mu\nu}$ at infinity represent something physical. It is exactly at this point that Einstein comes face-to-face with the true core of his problem, since the fundamental, theoretical difficulty which crops up is that a universe which is not static does not seem to make much sense: the increase of the stellar velocities, implied by the condition $A \to 0$ and $B \to \infty$, makes it impossible to find a reference system which could be adapted to the whole universe. Or, more accurately: with (2.12), not all reference systems would be equivalent in describing the matter distribution. The metric (2.13) is appropriate to the description of that part of the universe where the approximation $ds \approx \sqrt{g_{44}}\, dx_4$ is indeed observed. But beyond any given neighbourhood, where the stellar velocities would become significant, it would be necessary to choose a new reference system with respect to which the stars return to rest.

What is going wrong? Is the fault in the metric or in the boundary conditions? Einstein continues his cosmological memoir by sketching two alternative possibilities for the boundary conditions. Firstly the universe cannot be a problem because only systems within the universe are properly speaking the subject matter of physical science. Einstein views this as a mere clinging to the hitherto assumed validity of general relativity. His words are virtually a repetition of his November letter to De Sitter: "at the spatial limit of the domain under consideration we have to give the $g_{\mu\nu}$ separately in each individual case". In other words, because the proposed boundary conditions do not make sense, a metric for the whole universe cannot make sense either. The second way of evading the difficulties with the boundary conditions is to assume that the pseudo-Euclidean values (2.10), which can be taken as the boundary conditions of the system of planets around the Sun, are also applicable to the entire universe. Of course, as De Sitter had already argued, this presupposes a determined choice of the reference system. But Einstein now realizes that it is equally in contradiction with his own, Machian, conception of the relativity of inertia:

For the inertia of a material point of mass m (in natural measure) depends upon the $g_{\mu\nu}$; but these differ but little from their postulated values, as given above, for spatial infinity. Thus inertia would indeed be *influenced*, but would not be *conditioned* by matter (present in finite space).

Any single body would possess an amount of inertia that is virtually irrespective of the actual quantity of matter in the universe as a whole. This would boil down to a restricted version of Mach's principle, the inertia of a single point of mass being a limited variation around some otherwise determined value.

Einstein's obstinate wish to incorporate a complete relativity of rotation in the theory of general relativity thus results in him being left with a fairly unsatisfactory alternative. This was the situation he was in by the end of 1916; then on 2 February 1917 Einstein announces in a letter to De Sitter that he is doing fresh work on the problem of boundary conditions, in which he claims to have finally overcome the awkward degenerating field of $g_{\mu\nu}$: "I am curious to know what you will have to say about this somewhat fanciful conception". Two days later, he writes to Ehrenfest: "I have. . .again perpetrated something about gravitation theory which somewhat exposes me to the danger of being confined to a madhouse".

(f) "A method which does not itself claim to be taken seriously"

The essence of Einstein's new argument is quite the opposite of the point of view he expressed to Besso at the end of October: there *is* a way of bringing together the absence of boundary conditions with a consistent discourse about the universe envisaged as a totality. Einstein achieves this by taking a fresh look at his recent investigations of boundary conditions and their physical nature.

It had gradually become apparent that no boundary condition could be harmonized with a 'sensible' conception of the universe. If the universe was just a special 'extended' local system, then the boundary conditions (2.10) would tell us that any remainder could be seen as void; but there *is* in effect no remainder outside the universe. If this remainder is taken as being indeed physically nothing, yet distinct in character from mere emptiness (the conditions (2.12)), then no metric can be found which will satisfactorily describe the universe as a static one. Einstein's breakthrough was to think of a third possibility: "For if it were possible to regard the universe as a continuum which is *finite (closed) with respect to its spatial dimensions*, we should have no need at all for any such boundary conditions" (1917b, p. 183). This is the spark, the new way of looking at the same thing. Einstein's discovery was to see that there was an immense conceptual distinction between disregarding the boundary conditions or treating them with indifference and explicitly rejecting them; this was the needful leap that carried from a formally universal physics to the idea of a physics of the universe. The transition, although smooth, was very demanding. The example of the Schwarzschild exterior solution made it clear that the specification of boundary conditions was just as artificial, just as fictive, as the presumption that an isolated body could describe a physical reality. The infinite had long been acknowledged as a nonentity, both mathematically and physically. But the finiteness of space, if it is to have a counterpart in physics, implies that this nothing must be transformed in what was vividly described by Eddington as a space "boundless by re-entrant form, not by great extension. *That which is* is a shell floating in the infinitude of *that which is not*" (1928, p. 83).

Proceeding to his own interpretation of finite space, Einstein sends his reader back to the much earlier discussions of Riemann, Helmholtz, and Poincaré, in accordance with which the *non-infinity* of space does not necessarily stand in contradiction with the laws of thought, and even less with the facts of experience (1918c, p. 108). His argument in the popular exposition of relativity is essentially concerned with an 'intuitive' proof of the consistency of finite space. Of course, it is not really necessary to rephrase this type of proof here, since we have become very much accustomed to finite space. The fundamental feature to be noted is that we have to try to conceive the surface of an ordinary, two-dimensional sphere as existing with three dimensions, without an extra dimension in which it would be embedded. From a given point, draw a series of spheres with increasing radii. In the case of finite, spherical space, the strange property of the areas of these spheres is that they do not increase as quickly as the radii, that is, the surface of a sphere of radius r is no longer proportional to r^2. At a certain distance from the given point (the "circumference of the universe", as Einstein calls it), we have reached a sphere of maximum possible area. Beyond this, the areas decrease until the last sphere shrinks to a point. The point is the antipode of the starting point; there is simply no 'space' apart from these spheres.

More relevant is Einstein's implicit use of an idea from Riemann's famous 1854 lecture. The idea is that the homogeneity and isotropy are, in Riemannian manifolds, *global* properties of these manifolds that can be obtained only when the space under consideration is of constant curvature (Riemann 1854, p. 19. Some developments can be found in L.P. Eisenhart 1926, p. 86, and R. Torretti 1978, p. 100. Constant curvature is the only case in which homogeneity and isotropy mutually imply one another). Furthermore, Riemann expressed the view that only a manifold of constant curvature allows for the very physical existence of rigid bodies. As a result, Einstein shows that spherical space, because all its points are equivalent, is the most appropriate to the properties of a "world" in his general theory.

So much for the geometry. Setting about the task of making physical sense of the finiteness of space, Einstein's problem was to find out how the distribution of matter could be made responsible for space to be 're-entrant'. In other words, how is it possible to make sense of the apparently impossible, that is, a particular *solution* to the field equations which makes no use of boundary conditions? More specifically, what he had to find was a single reference system which could cover the whole of finite space in such a way as to overcome the prediction associated with a global requirement (Mach's principle) in infinite space, i.e. the non-static distribution of matter.

Two requirements must be set down (1917b, pp. 183-4). The first is a sort of principle, stating the possibility to disregard the local lack of uniformity in the matter distribution; over enormous spaces, it should be allowed to represent a uniform distribution. The second requirement is the fact of experience: "the relative velocities of the stars are very small as compared with the velocity of light". Einstein now makes clear that this fact of experience is a condition without which he would lose his foothold: "There is a system of reference relative to which matter may be looked upon as being permanently

THE INVENTION OF COSMOLOGY   157

at rest". From these two requirements, Einstein sets out to calculate the corresponding components of the field equations.

For the energy–momentum tensor, rest implies that only T_{44} is not vanishing: it is equal to the density of matter. The density, Einstein says, is not a function of the space coordinates *because* space is taken as being finite. In a sense, Einstein seeks to 'deduce' the uniform distribution from the finiteness of space. The determination of the purely spatial components of the metrical tensor (g_{11}, g_{12}, . . .g_{33}) is, in turn, based on this uniformity. From this assumption, "it follows that the curvature of the required space must be constant" (1917b, p. 185): this is the spherical space. The chain of reasoning is as follows: finiteness of space → uniform distribution of matter → sphericity (constant curvature) of space. The calculation of the temporal components amalgamates *all* of these assumptions. Matter at rest implies no acceleration for a test body, that is, the global contribution of the gravitational field is null. Like T_{44}, g_{44} is thus independent of locality. We can write $g_{44} = 1$ and, because the static field involves time to be orthogonal to space, $g_{14} = g_{24} = g_{34} = 0$ for any kind of spatial relations. The constancy of g_{44} means that there is a *cosmic time* superimposed on all proper times. Cosmic time is the new concept which translates Einstein's getting over his earlier problem with the boundary conditions. The direction of time is the same from all points. The paths of material particles or light rays are geodesics along the time axis of space–time; cosmic time means that, the 'length' of geodesics being measured along this direction, there can be neither convergence nor divergence of world-lines.

When the space and time components are taken together, the figure of the new universe can be drawn. This is what has become known as the *cylindrical universe*, a representation first introduced by mathematician Felix Klein soon after publication of Einstein's memoir (1918, p. 408). The universe is compared to a cylinder whose axis of symmetry is time. Reduce the three-dimensional Riemannian space to one dimension: this is the circular line in a Euclidean plane. If the lines representing time are all parallel to each other and independent of the space curvature, we have the cylinder drawn on Fig. 2.7. Each circular 'slice' of the cylinder represents a contemporaneous space, i.e., an instantaneous 'photograph' of the world.

In trying to bring about the connection between the geometrical and the physical, Einstein seems to make an implicit use of one of the theorems demonstrated by Schur (1886) stating that a Riemannian manifold which is isotropic about *each* point is also homogeneous. (The flaw of Newtonian cosmology was that, if the matter distribution around one point is isotropic, this point is unique, i.e. there is no homogeneity.) It is the mutual implication of large-scale homogeneity and large-scale isotropy that has become known as the assumption already briefly touched on above and formulated as the *cosmological principle*. That this idea remains almost entirely implicit at this

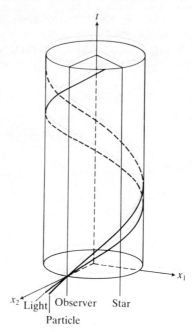

FIG. 2.7. World-lines in Einstein's cylindrical universe (from H. Roberston 1933, p. 70).

stage is witnessed by the fact that only Milne, in the mid-1930s, was the first to introduce the idea with this term; and Milne attributed the idea to Einstein's pioneering work. (Milne is often credited with the enunciation of the principle, although he repeatedly accords the honour to Einstein's 1917b. See Milne 1935, p. 24 and pp. 60 ff.)

Schur's theorem, it is to worth noting, says nothing about *constant* curvature. The actual progression of Einstein's demonstration can then be stated in the following way:

1. The very idea of the finiteness of space is to Einstein concomitant with isotropy about each point. In finite space, the existence of a centre has no *raison d'être*. Here is the crucial difference from Newton, since the Newtonian prediction of a finite distribution of matter *in* space implies isotropy about one point only.
2. The implicit use of Schur's theorem implies that homogeneity has to be added to isotropy for the distribution of matter.
3. Riemann's idea of global homogeneity and isotropy makes it possible to speak of constant curvature for the finite space.

What has been 'deduced' in 3. is *also* a fundamental requirement of general relativity, namely, the concept of ideal rigidity of physical bodies, in accordance with Riemann's own views.

And there is still more because the role of linear, uncurved time is justified in the same way: g_{44} is constant *because* it is independent of locality. In other words, if matter is to generate the *whole* of the metric field (both in its spatial and temporal components), the uniform distribution of this matter can only be thought of as permanent. In the Newtonian theory, a state of uniformity and of rest could still be conceived as the *final* phase of a general collapse of the island-like universe. But Einstein makes the state of uniformity a lasting thing and so rules out the idea of evolution altogether.

In consequence, only a spherical space could make the mutual implication of rest and isotropy valid for the whole of space, in accordance with the general assumption of uniformity. As we can see, this accord embodied an interaction between the three features which Einstein wanted to combine: Mach's principle, the possibility of one reference system covering the whole of space, and the fact of the weak stellar velocities. Rest itself is more than a fact: it shares in the theoretical framework because it cannot be removed without annihilating the entire interaction.

The mathematical representation of three-dimensional spherical space was not essentially different from that by which Schwarzschild could arrive at the interior metric. The similarity was of course limited to the spatial geometry, since neither the time nor the material tensor were identical. In fact, the eradication of the boundary conditions in the case of cosmology was the major difference. But it was also the most extraordinary stumbling-block of Einstein's innovation. We know that the interior metric implied a definite choice of boundary condition, related to the pressure. In cosmology, there would be an enormous price to pay for getting rid of boundary conditions, and this price was certainly connected to the apparently inoffensive question of the pressure.

The correct understanding of the status of boundary conditions implied a careful distinction between the *solutions* to the field equations and the equations themselves, a fact which De Sitter had stressed with considerable brilliance. Now, what about the consequences of finite space on the *equations* themselves? This was the price to pay:

. . . and if it were certain that the field equations which I have hitherto employed were the only ones compatible with the postulate of general relativity, we should probably have to conclude that the theory of general relativity does not admit the hypothesis of a spatially finite universe (1917b, p. 186).

Einstein actually obtained his metric for the universe with the assistance of calculations by no means dissimilar to those of the Schwarzschild interior metric (see Chapter 1). Inserting the expressions for the pressure and the density in (1.7a) and (1.7b), and taking into account the fundamental equation of perfect fluids (1.11), we are led to the following system of equations:

$$8\pi p = e^{-\lambda}(\mu'/r + 1/r^2) - 1/r^2 \quad \text{(a)}$$
$$8\pi p = e^{-\lambda}(\lambda'/r - 1/r^2) + 1/r^2 \quad \text{(b)} \quad (2.21)$$
$$dp/dr = -[(p + \rho)/2]\mu'. \quad \text{(c)}$$

The two metrics differ in the choice of boundary conditions. First, the cosmological metric demands that λ and μ tend to zero when r itself tends to zero, while the interior metric demands the same when r tends to infinity. Secondly, there is a definite option for the boundary condition of the pressure in the interior case (a decreasing pressure from centre to periphery), while the cosmological metric requires a constant pressure throughout space. While a null pressure at the periphery led Eqn (2.21c) to become (1.12), an overall constant pressure leads to

$$(p + \rho)\mu' = 0. \quad (2.22)$$

The density ρ must remain positive, and p is assumed to be constant. As a result of $\mu' = 0$, we have $\mu = cst$. In fact, μ is even zero, since $e^\mu = 1$ for $r = 0$. From Eqn (2.21a), we have $e^{-\lambda} = 1 - r^2(-8\pi\rho)$. With this relation, the metric of the cylindrical universe can be written:

$$ds^2 = \frac{dr^2}{1 - r^2(8\pi p)} - r^2 d\theta^2 - r^2 \sin^2\theta d\phi^2 + dt^2. \quad (2.23)$$

The coefficients of this metric will have the appropriate Minkowski signature $(+,-,-,-)$ only for r^2 less than $-1/(8\pi p)$. In other words, the radius of curvature R of the whole universe is

$$R^2 = \frac{1}{-8\pi p}. \quad (2.24)$$

Indeed, that is the only value which satisfies the general condition of a spatially finite universe. This is also the fullest expression of the problem, since the right-hand side of this equation should be positive like R^2. This is the reason why Einstein introduces the so-called *cosmological constant* Λ:

$$R^2 = \frac{1}{\Lambda - 8\pi p} \quad (2.25)$$

This enables him to make the right-hand side positive. As he puts it to Besso, "the closed universe is also suggested by the fact that *the curvature has the same sign everywhere, since experience teaches us that the energy-density does not become negative"* (Einstein and M. Besso 1972, p. 100, letter of December 1916. Einstein's emphasis.)

How is it possible to justify this new constant? At first sight, it seems that the problem of a negative pressure has been simply transferred to Λ. Einstein was immediately disturbed by that mathematical makeshift. Less than two years after the cosmological memoir, he expressed his doubts as to the physical nature of the constant. He tried to show that Λ is comparable to a mere constant of integration and that it has nothing to do with a constant of nature (Einstein 1919, p. 354. See also Chapter 3.) Already in the cosmological memoir, Einstein chooses to take the pressure to be zero. This is certainly a very good approximation of the observed state of stars, if at least the local motions are neglected. But, just as the low stellar velocities were a conceptual ingredient without which much of the whole construction would have been blocked, the fact that Einstein takes the pressure to be zero is not really an accident, since it bears a theoretical significance: the radius of curvature becomes

$$R^2 = \frac{1}{\Lambda}. \qquad (2.26)$$

With a zero pressure, the value of the cosmological constant is not confined a priori to a particular set of values. In doing so, Einstein is certainly following De Sitter's criterion of freedom for the values of such constants in general relativity. This tallies with De Sitter's troubles about the mathematics of general relativity. And furthermore, as Einstein made it clear in a 24 March 1918 letter to mathematician Felix Klein, the point of having zero pressure is that it makes cosmological considerations theoretically distinctive from local ones: in the case of the Schwarzschild sphere, "g_{44} is variable and equilibrium is possible only with spatially variable pressure", while in his solution, as Einstein went on to say, the cosmological constant makes two things possible at one stroke, namely, "the constance of g_{44} and the vanishing of the pressure". Yet, the relation of Λ to R remains, of course, very puzzling indeed. It is really unclear that any serious reason can be invoked why Λ should not be fixed, on account of its direct relation to R. It seems difficult to speak of a definite size for the universe if Λ is to remain undetermined. (The radius of curvature is not the only quantity to be directly related to Λ. Both the volume and the total mass are dependent on it: We have the relation $8\pi\rho = 2/R^2$, from which we find $R^2 = 1/\Lambda = 1/4\pi\rho$, and the volume is just $2\pi^2 R^3$ while the total mass is the product of volume and mean density.)

In fact, Einstein introduces the cosmological constant by again appealing to a parallel with the Newtonian theory. As he reminds us at the end of the analysis of Newtonian cosmology with which he opened his memoir, the strategy for overcoming the paradoxes of the island universe involves what looks like a piece of similar contrivance. It was in the third edition of his popular exposition of relativity that Einstein gave Seeliger credit for the modification of Newton's law, according to which "the force of attraction between two masses diminishes more rapidly than would result from the inverse square law" (1918c, p. 107). In fact, C. Neumann had reached similar conclusions (see Seeliger 1895 and Neumann 1896). Einstein went on to emphasize that neither a theoretical principle nor an observation would ever justify the proposed modification; any other convenient law would do the same job. Seeliger had based his calculations on the fact that static equilibrium with constant density up to infinity is not possible according to Newton's theory. The gravity at the surface of a homogeneous sphere of radius r increases with the radius, in the proportion given by $(4/3)\pi G\rho r$. An infinite radius leads to infinite gravity, and those solutions of Poisson's equation which imply a vanishing field at infinity are constructed so as to deprive the infinite of physical existence. Seeliger's modification of the inverse square law was $F = Gmm'e^{-\Lambda r}/r^2$. Of course, Λ can be so small as to have no bearing on empirical verification of the hitherto established law within the solar system. Seeliger maintained that the Newtonian law is taken by itself purely empirically, and nothing compels us to believe in its absolute accuracy. The cosmological constant really plays a very peculiar role, since it enables us to avoid the action of the infinite over the finite while at the same time rendering the material universe actually infinite. This proved to Seeliger that the construction was artificial: the constant seems to make the universe infinite, but he believed the universe really did have an island-like shape (see S.L. Jaki 1972, p. 278).

Einstein explains that the corresponding modification in Poisson's equation is

$$\Delta\phi - \lambda\phi = 4\pi G\rho_0 \tag{2.27}$$

where ρ_0 is the uniform density. The new solution,

$$\phi = -\frac{4\pi G}{\Lambda}\rho_0, \tag{2.28}$$

according to Einstein, "corresponds to an infinite extension of the central space, filled uniformly with matter" (1917b, p. 179). The conflict with statistical mechanics is removed, since the infinite is no longer a 'place' within reach, distinct from other places. An infinite extension of the centre is precisely the demand of Newton's law, since this law knows only of central forces. It can be said that the introduction of Λ in Newton's law is already a covert way of removing any necessity for the rigorous statement of boundary conditions, since no cosmic periphery need be considered: "A universe so constituted would have, with respect to its gravitational field, no centre".

The magic of Λ in the Newtonian theory is thus that it removes the ambiguity of a general arbitrariness of the periphery in this theory. The cosmological constant is really the heart of Einstein's new argument, since it has bearing on the most delicate of all questions, which is how the new theory stands comparison with the old. Einstein makes it clear that what he is going to unveil is "a method which does not in itself claim to be taken seriously; it merely serves as a foil for what is to follow". Beyond any surface similarity, there is indeed a fundamental contrast, because the Λ-term now fixes the periphery and removes all reference to the centre. Thus, the new constant is parallel to Newton insofar as the *form* of the laws is concerned, but the *interpretation* of it diverges sharply from formal analogy.

The introduction of Λ into the field equations transforms the original equations (1.3) in the following manner:

$$R_{\mu\nu} - \Lambda g_{\mu\nu} = - k \, (T_{\mu\nu} - \tfrac{1}{2} g_{\mu\nu} \, T). \qquad (2.29)$$

Amazingly, Einstein closes his memoir by emphasizing a completely unexpected function:

It is to be emphasized...that a positive curvature of space is given by our results, even if the supplementary term is not introduced. That term is necessary only for the purpose of making possible a quasi-static distribution of matter, as required by the fact of the small velocities of the stars (1917b, p. 188).

This testifies to Einstein's quite fantastic wavering over the true significance of his own labours. This statement is in explicit contradiction to his earlier view of Λ as concomitant with the hypothesis of finite space. On the one hand, it highlights the fact that the finiteness of space follows from the purely geometrical part of the new cosmological considerations: this is achieved by the decision, independent of Newton's theory, to incorporate non-Euclidean geometry in order to describe large-scale portions of space–time. On the other hand, it lends support to an interpretation of the constant which derives clearly enough from the Newtonian theory: quite apart from its mathematical role as a constant, Λ is the *physical* force of repulsion which countermands the effects of gravitation, and contrary to all known forces, the repulsion increases with distance so as to cancel the additive (and devastating) effects of gravitation. As Gamow suggested in one of his striking expositions, if Λ were a true force of repulsion, this force would have the peculiarity of affecting in exactly the same way such different objects as a common apple, the Moon, and the Sun! (see G. Gamow 1956, p. 319).

Because of the difficulty in correctly interpreting the new constant, Einstein was quick to warn against too serious a comparison between Newton's and his own theory, especially in their treatment of Λ. But, after all, is Λ itself to be taken seriously? Some physicists and commentators believe it should, while others try to minimize its theoretical significance. Einstein's own waverings on this score are in this respect characteristic of the ensuing disputes which spelled it out in all of its potential contradiction; this toing and froing has prevailed from the beginning until the present. Perhaps no topic has been as controversial in the whole development of twentieth century theoretical cosmology. (For a discussion of its impact today, see the important review article by W.H. McCrea 1971.) The problem is whether Λ is a true force or a simple additive constant in the pressure. In the years following the rediscovery of non-statical solutions, Einstein welcomed every suggestion which promised a way of avoiding what he saw as the unsatisfactory character of Λ. In the second edition of his *Meaning of Relativity* he wrote that the cosmological constant has lost "its sole original justification—that of leading to a natural solution of the cosmological problem" (1945, p. 127), but strikingly enough he had begun his argument by stating that Λ "can only be justified by the difficulty produced by the almost unavoidable introduction of a finite average density of matter" (pp. 111-12). Already in a paper of 1931 he attempted to justify the withdrawal of Λ by showing that it is not necessary once the 1917 hypothesis of spatial structure and density being constant in time is dropped in view of Hubble's empirical findings. Similarly, in replying to Georges Lemaître in his 'Autobiographical Notes', Einstein made it clear that "the introduction of such a constant implies a considerable renunciation of the logical simplicity of the theory, a renunciation which appeared to me unavoidable only so long as one had no reason to doubt the essentially static nature of space" (1949, pp. 684-5). And earlier in a letter to Lemaître which was a private reply to Lemaître's paper, he confessed straight out that "since I have introduced this term I had always a bad conscience" (26 September 1947). Criticizing all attempts to view Λ as a force of repulsion of the Newtonian kind, McVittie has argued quite convincingly that this constant does only manifest "itself by the appearance of a trivial additive constant in the pressure" (1954, p. 180). But it is probably Leopold Infeld who has best portrayed the disarming simplicity in the original purpose of Λ: in the equation (2.29), "if the geometry of the universe is known (that is, the $g_{\mu\nu}$ tensor), if Λ is known, then the physics of our universe (that is, the energy-momentum tensor) is know too" (Infeld 1949, p.484). In other words, Λ fills the gap between the *independent* positing of the finite geometry and the '*true*' physical aspects of the universe at large; the hope is that, if Λ is vindicated empirically, then the positing of the geometry will look less gratuitous, i.e. the rejection of a privileged centre in the universe will not seem to be resulting primarily from our own 'aesthetic' demands.

This brings us to those authors who have argued in favour of Λ, and therefore done more justice to Einstein's original plan. Apart from Lemaître (whose ideas we will examine in greater detail in Chapter 5), Eddington's arguments are worth attending to in this respect. Eddington is certainly the most eminent theorist after Einstein to think of the cosmological constant as forming an integral part of the relativity theory, and indeed of any consistent physical theory of nature. As he wrote in what is perhaps his best known book, *The Expanding Universe*, "I would as soon think of reverting to Newtonian theory as of dropping the cosmical constant" (1933, p. 24). To be sure, he is at some distance from Einstein since Einstein's "original reason was not very convincing, and for some years the cosmical term was looked on as a fancy addition rather than as an integral part of the theory". However, even in his first book on relativity, there is little doubt that Eddington had come to the most illuminating possible account of any physical meaning that could be attached to Λ. It is an account which certainly expresses with the greatest lucidity what Einstein might well have had in mind when he resorted to it.

Eddington sets out to discover the specific role played by the new constant in the cylindrical universe (1920, pp. 163-4). Because the total amount of matter in the universe depends on the constant, it can be said that this amount "is determined by the law of gravitation". The problem is to find "some plausible explanation of how the adjustment is brought about". The significance of Λ's value being *fixed* is that it means there is an upper limit to the amount of matter the whole universe may contain. By pouring more and more matter into the nothingness, into what is not (to use Eddington's expression), the space produced by this matter must curve around it until it closes, i.e. until "there is no more space". The moment when the process of adding more and more matter stops is precisely determined by the value of Λ. As Eddington puts it: "The more matter there is, the more space is created to contain it". Were the process endless, space would not necessarily be finite and Λ expresses the physical counterpart of the process. Of course, a further problem is that we are left in complete limbo as to what mechanism is operating, "whereby either gravitation creates matter, or all the matter in the universe conspires to define a law of gravitation". It is not until much later that Eddington found the satisfactory answer, that the cosmological constant can be regarded as that which defines an absolute and '*true*' zero condition for our reckoning the energy in the universe, as distinct from a merely *standard* zero condition if Λ is not taken into account. Eddington writes:

If we are contemplating a limited region of space, it is natural to take emptiness as the standard zero condition—the energy in a region is that which we should have to take away in order to leave it a complete vacuum. But a standard zero condition *for the whole universe* cannot be defined in that way. A region can be emptied by transferring the matter elsewhere; but to empty the universe would involve, not transference, but annihilation of matter and energy. It would be absurd to define our reckoning of energy by reference to a fictitious process which conflicts with the most important property of energy: its conservation" (1939, p.233).

The idea is that, if Λ can be viewed as a true force in the Newtonian sense, it is then the physical counterpart of the constant of integration in the law of potential, and that is why it may define the natural zero of energy. Thus, the spatially finite universe with Λ seems to proffer a very natural justification for the laws of energy conservation. In 1923 Eddington had another kind of conservation in mind; that of the distribution of matter. He pointed out that, when Λ has the value zero, the radius of curvature for the universe is infinite, which leads to the inadmissible conclusion that a *re-arrangement* of the matter can occur (1923, pp. 166-7). The spherical space of constant curvature has the property of ruling out any such re-arrangement; Λ removes, as it were, any possible contingency through which the distribution of matter would deviate from the natural geometry of the world. Even though Λ does not say anything about the mechanism whereby space is 'created' by matter, it does tell us that the constant curvature of space is the necessary consequence of this process.

Einstein's apparent confusion between two explanations for the occurrence of Λ (the finiteness of space, and the static distribution of matter) might well be clarified on this basis. He first of all conceived of the time linearity (g_{44} constant) as bridging the gap between the geometrical and the physical, but it is now clear that this linearity is far more an expression of the fundamental relation which Λ really seems to make possible. The privileged direction of time makes it *physically* impossible to re-arrange the distribution of matter but it is the closed nature of space which makes it *geometrically* impossible. In fact, the cosmological constant enables Einstein to link the physical and the geometrical in the finite universe, since it brings about a natural connection between spatial closedness and cosmic time.

It is therefore with some justification that he finally refers to his model as "logically consistent" (1917b, p. 188). The weakness of stellar velocities and the vanishing of the pressure are just approximations to the facts. But the invention of cosmology has made their insertion in the corresponding equations *structural*, so that without them the whole theoretical edifice would have fallen away. The new laws of gravitation had already succeeded in clinching the interdependence of previously unrelated terms, or even entire laws of nature, like the identity of inertial and gravitational mass or Poisson's equation and the conservation laws. It was no little satisfaction for Einstein to realize that his cosmological model achieved an interdependence between variables of an entirely different kind.

For all its blending of implicit background and explicit articulation, Einstein's cylindrical model emerged as a remarkable product of both theory and observation. In implementing a static metric, it did much more than tally with observation of small stellar velocities; it also reflected Minkowski's argument for the mathematical foundation of any possible field physics, namely, that a four-dimensional space–time would make it impossible to relate physical events to an independent time variable defining their 'history'. And constructing a homogeneous distribution of sources contributed to

Einstein's wish to remove all duality between source and field—the basic units of the universe are no singularities as they smoothly merge into the field at large. This alleged accord between mathematics and physics allowed for an extraordinary flexibility in reasoning. Indeed, in order to be solved, field equations need definite boundary conditions. To the technical sense of boundary conditions in the solutions of differential equations, the attempt to comply with Minkowski's pre-established harmony added the more general sense of necessary requirements allowing equations themselves to be set down in the first place. Minkowski's own requirement could be interpreted as general covariance, and Einstein wanted to translate it in physical terms by relying on Mach's principle. He could then hit on the cosmological constant as an extension of the original field equations in the process of working out their solution in the global static symmetric case.

However, the logical consistency of the new cosmology rested, in its turn, upon Λ. Because Einstein was loath to spell out a wholly convincing and unequivocal interpretation of it, he made the entire logical consistency of his system rest on the decision to incorporate it: "In order to arrive at this consistent view, we admittedly had to introduce an extension of the field equations of gravitation which is not justified by our actual knowledge of gravitation" (p. 188). What is exactly the ground for making Λ essential to the logical consistency of the new theory? This was the question that was left to confront the astronomer, Willem De Sitter.

References

Alexander, H.G., ed. (1956). *The Leibniz-Clarke correspondence.* Manchester University Press.
Born, M. (1962). *Einstein's theory of relativity*, Dover, New York.
Cassirer, E. (1921). *Einstein's theory of relativity.* (trans. W.C. Swabey and M.C. Swabey) 1953 edition. Dover, New York.
De Sitter, W. (1911). On the bearing of the principle of relativity on gravitational astronomy, *Monthly Not. Roy. Astr. Soc.* **71**, 388-415.
De Sitter, W. (1913a). A proof of the constancy of the velocity of light, *Proc. Kon. Akad. Wetensch. Amst.* **15**, 1297-8.
De Sitter, W. (1913b). On the constancy of the velocity of light, *Proc. Kon. Akad. Wetensch. Amst.* **16**, 395-6.
De Sitter, W. (1913c). Ein astronomischer Beweis für die Konstanz der Lichtgeschwindigkeit, *Phys. Zeitsch.* **14**, p. 429.
De Sitter, W. (1916a). On Einstein's theory of gravitation and its astronomical consequences. First paper, *Monthly Not. Roy. Astr. Soc.* **76**, 699-728.
De Sitter, W. (1916b). On the relativity of rotation in Einstein's theory, *Proc. Kon. Akad. Wetensch. Amst.* **19**, 527-32.
De Sitter, W. (1916c). On Einstein's theory of gravitation and its astronomical consequences. Second paper, *Monthly Not. Roy. Astr. Soc.* **77**, 155-83.

De Sitter, W. (1917). On the relativity of inertia, *Proc. Kon. Akad. Wetensch. Amst.* **19**, 1217-25.

De Sitter, W. (1932). *Kosmos*, Harvard University Press, Cambridge, Mass.

Earman, J. and Glymour, C. (1978). Lost in the tensors: Einstein's struggles with covariance principles, 1912-1916, *Studies in Hist. and Phil. of Sci.* **9**, 251-78.

Earman, J. and Glymour, C. (1980). The gravitational redshift as a test of general relativity: history and analysis, *Studies in Hist. and Phil. of Sci.* **11**, 175-214.

Eddington, A.S. (1917). Einstein's theory of gravitation, *The Observatory*, **40**, 93-5.

Eddington, A.S. (1920). *Space, time and gravitation*, Cambridge University Press.

Eddington, A.S. (1923). *The mathematical theory of relativity*, Cambridge University Press.

Eddington, A.S. (1928). *The nature of the physical world*, Cambridge University Press.

Eddington, A.S. (1933). *The expanding universe*, Cambridge University Press.

Eddington, A.S. (1939). The cosmological controversy, *Science Progress* **34**, 231-44.

Ehrenfest, P. (1913). Zur Krise der Lichäther-Hypothese, In *Paul Ehrenfest: the collected scientific papers* (ed. M.J. Klein) 1959, pp. 306-27. North-Holland, Amsterdam.

Einstein, A. (1911). Ueber den Einfluss der Schwerkraft auf die Ausbreitung des Lichtes, *Ann. Phys.* **23**, 898-909.

Einstein, A. (1912). Gibt es eine Gravitationswirkung die der elektrodynamisches Induktionswirkung analog ist? *Vierteljahrsschrift für gerichtliche Medizin*, Ser. 3, **44**, 37-40.

Einstein, A. (1913). Zum gegenwärtigen Stände des Gravitationsproblem, *Phys. Zeitschr.* **14**, 1249-66.

Einstein, A. (1914). Zum Relativitätsproblem, *Scientia*, **15**, 337-48.

Einstein, A. (1916). Die Grundlagen der allgemeinen Relativitätstheorie, *Ann. Phys.* **49**, 769-822, (trans. W. Perrett and G.B. Jeffery) In *The principle of relativity*. 1952, pp. 111-64. Dover, New York.

Einstein, A. (1917a). *Über die spezielle and die allgemeine Relativitätstheorie, gemeinverständlich*, Vieweg, Braunschweig. Trans. R.W. Lawson under the title *Relativity, the special and the general theory*. 1920. Methuen, London.

Einstein, A. (1917b). Kosmologische Betrachtungen zur allgemeinen Relativitätstheorie, *Sitz. Ber. Preuss. Akad. Wiss. Berlin*, 142-52. In *The principle of relativity*, pp. 177-88. Dover, New York.

Einstein, A. (1918a). Principles of research, (trans. A. Harris) In *A. Einstein: essays in science*, 1934, pp. 1-5. Philos. Library, New York.

Einstein, A. (1918b). Prinzipielles zur allgemeinen Relativitätstheorie, *Ann. Phys.* **55**, 241-4.

Einstein, A. (1918c). *Über die spezielle und die allgemeine Relativitätstheorie, gemeinverständlich*, Vieweg, Braunschweig. Trans. R.W. Lawson. Third edition of 1917a.

Einstein, A. (1919). Spielen Gravitationsfelder im Aufbau der materiellen Elementarteilchen eine wesentliche Rolle? *Sitz. Ber. Preuss. Akad. Wiss. Berlin*, 349-56.

Einstein, A. (1921). *The meaning of relativity*. Four Lectures delivered at Princeton University (trans. E.P. Adams). Methuen, London.

Einstein, A. (1931). Zum Kosmologischen Problem der allgemeinen Relativitätstheorie, *Sitz. Ber. Preuss. Akad. Wiss. Berlin*, 235-7.

Einstein, A. (1934). Notes on the origin of the general theory of relativity, (trans. A. Harris) In *A. Einstein: Essays in Science*, pp. 78-84. Philos. Library, New York.

Einstein, A. (1945). *The meaning of relativity*. Princeton University Press. Second edition of (1921).

Einstein, A. (1949). Reply to criticisms. In *Albert Einstein, Philosopher-Scientist* (ed. P.A. Schilpp) pp. 665-88. Open Court, Evanston.

Einstein, A. (1954). *Relativity, The Special and the General Theory*, trans. R.W. Lawson. Methuen, London. Fifteenth edition of 1917a.

Einstein, A. and Besso, M. (1972). *Correspondance 1903-1955*. (ed. P. Speziali) Hermann, Paris.

Eisenhart, L.P. (1926). *Riemannian geometry*, Princeton University Press.

Friedman, M. (1983). *Foundations of space-time theories*. Princeton University Press.

Gamow, G. (1956). The evolutionary universe, *Scientific American*. in *New Frontiers in Astronomy* ed., O. Gingerich, 1975, pp. 316-23. Freeman, San Francisco.

Hermann, A. ed. (1968). *A. Einstein-A. Sommerfeld, Briefwechsel*. Schwabe, Basle.

Hins, C.H. (1935). In memoriam Willem de Sitter, *Hemel en Dampkring*. **33**, 3-18.

Holton, G. (1973). Einstein, Mach and the search for reality. In *Thematic origins of scientific thought*. Harvard University Press, Cambridge Mass.

Infeld, L. (1949). General relativity and the structure of our universe, In *Albert Einstein, philosopher-scientist* (ed., P.A. Schilpp) pp. 475-499. Open Court, Evanston.

Jaki, S. (1972). *The milky way*. Science History Publications, New York.

Jeans, J. (1917). Einstein's theory of gravitation, *The Observatory* **40**, 57-8.

Kahn, C. and Kahn, F. (1975). Letters from Einstein to De Sitter on the nature of the universe, *Nature* **257**, 451-4.

Klein, F. (1918). Uber die Integralform der Erhaltungssätze und die Theorie der räumlich-geschlossenen Welt, *Nachr. Gesellsch. Wiss. Göttingen*, 394-423.

Klein, M.J. (1970). *Paul Ehrenfest*. North-Holland. Amsterdam.

Klein, O. (1962). Mach's principle and cosmology in their relation to general relativity, In *Recent developments in general relativity*, pp. 291-302. Pergamon Press, Oxford.

Kopff, A. (1921). Das Rotationsproblem in der Relavitätstheorie, *Die Naturwissenschaften* **9**, 9-15.

Kretschmann, E. (1917). Ueber den Physikalischen Sinn der Relativitätspostulaten, *Ann. Phys.* **53**, 575-614.

Lanczos, C. (1972). Einstein's path from special to general relativity, In *General relativity*. (ed., L. O'Raifeartaigh) pp. 5-19. Clarendon Press, Oxford.

Lorentz, H.A. (1910). Alte und Neue Fragen der Physik, *Phys. Zeitschr.* **11**, 1234-57.

Mach, E. (1883). *The science of mechanics*, (trans. T.J. McCormack) 4th ed. 1907, Open Court. Chicago.

Maxwell, J.C. (1887). *Matter and motion*, reprinted Dover, New York. 1952.

McCrea, W.H. (1971). The cosmical constant, *Qu. J. Roy. Astr. Soc.* **12**, 140-53.

McVittie, G.C. (1954). Relativistic and Newtonian cosmology, *Astron. J.* **60**, 173-80.

Mehra, J. (1973). Einstein, Hilbert and the theory of gravitation, In *The Physicist's Conception of Nature* (ed., J. Mehra) Reidel, Dordrecht.

Merleau-Ponty, J. (1965). *Cosmologie du XXème siècle*, Gallimard, Paris.

Miller, A.I. (1975). A study of Henri Poincaré's 'Sur la dynamique de l'electron', *Arch. for Hist. of Exact Sci.* **10**, 207–328.

Miller, A.I. (1981). *Albert Einstein's special theory of relativity*. Addison-Wesley, Reading, Mass.

Milne, E.A. (1935). *Relativity, gravitation and world-structure*. Clarendon Press, Oxford.

Minkowski, H. (1908). Raum und Zeit, In *Phys. Zeitschr.* **10** (1909), 104–11. In *The principle of relativity*, pp. 75–91. Dover, New York.

Misner, C., Thorne, K., and Wheeler, J.A. (1973). *Gravitation*. Freeman, San Francisco.

Neumann, C. (1896). *Allgemeine Untersuchungen über das Newtonsche Prinzip der Fernwirkungen*. Teubner, Leipzig.

Newton, I. (1729). *Mathematical principles of natural philosophy*, (trans. A. Motte) University of California Press, Berkeley. 1934.

North, J. (1965). *The measure of the universe*. Clarendon Press, Oxford.

Oort, J.H. (1935). Willem de Sitter, Obituary, *The Observatory* **58**, 22–8.

Pais, A. (1982). *Subtle is the Lord: the science and life of Albert Einstein*, Oxford University Press.

Pauli, W. (1958). *The theory of relativity*. Pergamon Press, Oxford.

Peebles, P.J.E. (1980a). *The large-scale structure of the universe*. Princeton University Press.

Peebles, P.J.E. (1980b). Comment on 'The size and shape of the universe' by M. Rees. In ed., H. Woolf, *Some strangeness in the proportion*. pp. 302–5. Addison-Wesley, Reading, Mass.

Plummer, H.C. (1910). On the theory of aberration and the principle of relativity, *Monthly Not. Roy. Astr. Soc.* **70**, 252–75.

Poincaré, H. (1898). De la mesure du temps, *Rev. Métaph. Morale* **6**, 1–13 reprinted in his (1905b), pp. 26–36.

Poincaré, H. (1902). *La science et l'hypothèse*. Flammarion, Paris.

Poincaré, H. (1905a). Sur la dynamique de l'électron, *Rendiconti del circolo matematico di Palermo* **21**, 129–175.

Poincaré, H. (1905b). *La valeur de la science*, Flammarion. Paris.

Poincaré, H. (1912). Space and time. In *Mathematics and Science: last essays*. (Trans. J.W. Bolduc), pp. 15–24. Dover, New York.

Riemann, B. (1854). Ueber die Hypothesen welche der Geometrie zugrunde liegen, *Abhandl. der Gesellsch. der Wiss. zu Göttingen*, Bd. 13. English trans. On the hypotheses which lie at the foundations of geometry, In *A Source Book in Mathematics* (ed., D.E. Smith) 1929, pp. 423–34. McGraw-Hill, New York.

Robertson, H.P. (1933). Relativistic cosmology, *Rev. Mod. Phys.* **5**, 63–90.

Schur, M. (1886). Ueber den Zusammenhang der Räume constante Riemannschen Krümmungsmasses mit den projectiven Raumen, *Math. Ann.* **27**, 537–67.

Seeliger, H. von (1895). Ueber das Newton sche Gravitationsgesetz, *Astron. Nachr.* **137**, 129–33.

Sklar, L. (1974). *Space, time, and spacetime.* University of California Press, Berkeley.
Spencer-Jones, H. (1935). Willem de Sitter, Obituary, *Monthly Not. Roy. Astr. Soc.* **95**, 343-7.
Stachel, J. (1980). Einstein and the rigidly rotating disk. In *General relativity and gravitation.* (ed., A. Held) Vol.1, pp. 1-15. Plenum, New York.
Synge, J.L. (1937). On the concept of gravitational force and Gauss's theorem in general relativity, *Proc. Edin. Math. Soc.* 2nd ser., **5**, 93-102.
Synge, J.L. (1960). *Relativity: The general theory*, North-Holland, Amsterdam.
Tolman, R.C. (1910). The second postulate of relativity. *Phys. Rev.* **31**, 26-40.
Tonnelat, M.A. (1964). *Les vérifications expérimentales de la relativité générale*, Masson, Paris.
Torretti, R. (1978). *Philosophy of geometry from Riemann to Poincaré*, Reidel, Dordrecht.
Torretti, R. (1983). *Relativity and geometry.* Pergamon Press, Oxford.
Weyl, H. (1918). *Space, time, matter*, (trans. H.L. Brose of the fourth German ed.) 1952. Dover, New York.
Weyl, H. (1949). *Philosophy of mathematics and natural science.* Princeton University Press.
Whitrow, G.J. (1980). *The natural philosophy of time*, Oxford University Press.
Whittaker, E.T. (1910). Recent researches on space, time and force, *Monthly Not. Roy. Astr. Soc.* **70**, 363-6.

3

The almost full and the almost empty

1. The changing picture of general relativity from September 1916 to February 1917. First reactions to the cosmological considerations

Einstein's gradual recognition of the need for a concrete model of the universe may be seen as the key to an understanding of the foundational debate between him and De Sitter as to the nature of relative rotation in general relativity. Certainly it is true that Einstein's invention of the cosmological model was derived from a new, totally reshaped response to De Sitter's epistemological queries. In the first instance, following Mach's objections to Newton, Einstein had tried to reject any principle of unconditionality in physical science by denying in his thought experiment of rotating spheres any possibility of a difference between an 'idealized' and a 'real' universe. As De Sitter was to demonstrate, this in turn involved a new, disguised form of unconditional—that of distant matter. Over and above Einstein's intention, De Sitter tried to show that the notion of distant matter depended upon an implicit argument of pure intelligibility, rather similar in kind to that which allowed Newton to conceive of the physical centre of the world or even of absolute space at absolute rest. However, were we able to make experiments on the *global* distribution of matter and motion in the universe, the awkward status of distant matter, as an unconditional, would disappear taking with it the privilege given to a special type of solution to the problem of dynamic relativity. The weight of tradition might well have forced Einstein to overlook cosmology as a new physical problem: at first he tried various sorts of physics of the infinite, failing to perceive the shrewdness of De Sitter's observations about the as yet excessively 'formal' character of general relativity. The drastic step of abolishing any physics of the infinite, concomitant with the introduction of the cosmological constant, was to be Einstein's ultimate answer to De Sitter. It marked the emergence of relativistic cosmology, by providing science with a law which enabled the cutting off of the indefinitely extended chain of distant masses. Einstein's closed universe brought to an end the impossible search for the physical consequences of a mathematical makeshift, a makeshift by which the field equations were solved on a local scale, with a zero field at infinity. A closed universe can also be seen as the first global solution to the conflict between Mach's principle and the general relativity principle; the original thought experiment could offer only a local solution.

In his *Four Lectures* of 1921, Einstein had arrived at the position of understanding Mach's 'empirical' method of defining inertia (a point of mass unaccelerated with respect to the entire universe) as *implying* that the series of mechanical causes were *closed* (1921b, p. 62). Mach, however, never went this far, and Einstein was in fact moving a long way from the strict obedience to Mach's behests. Mach assumed almost implicitly that the distribution of matter was uniform and infinite. Einstein's cosmology activated Mach's notion that, even in the case of only two masses, the relation of this system to the whole universe could not be ignored. But it was primarily De Sitter's critique which made Einstein begin to realize that his new thought experiment had brought this relation into question.

Einstein sent a letter to De Sitter on the 12 March 1917 (probably along with his cosmological memoir), which revealed much of his drift. He confessed how scorching the question had become for him either way: whether the construction of a model for the whole universe "offers an extension of the relativistic way of thinking, or whether it leads to contradiction". It was a relief, he concluded, that all obstacles had been overcome. From the standpoint of astronomy, Einstein surmised that what he had built was of course "a spacious castle in the air". But the more important thing was the theoretical background. To Besso, he spoke of his model as "a proof that general relativity can lead to a non-contradictory system" (A. Einstein and M. Besso (1972 p. 102), a statement which underlines the fact that he regarded cosmology as an almost transparently logical translation of the physical sense of his physical theory. It was also in this letter to De Sitter that Einstein went on to insist that the model was no violation against the principle of relativity in as much as only a *statical* conception was possible. He thus made quasi explicit the view that the choice of a statical metric involved much more than mere compliance with the facts. It was the theoretical proof that elimination of boundary conditions did away with their inevitable but unacceptable physical consequences.

In what sense can we see Einstein's model as theoretically progressive? In what sense is it satisfactory from the theoretical, as *distinct* from the empirical point of view? Let us consider the two aspects in succession, first, spherical geometry, and then, linear time. By recalling the substance of his "Cosmological Considerations" in the 1921 Address to the Prussian Academy of Sciences, Berlin, Einstein (1921a, pp. 40ff.) was now in a position to discard the vary basis of Poincaré's epistemological argument. The *theoretical* possibility of a spherical universe overcomes the apparent independence of Euclidean geometry from decisive observations, for the distances postulated by this theory exceed all detectable parallaxes. It is precisely *because* the new model is independent of parallaxes that it overcomes Poincaré's doubts. And as we will see, an important development due to De Sitter was a demonstration of the total irrelevance of parallaxes in estimating the size of Einstein's universe. What Einstein could not suspect in 1917 was that new observational means would evolve rapidly, which increased in a really dramatic way the capacity for scrutinizing the depths of the universe. In the early 1920s, these new means did have sufficient impact to make Einstein declare: "I do not even consider it impossible that this question will be answered before long by astronomy" (p. 41). But once again, no progress was made in this direction until De Sitter came to explore the entirely new range of possibilities opened up by Einstein's cosmology. So we come back to De Sitter's earlier reaction, when the connection between the observational and the theoretical was first delineated.

Note that on 15 March 1917, De Sitter wrote a brief postcard, thanking Einstein for his explanations. It is his first recorded reaction to Einstein's model. He agreed with Einstein's opinion to the extent of thinking the model should not be seen as an attempt to "force" reality. He had no objection on the grounds that it was an "uncontradicted train of thought". In this postcard, De Sitter also alluded to the final printing of his second paper for *Monthly Notices*. He apologized for not sending an offprint, but added that Einstein would know the basic contents from their earlier discussions in Leiden. One point in particular should be stressed, the postcard highlights De Sitter's strictly astronomical views on the problem of the total mass of the stars. He claimed that "the simple fact that we can identify spectral lines proves the potentials among all stars and nebulae. . .to be of the same order as here. This proof is stronger than that of the small stellar velocities". Since no violetshift is observed, an upper limit for the mass can be known within the distance allowed by observation of spectral lines.

This insistence was to be the prelude to a most baffling intellectual journey for De Sitter. The point is that he is inclined to believe that this knowledge of the total mass of the stars confined within the limits of observation is the *only* knowledge we have of what lies outside the solar system. So, in Einstein's new model, the theoretical determination is exposed to the danger of extrapolation. And because of the peculiar status of time in this model, De Sitter is now eager to emphasize the danger of extrapolation into both space and time from the 'photographic image' we have of the universe. Indeed, the image is a static one: in a letter to Einstein (1 April 1917), De Sitter said that "all extrapolations beyond the region of observation are insecure. . .Your premise that the world is quasi-static mechanically I contest with the utmost energy. We have of the world only a snapshot, and we must not conclude from the fact that the picture shows no great transformations that everything will always remain as it was when the photo was made": one should not be misled by the observation that the universe *appears* static. The argument is really far-reaching. For by contrast, Einstein was obviously willing to take into account physical phenomena which contradicted the *apparent* observations on other occasions, such as relative time in special relativity (which defies our 'natural' sense of simultaneity), or even the homogeneous distribution of matter in the cylindrical model (which defies every observation of the sky at night). The whole problem can be given a striking form by referring to general relativity incorporating the 'classical' potentials of gravitation in the space-time geometry. For if only the *facts* require a return to linear, 'classical' time, then general relativity (when applied to large-scale portions of space-time) would not be consistent in the *mathematical* sense outlined by De Sitter in his earlier discussion of the relativity of rotation. That is why De Sitter will strive to extend the basis of his critique in discussing the logical consistency postulated by Einstein at the end of his 'Cosmological Considerations'. This is quite a jump, from admitting that the model does not contradict the facts to actually challenging that the model may be unique.

From this critique, De Sitter was led to formulate his own model of the universe. It is almost as if he became a cosmologist against his own will, finding no better way of criticizing Einstein's new hopes than constructing a counter-example; this unexpectedly became an alternative model of the universe. His first discussion of it was in a letter to Einstein dated 20 March. Einstein replied on the 24th, and De Sitter communicated his model at a meeting of the Amsterdam Royal Academy of Sciences on 31 March (see De Sitter 1917a). This printed paper was not the most accomplished of De Sitter's various technical grapplings with his new problem. In fact, De Sitter published a detailed paper in *Monthly Notices* later in the year, which was to be the third in his series on astronomical consequences of Einstein's theory (1917d). Nevertheless, the March article was in itself such a giant step, that it is certainly more profitable to begin with an analysis of its revolutionary ideas as they first sprang from De Sitter's mind. The continuing correspondence with Einstein, wherein De Sitter was obliged to rebut some thoroughgoing objections, may act as a stimulus as we attempt simultaneously to assess the philosophical basis of the new ideas as well.

2. Cosmological consequences of the relativity of inertia

There is perhaps only one clear-cut and decisive argument which in Einstein's mind justified the cylindrical model being regarded as the only possible model for the universe: that the model cannot really be regarded as a particular *solution* to the field equations in the usual sense, in that a solution like the Schwarzschild exterior metric makes use of boundary conditions and obviously does not satisfy Mach's principle. Rather than a solution, Einstein seems to see cosmology as being in keeping with the *principles* of general relativity, those principles (such as the equivalence principle) which were until then based on local considerations and in which boundary conditions were neglected. Hence, Einstein's use of all-embracing principles in discussing the merits of his hypothesis, and the apparent facility with which he can discard objections drawn from the limits of observation. This aspect of Einstein's cosmology must be borne in mind throughout, and De Sitter himself will repeatedly discuss it; its full implications will be addressed in the last section of this chapter.

De Sitter's approach to the problems of cosmology is strikingly reminiscent of his earlier criticisms, in which the concept of distant matter was opposed to the real premises of general relativity. He conceived the consistency of general relativity in terms of a theory which dispensed with the need to explain inertia, and this general epistemological standpoint was now used to attack Einstein's cosmology. As De Sitter said to Einstein in his 20 March letter, even before offering a full proof of his new conjectures, he simply did not know whether they could be said to help explain inertia. In fact he is still not concerned with explanations. And a few days later (1 April), De Sitter was almost apologetic for having dared build a theoretical construction which Einstein ventured to call a "world". All extrapolations remained dangerous, and De Sitter found it difficult to take Einstein at his word when he spoke of "your" model and "my" model.

The attack is first and foremost a philosophical one. De Sitter objects to the alleged non-contradiction in the relativity postulate: "From the point of view of the theory of relativity, it appears at first sight incorrect to say: the world *is* spherical" (1917a, p. 1218). De Sitter is eager to underline the word "is". He does not appear to baulk at the idea of finiteness, provided it is used for purely geometrical purposes. With Einstein's extension of it so that it takes on a physical application, however, De Sitter points to the residual distinction between the strictly formal requirement of the principle of relativity and any assertion as to the actual existence of postulated entities. His trouble lies in the relation of generalized mathematical transformations to the existence of some unique, world-wide space-time. The connection between homogeneous distribution of matter and identity of all proper times, on the one hand, and constancy of space curvature, on the other, may well be compatible with the formal invariance of the laws, but this formal invariance should in turn remain compatible with other scenarios as well.

In order to explore *general* implications of the relation of the formal to the objectual, and before drawing his own conclusions, De Sitter, initially, searches for possible motivations which may account for Einstein's strategy. This is a crucial task, since it really amounts to 'testing' the claim that a physics of the whole universe could not be different from the proposed one. De Sitter's own view of the true consistency of general relativity fits in very well with this schema. It should be remembered that, with respect to the problem of rotation, De Sitter had rejected any space–time which could have different dynamic effects depending on the coordinate system chosen. But his solution remained at every point compatible with the existence of a space–time entirely deprived of dynamic effect and independent of matter. Applying this conclusion to the new problem in question, it appears at least possible to explore the consequences of an appropriate geometrical transformation, leaving the relevant invariants unaltered, whereby we should see the relation of the physically finite model to some other kind of space. Because general relativity has no physical content other than the coincidences between events, the 'absolute' properties of space–time should be left unchanged under any transformation of coordinates. It is De Sitter's contention that, if Einstein is correct in the belief that his physics of the whole universe is consistent with local physics, this invariance can also be applied to such an object as the whole universe. De Sitter chooses the stereographic projection, which maps the Einstein universe onto Euclidean space—he could have chosen hyperbolic space as well as he did a little later.

What is a stereographic projection? Some of its properties important to a cosmological context will be explained here. It is one of the projections of a sphere onto a Euclidean, infinite plane (see Fig. 3.1). Under this transformation, circles remain circles, and all finite figures remain finite figures. Consequently, the projection functions as a mathematical device by which it is possible to represent the new finite universe in terms of the old, island-like one. This, in turn, implies that boundary conditions can be written. The values for the $g_{\mu\nu}$ at infinity are:

$$\begin{matrix} 0 & 0 & 0 & 0 \\ 0 & 0 & 0 & 0 \\ 0 & 0 & 0 & 0 \\ 0 & 0 & 0 & 1 \end{matrix} \quad (3.1)$$

The meaning of these boundary conditions should not be misunderstood. Stated in *natural* rather than in *coordinate* measures, it is clear that the universe is finite. In De Sitter's words: "It is, in fact, evident that, if the universe in natural measure be finite, then, if Euclidean coordinates are introduced the $g_{\mu\nu}$ must necessarily be zero at infinity, and inversely, if the $g_{\mu\nu}$ at infinity are zero of a sufficiently large order, then the universe is finite in natural measure". What comes immediately after that statement is the crucial step: De Sitter points to the ambiguity of the set (3.1):

Einstein only assumes *three*-dimensional space to be finite. It is in consequence of this assumption that in (3.1) g_{44} remains 1, instead of becoming zero with the other $g_{\mu\nu}$. This has suggested the idea to extend Einstein's hypothesis to the *four*-dimensional space–time (1917a, p. 1219).

In accordance with this extension, the potentials at infinity would now degenerate to the values:

$$\begin{matrix} 0 & 0 & 0 & 0 \\ 0 & 0 & 0 & 0 \\ 0 & 0 & 0 & 0 \\ 0 & 0 & 0 & 0 \end{matrix} \quad (3.2)$$

It was at this stage that De Sitter acknowledged his debt to Ehrenfest, who had suggested the idea several months earlier. Ehrenfest himself had dared not develop this idea, even though it had looked so natural, fearing the concept of time would become unintelligible. Having gone so far in his attempt to discover the essence of Einstein's strategy, De Sitter could no longer recoil before the formidable idea of spelling out the actual meaning of a fully relative time.

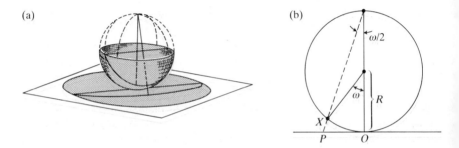

FIG. 3.1. (a) Stereographic projection of the sphere onto the plane. (From F. Klein 1928, p. 296.) (b) Projection of a circle. Each point of the circle is represented by its projection P onto the tangent in O. To the angular coordinate ω on the circle corresponds the distance OP = $2R\tan(\omega/2)$.

Setting about the task of developing Ehrenfest's idea, De Sitter emphasizes the fact that, in accordance with Minkowski's notion of continuum, space and time should be considered on the same footing a priori. The set (3.2) allows some sort of interchange between space and time since, as De Sitter said to Einstein (in his 20 March letter), these $g_{\mu\nu}$ make it possible to consider the infinite as "either spatial, or temporal, or both". The system is indeed invariant for all transformations. In contrast, the link between space and time in Einstein's model was very different, being dependent upon the assumption of a uniform distribution of matter. Thus, in trying to uncover the nature of Einstein's model, De Sitter unearths a problem which actually goes beyond his original purpose. Einstein conceived the cosmological problem as involving the means whereby Riemannian space could be made the *only* possible *physical* space. But De Sitter holds in reserve some sort of entirely amorphous space, which enables him to face the question of the relation of this physical space to the a priori possibility of a physical continuum. Following on a mathematically extended analysis of Einstein's original hypothesis alone De Sitter was to hit on an entirely new model, although at this stage he considered it primarily in terms of a reflection on what actually constitutes the 'materiality' of the postulated continuum. In fact, in his letters to Einstein, this position was to cause him considerable embarrassment, since he was reluctant to call the new hypothesis a model of the universe. That he would finally assent to the identification, under pressure from Einstein himself, is quite another part of the story to which we will turn later.

De Sitter distinguishes Einstein's original model from his own interpretation by labelling them A and B respectively. The equations are written so as to face each other, a procedure which makes their opposition look all the more relentless. (J. Merleau-Ponty 1965, p. 55, speaks of this procedure as being quite reminiscent of Kant's antinomies. The comparison is certainly right, and even more than formal: We will see that De Sitter was to discover many antinomical facets in cosmology.) To begin with, a full symmetry between the two models requires model B to be a four-dimensional sphere, with the time component coiling around itself like space. The $g_{\mu\nu}$ are:

$$g_{ij} = -\delta_{ij} - \frac{x_i x_j}{R^2 - \Sigma' x_i^2} \quad \Big| \quad g_{\mu\nu} = -\delta_{\mu\nu} - \frac{x_\mu x_\nu}{R^2 - \Sigma x_\mu^2} \quad (3.3)$$

$$g_{44} = 1$$

where i and j run from 1 to 3, while μ and ν run from 1 to 4; Σ is the sum from 1 to 4, and Σ' from 1 to 3; $\delta_{\mu\nu} = 0$ if $\mu \neq \nu$, and $\delta_{\mu\nu} = 1$ if $\mu = \nu$ (the same holds for δ_{ij}). Symmetry implies $x_4 = ict$, that is *imaginary time* is homogeneous with spatial coordinates. Model B is a four-dimensional, spherical space–time, while A is spherical only with respect to the three-dimensional space. De Sitter will be troubled only later by the question of whether imaginary or real time should be used.

In order to highlight further the differencce between A and B, De Sitter introduces a change of variables which enables him to write the two metrics in polar coordinates:

$$\begin{array}{l|l}
x_1 = R \sin\chi \sin\psi \sin\theta & x_1 = R \sin\omega \sin\chi \sin\psi \sin\theta \\
x_2 = R \sin\chi \sin\psi \cos\theta & x_2 = R \sin\omega \sin\chi \sin\psi \cos\theta \\
x_3 = R \sin\chi \cos\psi & x_3 = R \sin\omega \sin\chi \cos\psi \\
 & x_4 = R \sin\omega \cos\chi
\end{array} \quad (3.4)$$

where the additional curvature in B is represented by ω. In A, linear time forbids x_4 to be related to R. The metric for A is thus

$$ds^2 = -R^2[d\chi^2 + \sin^2\chi\,(d\psi^2 + \sin^2\psi\,d\theta^2)] + c^2\,dt^2, \quad (3.5a)$$

with $0 < \theta < 2\pi$ and $0 \le \chi,\psi < \pi$. This is, in fact, the metric representing the three-dimensional spherical surface embedded in a four-dimensional Euclidean space. It is immediately apparent that this is a generalization of the more familiar metric of a two-dimensional spherical surface: $ds^2 = R^2\,d\phi^2 + \sin^2\phi\,d\theta^2$. The metric for B is obtained by adding the dimension ω:

$$ds^2 = R^2\{d\omega^2 + \sin^2\omega\,[d\chi^2 + \sin^2\chi\,(d\psi^2 + \sin^2\psi\,d\theta^2)]\}, \quad (3.5b)$$

with $0 < \theta < 2\pi$ and $0 < \omega,\chi,\psi < \pi$. This is a four-dimensional spherical surface embedded in a five-dimensional Euclidean space.

The next step is the stereographic projection of A and B, in order to find the $g_{\mu\nu}$ of the two universes in Euclidean space. This requires Cartesian coordinates, i.e. $x^2 + y^2 + z^2 = r^2$, $x^2 + y^2 + z^2 - c^2t^2 = h^2$ which are obtained by:

$$\begin{aligned} x &= r\sin\psi\cos\theta & x &= h\sin\chi\sin\psi\sin\theta \\ y &= r\sin\psi\cos\theta & y &= h\sin\chi\sin\psi\cos\theta, \\ z &= r\cos\psi & z &= h\sin\chi\cos\psi \\ & & ict &= h\cos\chi \end{aligned} \quad (3.6)$$

with $r = 2R\tan\tfrac{1}{2}\chi$, $h = 2R\tan\tfrac{1}{2}\omega$.

Comparing the changes of variables (3.4) and (3.6), we can see that the transformation used by Einstein is equivalent to $r = R\sin\chi$. The transformation proposed by the De Sitter, $r = 2R\tan\tfrac{1}{2}\chi$, leads to the new form of the metric:

$$ds^2 = -\frac{dx^2 + dy^2 + dz^2}{[1 + 1/\sigma\,(x^2 + y^2 + z^2)]^2} + c^2dt^2, \quad (3.7a)$$

where $\sigma = 1/4R^2$, or

$$ds^2 = -\frac{dr^2 + r^2(d\psi^2 + \sin^2\psi d\theta^2)}{(1 + \sigma r^2)^2} + c^2dt^2. \quad (3.8a)$$

The metric for B, with $h = 2R\tan\tfrac{1}{2}\omega$, becomes

$$ds^2 = -\frac{dx^2 + dy^2 + dz^2 - c^2\,dt^2}{[1 + 1/\sigma\,(x^2 + y^2 + z^2 - c^2dt^2)]^2} \quad (3.7b)$$

or

$$ds^2 = -\frac{dh^2 + h^2(d\omega^2 + \sin^2\omega d\theta^2)}{(1 + \sigma h^2)^2} + \frac{c^2dt^2}{(1 + \sigma h^2)^2}. \quad (3.8b)$$

Calculation of the $g_{\mu\nu}$ yields

$$g_{ij} = -\frac{\delta_{ij}}{(1+\sigma r^2)^2} \quad \bigg| \quad g_{ij} = -\frac{\delta_{ij}}{(1+\sigma h^2)^2}$$

$$g_{44} = 1 \quad \bigg| \quad g_{44} = \frac{1}{(1+\sigma h^2)^2} \, . \quad (3.9)$$

These components take the pseudo-Euclidean values for $r = 0$, and they also degenerate to the values (3.1) and (3.2) when $r = \infty$.

Passing now to the *physical* difference between A and B, it appears that such an assessment is dependent upon a comparison between δ and Λ. In other words, the above-mentioned $g_{\mu\nu}$ have to be inserted into the field equations with the cosmological constant. In his letter of 20 March, De Sitter insisted that his considerations had been up until then quite independent of all physical masses (like the Sun): "*physically* I *only* know that at finite distance the $g_{\mu\nu}$ must not be very different from the old theory of relativity". A definite matter distribution is required, however, otherwise a comparison between δ and Λ in the two models is not possible. In view of the smallness of Λ (as indicated by our knowledge of the perturbations in the solar system), the predicted quantity of matter in the universe is absolutely enormous. In fact, as we have seen, the rough estimation amounted to something like ten thousand Milky Ways. An astronomer like De Sitter would have found this quite unacceptable, given that at the time the very question of whether at least one nebula was actually extra-galactic had not been settled. De Sitter stated this problem of the quantity of matter in A in the following terms: "It is found necessary to suppose the whole three-dimensional space to be filled with matter, of which the total mass is so enormously great, that compared with it all matter known to us is utterly negligible. This hypothetical matter I will call the 'world-matter' " (1917a, pp. 1218-19). This world-matter is of course quite reminiscent of the earlier distant masses, which are always found to lie beyond any given field of observation. However, for the sake of comparison, De Sitter selects the hypothesis that Einstein had already chosen: the universe is a smoothed-out system, quite similar to a perfect fluid at rest. This implies, as before, $T_{44} = \rho.g_{44}$ and all other $T_{\mu\nu} = 0$. The field equations are, in both models: $R_{ij} - (\Lambda + \frac{1}{2}k\rho)g_{ij} = 0; R_{44} - (\Lambda + \frac{1}{2}k\rho)g_{44} = -\kappa\rho$. The $R_{\mu\nu}$ in the two systems are $R_{ij} = 8\sigma g_{ij}$ and $R_{44} = 0; R_{\mu\nu} = 12\sigma g_{\mu\nu}$; which leads to $\Lambda = 4\sigma$ and $\Lambda = 12\sigma$ respectively. In other words, the radius of curvature in A is $R = \sqrt{1/\Lambda}$, while in B it is $R = \sqrt{3/\Lambda}$. These values for Λ lead to the following values for ρ: in A, $\rho = 8\sigma/\kappa$ while in B, $\rho = 0$. In this way, "the result for A is the same as found by Einstein. For B we have $\rho = 0$: the hypothetical world-matter does not exist" (De Sitter 1917a, p. 1221). This indicates that there is in effect no necessary relation (in a logical sense) between the curvature of space and the particular form of the material tensor chosen by Einstein.

De Sitter's first letter to Einstein (20 March) on these cosmological issues stops here. I have followed his argument through step by step, inserting from the printed article of 31 March only those parts which appeared crucial to his exposition. De Sitter is reluctant to base physical arguments on cosmology. He expresses his preference for the four-dimensional system without "supernatural" masses, and he would like even more the original theory without invariant $g_{\mu\nu}$ at infinity and without the "indeterminable" quantity Λ. In De Sitter's view, the new construction succeeds in showing that a test body in an otherwise empty universe is still endowed with inertia. Remove the *physical*, actually observed masses from your universe, De Sitter says to Einstein, and the result is the same: a test body which still has inertia. Only the *reason* for this assertion differs in the two models: in A, the neglect of distant masses is as impossible as a physical proof which began with the statement: "If the world were not there" (De Sitter's words). Model B has no meaning of its own; at this stage, its only purpose is to show that, in A, the argument for a causal connection between the spatial curvature and the uniform distribution of matter rests upon this type of reasoning. De Sitter makes a tentative suggestion that metric B could also be interpreted as a spatially finite world, with a radius of curvature $R = \sqrt{3/\Lambda}$. But he is evidently uneasy about this. He simply claims that, without supernatural masses and only by keeping Λ, he has reached a conclusion analogous to Einstein's as to the inertia of a test body.

Before skipping to Einstein's immediate counter-argument, let us see how the mathematics of De Sitter's results differs from Einstein's. From Eqns (2.21a), (2.21b) and (2.22) in chapter 2, we have $8\pi (p + \rho) = e^{-\lambda}[(\lambda' + \mu')/r] = 0$. The condition $p + \rho = 0$ is indeed an alternative to Einstein's interpretation. Einstein had put $\mu' = 0$ because he thought, uncritically, that the density would necessarily be different from zero. The condition $p + \rho = 0$, on the contrary, leads to $p = 0$ and $\rho = 0$ since neither the pressure nor the density could be negative. This implies $\lambda' + \mu' = 0$. With this, the general solution to the differential equation is: $-e^{-\lambda} = -e^{\mu} = -[1 - (8\pi\rho + \Lambda/3)r^2]$. Taking $R^2 = 3/(8\pi\rho + \Lambda)$ which is, once again, the only way of preserving the signature, De Sitter's metric is obtained with $\rho = 0$ and $R^2 = 3/\Lambda$. Both the pressure and the density vanish, while the value of the cosmological constant (as in Einstein's model) is free of all restrictions.

3. Paradoxes with time. First hints of the geometry of De Sitter's model

Four days after De Sitter's letter, Einstein did not show as much cautiousness as De Sitter on the physical meaning of the concepts, for, taking De Sitter at his word, Einstein investigated his result as if it were a new form of the spatially closed and finite universe.

The difficulty lay, of course, in the theory of time. In a new letter, Einstein pointed out a cumbersome singularity occurring at finite distance in model B. Let us try to see what this means.

Until now, De Sitter has used the coordinate x_4 as imaginary time for the sake of symmetry between A and B. Because Einstein tries to make physical sense of what De Sitter has found, he uses a *real* time, that is, $x_4 = ct$. He does so almost inadvertently, without explicitly justifying the change. Instead of an imaginary hypersphere, we now have the *real* figure of the space–time geometry. The four-dimensional sphere had some curious features, since the coiling of time around itself leads to the theoretical possibility of an observer experiencing his own past. The real figure avoids this: instead of a sphere, we have an hyperboloid. The hyperboloid is just Einstein's original cylinder, twisted about its axis because of the curvature of time. De Sitter's universe is now a one sheet hyperboloid, with its two ends open in the time direction. The canonical equation of an easily intuitable three-dimensional hyperboloid is either

(a) $x^2 + y^2 - z^2 = 1$ or (b) $x^2 + y^2 - z^2 = -1$.

The first equation is that of a one-sheet hyperboloid, while the second is that of a two-sheet hyperboloid (see Fig. 3.2). The intersection of curve (b) with a plane $z = z_0$ gives $x^2 + y^2 = -1 + z_0^2$, so that curve (b) is a figure whose successive intersections by planes of constant z are circles of radius $R = \sqrt{-1 + z_0^2}$. It is clear that this is real only for $z_0 > 1$: so the hyperboloid is separated into two sheets disconnected from one another, with $z \geq 1$ or $z \leq -1$. In De Sitter's terminology, as we will see, equations (a) and (b) define *conjugated hyperboloids*.

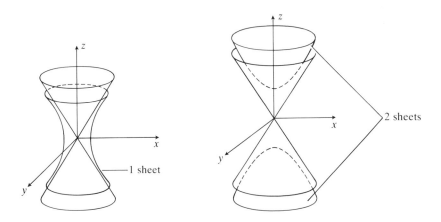

FIG. 3.2. De Sitter's conjugated hyperboloids: the difference between a one-sheet and a two-sheet hyperboloid.

More generally, the equation of a one-sheet hyperboloid is

(c) $x_1^2 + x_2^2 + \ldots + x_n^2 - x_{n+1}^2 = 1.$

With the change of variable $x_{n+1} = ix'_{n+1}$, equation (c) becomes:

(d) $x_1^2 + x_2^2 + \ldots + x_n^2 + x'^2_{n+1} = 1,$

that is to say, the equation of an n-dimensional sphere of radius 1. Thus, an n-dimensional hyperboloid with one sheet embedded in the $(n + 1)$-dimensional Euclidean space is equivalent to an n-dimensional sphere embedded in an $(n + 1)$-dimensional Minkowskian space. While x'_{n+1} is imaginary, x_{n+1} is real.

Let us now examine De Sitter's hyperboloid with stereographic coordinates. It is desirable to start with the two-dimensional case, which is quite easy to picture intuitively, as in Fig. 3.3(a). A point X is projected by the line

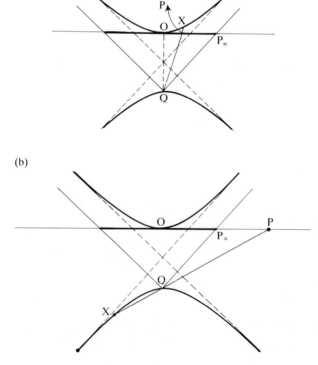

FIG. 3.3. De Sitter's hyperboloid with stereographic coordinates in two dimensions: (a) upper sheet; (b) lower sheet.

XQ onto the point P on the tangent to the hyperbola at O. The upper sheet of the hyperbola is entirely represented by the segment OP. As X moves away to infinity, the line QPX becomes parallel to the asymptote (indicated by the dotted line); QPX tends to coincide with the line QP_∞, parallel to the asymptote. A small distance on OP_∞ can be a very large distance on the 'real' space (the hyperbola); in fact, the $g_{\mu\nu}$ become infinite as $P \to P_\infty$. Thus, the fact that the $g_{\mu\nu}$ tend to infinity as $P \to P_\infty$ expresses the idea that P_∞ is "at infinity". As far as the lower part of the hyperbola is concerned, it is represented, in the projection, by that part of the infinite line OP which is not thick (Fig. 3.3(b)). Consider now the one sheet hyperboloid, drawn on Fig. 3.4(a). The stereographic plane is the tangent to the hyperboloid at O. It intersects the hyperboloid in a pair of straight lines Oa and Ob on the hyperboloid. The asymptotic cone, with vertex S, is also drawn. The projection of a

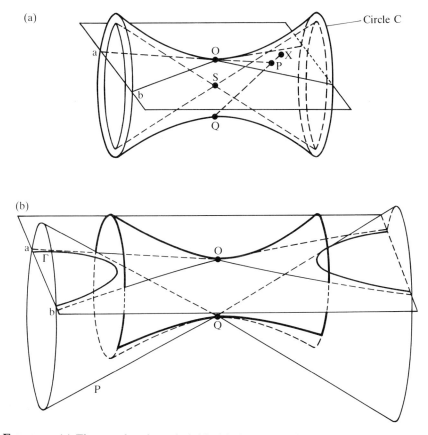

FIG. 3.4. (a) The one-sheet hyperboloid with the stereographic plane. (b) The same hyperboloid as the circle C tends towards infinity.

point X, located 'above' the plane, is P (that is, the intersection of the plane with QX). The question arises: does the projection of the hyperboloid cover the whole of the plane? Let us imagine the cone of lines which connect the point Q to the circle C on the hyperboloid. When C tends towards infinity, this cone tends to become cone P of Fig. 3.4(b): this is the cone parallel to the asymptotic cone, with vertex Q. The lines QX are always exterior to the cone P. Thus, the projected hyperboloid gives rise to points which are always exterior to the projection of P. On the plane, the hyperboloid is projected onto the hatched area of Fig. 3.5. The hyperbola Γ, intersection of the cone and the stereographic plane, is also represented on Fig. 3.4(b); in fact, this represents the points on the hyperboloid which are "at infinity".

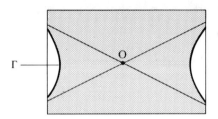

FIG. 3.5.   A view of Fig. 3.4(b) from the stereographic plane.

De Sitter was prompted by Einstein's 24 March letter to reconsider the implications of model B, since Einstein believed that the hyperboloid is affected by a 'singularity'. Einstein found that, in model B, "the surface with singular properties is. . .in physical finite space. Therefore it seems to me that your solution corresponds to no physical possibility". The surface in question is hyperboloidic: it has the equation

$$1 + \sigma h^2 = 0 \text{ or } 4R^2 + x^2 + y^2 + z^2 - c^2 t^2 = 0. \tag{3.10}$$

This hyperboloid will be called H. By a singularity, Einstein means that the values (3.10) for the $g_{\mu\nu}$ (3.9) make these $g_{\mu\nu}$ infinite.

We are now coming to the critical stage of the whole Einstein–De Sitter controversy. Einstein felt that the singularity discredited the physical meaning of model B, but De Sitter was convinced from the outset that the meaning was frail in any case. Einstein's argument would simply have tended to reinforce De Sitter's conviction, were it not for the fact that De Sitter discovered the fallacy of the argument. Because Einstein is wrong in his view of the nature of the singularity, it is ironic that De Sitter's reply is concerned with an attempt to promote the physical meaning of B. From then on, De Sitter began to suspect that his model could sustain such a meaning. Surprisingly, in the following months, a whole series of different kinds of correlation had the effect of confirming his suspicion. In his two letters of the 1 and 18 April, De Sitter was so cautious that he used inverted commas when speaking of "my" world. By the end of June, he was game enough to start exploring those indications "that the system B, and not A, would correspond to the truth" (1917b, p. 236).

The fallacy in Einstein's argument is that it confuses two types of singularity, one of which is produced by the coordinates, the other by the intrinsic features of the world. De Sitter claims that the singularity pointed out by Einstein is coordinate in kind. In his article on the relativity of inertia, he adds an important footnote (1917a, p. 1220) in which he characterizes the discontinuity produced by (3.10) as "apparent only". Einstein is not to be won over: "I have not yet understood your remark about the singularity at finite distance as being only apparent, due to the choice of the coordinates...I am waiting for clarification" (14 April). To which De Sitter replied: "The question of the discontinuity...is properly speaking not very interesting...". Einstein did not give up so quickly, and came back with new arguments some two months later: "Your four dimensional continuum lacks the property of having all its points similar..." (14 June). The dispute then takes another form, which cannot be appraised without a close analysis of the arguments used here.

The stereographic projection of the original spherical form of model B is given by Eqns (3.6). The parameter h in this projection corresponds to the distance OP of Fig. 3.4(a). This distance, in turn, is projected on the four dimensions x, y, z, ict of the stereographic plane. The $g_{\mu\nu}$ in these new coordinates are given by Eqn (3.9): they are infinite on H. Thus, De Sitter explains in his 1 April letter: the singularity "rests on the fact that the real, hyperbolic world has been represented as spherical. The infinite on the hyperboloid is brought back at finite distance, where it becomes a singularity". In the article itself, De Sitter commits a mistake which does not help clarify the situation: "The four-dimensional world, which we have for the sake of symmetry represented as spherical, is in reality hyperbolical, and consists of two sheets, which are only connected with each other at infinity" (1917a, p. 1220). We know that an n-dimensional hyperboloid with *one* sheet embedded in an $(n + 1)$-Euclidean space is equivalent to an n-dimensional sphere embedded in an $(n + 1)$-Minkowskian space, on condition that one of the real variables will be transformed into an imaginary one. In the transformations (3.4) right-hand column, we can verify (by adding $x_5 = R \cos\omega$) that $x_1^2 + x_2^2 + x_3^2 + x_4^2 + x_5^2 = R^2$. The four-dimensional spherical world is a sphere embedded in Minkowskian space. Thus, one of these five coordinates must be imaginary. For instance, by making $x_4' = ix_4$, x_4' becomes real and $x_1^2 + x_2^2 + x_3^2 - x_4'^2 + x_5^2 = R^2$, that is to say, we have "in reality" (to quote De Sitter) a four-dimensional hyperboloid with *one* sheet. But De Sitter spoke of *two* sheets. The mistake is corrected in the next article: "The hyperboloid has only *one* sheet. Its projection fills only part of the Euclidean space of four-dimensions: the part outside the limiting hyperboloid H is the projection of the conjugated hyperboloid (which is of two sheets)" (1917b, p. 229). Perhaps the mistake has arisen because De Sitter had first before his eyes his equation H, which is the equation of a three-dimensional hyperboloid with two sheets. If we look at Figs. 3.4(a) and (b), we see that, in the stereographic plane, H exists with infinite $g_{\mu\nu}$, because the projection of the world is entirely outside the sheets of this three-dimensional hyperboloid. This is what the correction tends to show.

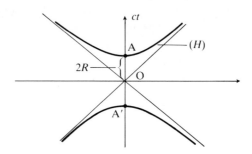

FIG. 3.6. Calculating distances in the De Sitter universe.

Bearing this mistake in mind, let us return to the original footnote of the first article. Take the De Sitter world with its time dimension and keep only one space dimension (Fig. 3.6). De Sitter believes that the results remain the same, whether the universe is a hyperboloid or a sphere (1917a, p. 1219). In particular, he is not really bothered by the two sheets, since "the formulae embrace both sheets, but only one of them represents the actual universe" (p. 1220). Einstein will be very puzzled by this curious and unjustified argument. But let us not anticipate. De Sitter explains the occurrence of the singularity in the following way:

The hyperboloid H is the limit between the two parts of the Euclidean space x, y, z, ct corresponding to these two sheets. It is intersected by the axis of t at the points $ct = 2R$, the distance of which from the origin is, in natural measure, $\int_0^{2R} c dt(1 - \sigma c^2 t^2)$ $= \infty$. The length in natural measure of the half-axis of x is, in both systems, $\int_0^\infty dx(1 + \sigma x^2) = \pi R$.

Points A and A' on Fig. 3.6 represent all the points of the three-dimensional hyperboloid Γ. In order to show that, De Sitter uses the $g_{\mu\nu}$ of system B. The distance AA' is $\int_0^{2R} ds$. On the ct axis, we have $\int_0^{2R} \sqrt{g_{44}} d(ct)$, that is, $\int_0^{2R} [1/(1 + \sigma h^2)] c dt$. As $\cos \chi = 1$ and $\sin \chi = 0$ on the ct axis, we have $h = ict$ and $\int_0^{2R} 1/(1 - \sigma^2 c^2 t^2) = \infty$. The distance AA' is thus infinite in natural measure; the $g_{\mu\nu}$ are infinite on those points which are impossible to reach by physical means. Through the projection, the points at infinity are brought back at finite distance, but this has nothing to do with a place where space would be so distorted that its geometrical properties break down. As De Sitter says, "the natural distance of those points from the origin is. . .*not* finite, contrary to what you have thought at first" (18 April). Einstein had calculated that $\int_0^{2R} ds$ is finite, which is an elementary mistake in calculus. De Sitter did not see this himself. In fact, his assistant, Kluyter, saw the mistake, and De Sitter crossed out Einstein's result.

Such a conclusive argument would seem to settle once and for all the question of the geometrical properties of the singularity. Not for Einstein. There is yet another way of seeing that something is wrong with the geometry: not by looking at the periphery, but rather at the *centre*. Einstein believes his argument transcends the as yet loose distinction between coordinates and structure. Instead of a singularity related to space, he now discovers a paradox with time: "The spatial extent of your world depends on t in a peculiar manner. For sufficiently early times one can put a rigid circular hoop into your world which has no place in it at time $t = 0$" (14 June). This is due to point O on Fig. 3.6: if the world is only one of the two sheets, it is difficult to find a justification for the 'privilege' of either one; the two sheets must be divided by a puzzling line $t = 0$ if they are to be symmetrical. That a rigid body will remain rigid at any place and at any time is of course one of the essential premises by which Einstein could give the Riemannian geometry a physical interpretation; but nothing seems to exist at $t = 0$, since the line divides two equally possible worlds.

At this stage, it becomes extremely difficult to reconstruct the actual significance of the arguments, since a certain number of gaps appear in the correspondence available to us. In another letter (22 June 1917), Einstein discusses the problem anew. Perhaps we should understand this discussion as a reaction to De Sitter's correction—model B is not in fact a two-sheet, but a one-sheet hyperboloid. De Sitter himself, it should be noted, realized the mistake only when Ehrenfest pointed it out in his letter of 14 April (the most significant passages are quoted in Chapter 2). At any rate, Einstein is now eager to make it clear that the privilege of the central point is certainly independent of all coordinates. Consider the light-cone originating from any point on the hyperboloid, and let H stand for "the infinitely remote (in natural measure)". All light-cones form "light-surfaces" (*Lichtfläche*), that is, they follow the hyperbolic shape of the four-dimensional world. The privilege of the central, null point comes from the fact that the light-surface originating from that point does *not* intersect H; it approaches it only asymptotically. As a result, Einstein says, "this point is *de facto* preferred. . . This is of course no disproof, but the circumstance irritates me".

What Einstein has discovered, is the existence of a so-called *event horizon* in De Sitter's universe. The phrasing is borrowed from a much later clarification by W. Rindler (1956). It refers to the existence of a light front, separating two classes of photons: those that reach a given observer within a finite time and those that do not; the latter are the photons that never get to the observer. Of course, when we are armed with the conception of an expanding universe, this can be easily understood in terms of a 'balloon' which is being blown up at a very rapid rate. An event horizon is to be distinguished from a *particle horizon*, one that divides all actually observable particles for a given observer into two classes: one containing the already observed particles, and the other all those yet to come within reach. Einstein was irritated by the fact that the particle horizon from O was *also* an event horizon.

In fact, Einstein began by confusing coordinates and structure, while the event/particle horizon distinction is entirely structural. At the time, such distinctions were largely ignored in the literature on relativity and various writers were to perpetuate the confusion for many years. (For a historical review of the problems and the influence of the early discussions on the present debates, see F.J. Tipler, C.J.S. Clarke and G.F.R. Ellis 1980.) Yet in spite of its clumsy terminology, this debate between Einstein and De Sitter certainly represents the first explicit recognition of the problem of horizons in modern cosmology. The various interpretations of horizons would henceforth constitute one of the vexing questions which would do much to determine the course of latter day cosmology. We will come to some of these interpretations in due course.

What is important in this first, decisive stage of the debate, is that according to Einstein, there seems to be some evidence for a natural origin of time in the De Sitter model. There is, of course, no serious anticipation of the 'big bang' theory in this remark, yet it must be emphasized that Einstein was never deterred from his belief that true singularities are inadmissible, whether they be in the form of the Schwarzschild singularity or the occurrence of event horizons in cosmology. As to the big bang, Einstein remained critical until the end of his life: when he began to consider the issue of unification of gravitation and electromagnetism in terms of large densities, he wrote that "one may not conclude that the 'beginning of expansion' must mean a singularity in the mathematical sense" (these are his words in the appendix on the cosmologic problem which he added for the second edition of *The Meaning of Relativity* 1945, p. 129). And ultimately he went as far as thinking that those cosmological theories which embarked on understanding the origin of time were as much "scientific" as biblical accounts of creation could be (see Einstein–Besso 1972, p. 500). What bothers him in this early discussion with De Sitter, is that the real figure of the De Sitter universe revives the old problems of boundary conditions. Indeed, whatever the justification for the change from set (3.1) to set (3.2), the curvature of time appears to keep pace with the positing of boundary conditions for time. While the difficulty with a natural limit in space could be overcome by means of non-Euclidean geometry, the temptation would be to identify these new boundary conditions

MATHEMATICAL AND PHYSICAL POSTULATES OF COSMOLOGY   191

with a natural origin of time. Yet the laws of mechanics, in whatever context, are by themselves impervious to the direction of time (the content of the equations is not changed if $+t$ is replaced by t). Thus, far from representing the last remnant of a classical conception of the world, Einstein may well have seen that linear time, in model A, is the only property of time that *completely* does away with boundary conditions. As he wrote it to Ehrenfest as early as 14th February 1917, it was quite ironical to realize that finally all the requirements of relativity would be satisfied with "a new quasi-absolute time and a preferred coordinate system".

In fact, Einstein's discussion of singularity was simply one aspect of a broader problem. He understood that the singularity was yet another issue which demonstrated how model B failed to meet the general requirement of *homogeneity*. Homogeneity on both the geometrical and the physical sides of the equations, was the great issue which loomed from the start. Let us, therefore, go back to late March 1917. Here again, De Sitter was to raise questions of a genuine epistemological significance.

4. Mathematical and physical postulates of cosmology

There is indeed another part of Einstein's first (24 March) reaction to model B, one which deals with the implications of Mach's principle. This part of the letter is so important that De Sitter asked Einstein's permission to quote an excerpt as a postscript to his article on the relativity of inertia (1917a, p. 1225). Einstein had written: "In my opinion, it would be unsatisfactory to think of the possibility of a world without matter. The field of $g_{\mu\nu}$ ought to be conditioned by matter, otherwise it would not exist at all." He then goes on to depict the material requirement of relative inertia as something quite radical, because it extends beyond any strictly Machian conception: he speaks of this as "the material property of geometry" in a striking return to the original foundations of Riemannian geometry and their actual incorporation in general relativity. Riemannian geometry is implicitly a kind of 'materialization' of geometry: the very act of surveying anything material with measuring instruments produces a ds^2, and the resulting metric is intrinsic, that is, our measurements would always remain arbitrary were it not for the metric which actually conforms to the shape of physically existent entities. Because of the connection between its conceptual foundations and the *large-scale* properties of the new geometry, Einstein adds that the cosmological constant for the first time meets all demands of general relativity.

This line of reasoning is called by De Sitter the "*material postulate* of the relativity of inertia". But model B shows that the relation of Λ to the world-matter is *not* a necessary one (postcard of 18 April). The very existence of a globally curved space-time, devoid of world-matter, shows that the large-scale connection between physics and mathematics cannot tally with the form envisaged by Einstein. The connection is certainly not secured by Λ, since Λ cannot lend itself to the interpretation Einstein sought. Once again, De Sitter is from the outset not concerned with the interpretation of his own model. He comes to the conclusion that *any* connection on a large scale between physics and mathematics has to be undermined: "We can also abandon the postulate of Mach, and replace it by the postulate that at infinity the $g_{\mu\nu}$, or only the g_{ij} of the three-dimensional space, shall be zero, or at least invariant for all transformations" (1917a, p. 1222). In contrast to the material postulate, De Sitter calls this condition of invariance the "*mathematical postulate* of the relativity of inertia". Note that models A and B do not form a physics/mathematics opposition: De Sitter maintains that *any* cosmological construct is doomed to be affected by this distinction. It is also clear that the condition of invariance provided De Sitter with the method by which he succeeded in formulating the boundary conditions for the finite universe.

The distinction contemplated by De Sitter between the two types of postulates comes as the first firm result yielded by his investigations. Provided Λ is maintained and all world-matter dispensed with, model B does give the generally covariant boundary conditions that Einstein was so eager to establish. It should be remembered that, in the calculation of these conditions (Chapter 2), the quantity A in the expression (2.13) of the ds^2 (isotropic and homogeneous gravitational field) had to tend to zero for a particle sufficiently remote from all other masses. In order for ds^2 to remain an invariant, the quantity B had to tend to infinity while A tended to zero. But De Sitter's new reflections make the differing behaviour of the two quantities unacceptable. Instead of time accelerating as the distance is increased, the new conditions $A \to 0$ *and* $B \to 0$ imply the deceleration of time, an infinite 'contraction' at infinity. At infinity time ceases to dilate in order to compensate for the 'dissolution' of space.

There was indeed something quite paradoxical about Einstein's earlier boundary conditions. The notion that clocks would run 'infinitely fast' at infinity is a direct consequence of the prediction which Einstein made as early as 1911, that clocks would slow down in the vicinity of ponderable masses. Clearly, the paradox resulted from the kind of boundary conditions under consideration: how will a test body behave when very remote from all other masses? The correct interpretation of the boundary conditions was vitiated from the start, since this type of boundary condition was already utterly 'classical' in the Newtonian sense, absolutely without relation to a consistent Machian concept which would have invalidated the question in the first place. The impossibility of predicting the behaviour of a supposedly remote system was already the subject of discussion in the thought experiment of two rotating bodies. In the present case, De Sitter does not pretend to uncover the true formulation of neo-Machian boundary conditions. Rather, the new boundary conditions are an indirect proof that the Machian concept is untenable: clocks now stand still where there is no matter at all. Why does this happen? The general condition of invariance at infinity simply reflects our inability to think of the infinite (whether to maintain it or to abolish it) in terms other than the purely mathematical.

Does this tend to show that no progress has been made in the problem of finding the origin of inertia? Does De Sitter's scepticism block the way to a better understanding of this origin? Einstein seems to think so. Replying to De Sitter on 14 April, Einstein asks that all degrees of belief be suspended: "Conviction is a good mainspring, but a bad regulator". Reflection should be confined to the actually available possibilities—an exceedingly harsh stricture since Einstein himself remains just as committed to his Machian principles. Progress has been made in that a new problem has been perceived, for it was Einstein's innovation to create a cosmology which would lead De Sitter to make the distinction between two types of postulate. Only because the finiteness of the world was conjectured as more than geometrical, could De Sitter say that little had changed in the pre-relativistic representation of the infinite as a 'nothing', i.e. that the infinite becomes again what it had always been—a mathematical fiction. This is much more than a return to square one, because for De Sitter the distinction allows the appropriate questions to be raised. And by addressing them, De Sitter was to discover gradually that what he had done was a good deal more than he first imagined.

The question which, from this point on, preoccupied De Sitter was this: "Which of the three systems is to be preferred: A with world-matter, B without it, both with the field equations (2.29) and at infinity the $g_{\mu\nu}$ (3.1) or (3.2); or the original system without world-matter, with the field equations (1.3) and the $g_{\mu\nu}$ (2.10) which retain the same values at infinity?" (1917a, p. 1221). Observation cannot settle the question, since the observed $g_{\mu\nu}$ are virtually pseudo-Euclidean everywhere. As De Sitter says: "The question thus really is: how are we to extrapolate outside our neighbourhood? The choice can thus not be decided by physical arguments, but must depend on metaphysical or philosophical considerations, in which of course also personal judgement or predilections will have some influence" (p. 1222). As we know, in the earlier discussions on relative rotation, De Sitter had strongly opposed philosophical concerns getting mixed up with particular solutions to the generally covariant field equations. But we should not be misled into thinking that our two authors have somehow exchanged positions. The problem is not that Einstein and De Sitter disagree about the reality of philosophical concerns. In De Sitter's mind it is all a matter of what changes need to be made to our basic concepts of physics, when we pass from local questions to the question of cosmology.

5. Inertia and gravitation; ordinary and world matter

First of all, it should be borne in mind that the De Sitter metric is formally very analogous to the Schwarzschild interior metric. The radius of curvature of a spherical mass had been calculated to be $R = \sqrt{3/\kappa\rho}$, but De Sitter obtains something very similar for his model, $R = \sqrt{3/\Lambda}$. Moreover, there is curvature of time in each case. Thus, *the De Sitter metric appears to be based on some sort of substitution of ρ for Λ in the interior of a homogeneous mass.* The cosmological constant now plays the role of the matter density in this homogeneous sphere. This has far-reaching consequences, since it suggests that the interior (local) metric and the De Sitter (cosmological) metric are formally identical, and differ only in that the former requires $\Lambda = 0$ and $\rho \neq 0$, the latter $\Lambda \neq 0$ and $\rho = 0$. So the relation between them is one of opposition: it would seem impossible to have both metrics at the same time. Even though a discussion of the interior metric is not explicit in De Sitter's writings, it certainly forms an important conceptual subtext, since the rest of his contribution is dominated by a reflection on what actually constitutes the 'cosmological' aspect of some given relativistic metric. As we will see in Chapter 4, Weyl shortly after became the first person to elucidate De Sitter's problem explicitly in terms of an antinomy.

This opposition between a local perspective and the cosmic viewpoint is first made clear in a note which Einstein added to his letter of 24 March. Here he claims that the argument of weak stellar velocities is still to be preferred to the absence of violetshift in the spectral lines, since it allows the static nature of the universe to be established. Indeed, as he pointed out, the weak stellar velocities are not limited to the observed facts within the visible part of the universe. On the contrary, "if we come up with the hypothesis of a mechanical, quasi-stationary behaviour for the matter", the weak velocities *prove* that large differences of potential cannot occur at all. According to Einstein, then, observation of spectral lines may play some role in determining the value of the cosmological constant. For if Λ were zero, the mean density of matter would become such that considerable violetshifts would occur. De Sitter was totally dissatisfied with these explanations. In the margin of Einstein's letter, he wrote: "This train of reasoning is completely erroneous...it supposes that the average density here remains the same till infinity. This is certainly not true". He refers to his earlier calculations, whereby the absence of violetshift provided only one clue to the relation between the quantity of matter and the distance: "This is the correct conclusion to draw from the observations, *not* something about Λ". He points to the error as lying in the hypothesis of quasi-stationarity. In his reply to Einstein (1 April) he makes this clear: "Of the world we have only a snapshot (*Momentphotographie*). Because we do not see many changes in the photographs, it would be too easy to conclude that it will always remain the same. Even the Milky Way does not seem to be a stable system, for the stars [he means the visible ones, not the world matter] are obviously *not* distributed homogeneously." All observations, he adds, speak against such a mean density that is everywhere the same. The next exchange of letters (14 and 18 April 1917) brought to light what De Sitter called the "difference in belief" between Einstein and himself. Einstein repeats that the constant Λ is offered only as a tentative extension of general relativity, observation having the last word on whether it should disappear or not. And among these observations, the fact that the spectral lines are a function of the distance from us has a decisive significance. De Sitter, on the other hand, speaks of the actual *value* of Λ as being beyond the possibilities of observation. He is not concerned with its *existence*, which he seems to take as part and parcel of Einstein's cosmological considerations. The divergence between the two authors is now complete, since De Sitter is willing to extrapolate in time but not in space, whereas Einstein extrapolates in space but not in time; De Sitter assumes Λ to be part of the theory of the universe, whereas Einstein expects observation to settle the question of its existence. In fact, De Sitter reserves the term "existence" or "non-existence" to the world-matter as distinct from what he terms "ordinary matter".

This is how he understands the opposition between the local and the cosmic. The earlier distinction between physical and mathematical postulates is now translated into this opposition:

To the question: If all matter is supposed not to exist, with the exception of one material point which is to be used as a test-body, has this test-body inertia or not? The school of Mach requires the answer *No*. Our experience however very decidedly gives the answer *Yes*, if by 'all matter' is meant all ordinary physical matter: stars, nebulae, clusters, etc. The followers of Mach are thus compelled to assume the existence of still more matter: the world-matter (1917a, p. 1222)

Until this point, the hypotheses on the material tensor of system B had represented an ideally uniform, homogeneous world-matter. But nothing had been said about ordinary matter, which was deliberately left out. Were the world-matter some kind of ideal arrangement of the ordinary matter, it is clear that the De Sitter universe would be *completely empty* by virtue of its equations. What De Sitter wants to prove here is that: if the world-matter is to exist at all, it *cannot* be a particular arrangement of the ordinary matter, i.e. the former is necessarily *added* to the latter. Suppose that the only existing matter is the ordinary kind. The circumstances under which this ordinary matter is supposed not to exist are identical with the situation of the exterior metric, where the matter surrounding a unique body such as the Sun is neglected. The inertial field created by the Sun exists, even though the Sun is artifically isolated from the planets. A Machian conception is thus forced to include *another* remoter matter: were this distant matter done away with, the inertial field around the Sun could not exist at all. De Sitter underlines, not without irony, that "this world-matter. . .serves no other purpose than to enable us to suppose it not to exist" (1917a, p. 1222). (In 1917d, p. 5, De Sitter adds: "and to assert that in that case there would be no inertia".) According to Mach's theory, then, the fiction by which we suppress all ordinary matter *is not the same* as the fiction by which the world-matter is supposed not to exist. However, there is nothing that allows us to believe that the two fictions differ, save the belief that this world-matter is more than mere idealization.

Einstein did not like this reasoning. He threw De Sitter's objection back at him, by showing that there was no world-matter outside the stars (14 June 1917). He claimed that ρ was, as far as he could see, nothing more than a uniform distribution of the existing stars. De Sitter was of course unconvinced. First in a letter of 20 June, and then with additional detail (1917b), he offered an in-depth analysis of the true difference between A and B, based on the implications arising from their physical properties rather than from "philosophical predilections".

De Sitter had received an interesting letter from Lorentz, dated 23 June, which seems to be the only record we have of Lorentz's reaction to the new cosmological developments. Lorentz thought he had hit on a way of smoothing out the difference between the systems A and B. He argued that non-vanishing internal stress and pressure in *Einstein's* model would yield a general case from which Einstein's actual solution could be deduced along with De Sitter's as particular cases. Lorentz's motivation lay in his conviction that the values of internal stress and pressure in the field equations should be naturally predictable as implications of gravitational effects. But De Sitter's own conviction was that such general analysis fell short of the essence of cosmological arguments. At the bottom of Lorentz's letter, he wrote this simple comment in the margin. "But precisely Einstein postulates that there is *no* internal stress." And he set out to develop his own general analysis, from which the difference between A and B would be maintained.

The problem was to specify the field outside some small sphere of matter in the two systems. In system A, it would not be possible to concentrate the whole of the world-matter into one point like the Sun. The density could then be written $\rho = \rho_0 + \rho_1$, where ρ_0 is the density of the world-matter and ρ_1 is the density of the ordinary matter. System A requires the density within the sun to be $\rho_0 + \rho_1$, and outside the sun, ρ_0. In system B, $\rho_0 = 0$ everywhere, except within the sun where the density is $\rho_0 = \rho_1$. In the calculations of the resulting fields in these two cases, De Sitter makes use of an entirely new conceptual framework. He explains it in the following terms.

In the theory of general relativity, there is no essential difference between inertia and gravitation. It will, however, be convenient to continue to make this difference (1917b, p. 230).

What is at stake is, in fact, much more than a matter of convenience. The distinction is required by the extension of general relativity to cosmology. Calling C the system of the special theory of relativity,

A field in which the line-element can be brought in one of the forms A, B or C. . .will be called a field of pure inertia, without gravitation. If the $g_{\mu\nu}$ deviate from these values we will say that there is gravitation. This is produced by matter, which I call 'ordinary' or 'gravitating' matter. . .

When this terminology is fixed, it appears that systems A an B are opposed to one another in the following respects:

In the systems B and C there is no other matter than this ordinary matter. In the system A the whole of space is filled with matter, which, in the simple case that the line-element is represented by A, produces no 'gravitation', but only 'inertia'. This matter I have called 'world-matter'.

In Einstein's universe, the world-matter must contribute to the totality of the inertial field. De Sitter interprets this requirement as meaning that the $g_{\mu\nu}$ cannot deviate from the values postulated for them in this universe. How, then, can we understand the connection between the world-matter and the ordinary matter in system A? The calculations which follow (developed in 1917b, pp. 237–42, more systematically in 1917d, pp. 19–23) are an attempt to explore the consequences of Einstein's suggestion that ordinary matter might be some kind of local condensation of the world-matter. De Sitter proceeds in three stages: first that of a purely inertial field in the three systems, then of ordinary matter in system A, and finally of the *same* ordinary matter in system B.

Taking first $\rho_0 = \rho_1$, that is, matter which is at rest and in which all kinds of internal forces (such as pressure and stresses) are neglected, the values of the material tensor can be designated by $T_{\mu\nu} = T^0_{\mu\nu}$. The result is predictable: "It is easily verified that all the different sets of $g_{\mu\nu}$ which have been given above for the inertial field satisfy these equations, if the appropriate values are taken for Λ and ρ_0" (1917d, p. 20). Indeed, in a purely inertial field with $\rho = \rho_0$ in A and $\rho = \rho_0 = \rho_1$ in B (and in C), the field equations with the cosmological constant are easily verifiable. It is quite remarkable that the equations remain verified for any constant value $T^0_{\mu\nu}$, in particular for $T^0_{\mu\nu} = 0$. In the verification of these equations, it is a matter of indifference whether we are dealing with world-matter or with an entirely empty universe. In other words, the action of inertia alone, whether or not it is linked to the existence of matter, is incapable of exerting a pressure or any other internal force from within the world-matter. From this point on, De Sitter takes the absence of pressure or stress in the world-matter as a sign of the non-necessary existence of this matter. For that very reason it becomes important to understand how ordinary matter can be superimposed on the world-matter of system A.

Let us start with the most general form of the metric around the sun:

$$ds^2 = -a dr^2 - b(d\psi^2 + \sin^2\psi d\theta^2) + f^2 c^2 dt^2. \tag{3.11}$$

If the mass of the Sun is small in comparison with the world-matter, the terms a, b, and f will not be very different from what they are in the purely inertial field. The corrections will be of the order of α, β, and γ:

$$a = 1 + \alpha, \ b = R^2 \sin^2\chi(1+\beta), f = 1 + \gamma. \tag{3.12}$$

De Sitter immediately finds that the influence of the Sun makes γ non-constant; γ vanishes only at the largest possible distance in Einstein's universe. As far as the material tensor is concerned, the effect of 'gravitation' is to account for the difference between $T^0_{\mu\nu}$ and a new $T_{\mu\nu}$. This difference only vanishes with ρ, for if the density of matter is zero, both $T^0_{\mu\nu}$ and $T_{\mu\nu}$ vanish. The corrections to be introduced in the material tensor are of the order $\epsilon\kappa\rho$, ϵ itself being of the order of α, β, γ. If the corrections are neglected, De Sitter notes, the system of equations which can be extracted from (3.11) and (3.12) is one where equations do not depend on each other, i.e., they are not exact. However, the corrections are quite negligible if, and only if, Λ is itself of the order of ϵ—in which case it then becomes possible to write the expressions for α, β, γ outside the Sun in system A.

Now, the system of equations which can be deduced from (3.11) *without* the corrections (3.12) is not a system of mutually dependent equations either. Indeed, the pressure being supposed to be the same everywhere, an equation like (2.22) can be derived. So, as De Sitter says: "the equations are dependent on each other, i.e. a stationary equilibrium, all matter being at rest without internal forces, is only possible when either $\rho = 0$ or $\mu' = 0$, i.e. $g_{44} = $ constant". On the one hand, supposing $\rho = 0$ amounts to making both $T_{\mu\nu}$ and $T^0_{\mu\nu}$ equal to zero, which means that both the world- and the ordinary matter do not exist. On the other hand, the presence of ordinary matter leads to an additional term γ in g_{44}, and g_{44} is thereby deprived of its constancy. In consequence: "If ordinary or gravitating matter does exist, then not only in those portions of space which are occupied by it, but throughout the whole of the world-matter $T_{\mu\nu}$ will differ from $T^0_{\mu\nu}$". To demonstrate this, De Sitter develops the mathematical expressions which include second-order effects, such as the motion of perihelion for the planets. This motion disappears when the values $T^0_{\mu\nu}$ are used. So we can say it is produced by the pressure of the world-matter; it disappears if it is supposed that no world-matter is present in the vicinity of the Sun.

INERTIA AND GRAVITATION; ORDINARY AND WORLD MATTER   199

The result derived by De Sitter is of the utmost importance. It is essentially a feature of system A which had not been realized. De Sitter's starting point was the necessity of overcoming the contradiction that, in A, we have *either* $\rho = 0$ or $g_{44} = cst$. How is it possible to have, at the same time, a stationary equilibrium and ordinary matter, since ordinary matter is primarily responsible for deviations from $g_{44} = cst$? This is possible, De Sitter argues, only if the material tensor is modified in accordance with the reality of internal pressure and stress. The pressure and stress would compensate for the deviations from $g_{44} = cst$. But this is only valid if the world-matter is identified with a continuous fluid; were the world-matter to be compared with separated material points, that is, were it to consist of discrete entities, the internal forces would have no effect and would prevent this world-matter from remaining at rest. In either case, the metric A is inappropriate, since it is supposed to describe a pressureless, stationary world-matter. In fact, by raising the problem of ordinary matter versus world-matter, the principal virtue of system B is to reveal system A as already affected by an internal contradiction between the local and the cosmological. Finally, how can system B itself account for the existence of ordinary matter? In this system, g_{44} is not constant. But does this imply $\rho = 0$ everywhere? De Sitter shows that in B, outside the Sun, the density of matter vanishes, so that the equations derivable from (3.11) are integrable since they depend on each other. An empty space-time around the Sun, capable of exerting inertial effects, has been substituted for the world-matter. In other words, the solution to the difficulty with ordinary matter in system A lies in the transition to system B, in which one *begins* by positing the actual existence of ordinary matter.

Of course, ordinary matter consists of discrete entities. In his synthesis prepared for the end of 1917, De Sitter sums the situation up in the following terms:

In A there is a world-matter, with which the whole world is filled, and this can be in a state of equilibrium without any internal stresses or pressures, if it is entirely homogeneous and at rest (1917d, p. 20).

In this case, when ordinary matter is dispensed with, it is impossible to *go back* to ordinary matter. In contrast, the occurrence of ordinary matter in system B makes it impossible to go back to the world-matter:

In B there may, or may not, be matter, but if there is more than one material particle these cannot be at rest, and if the whole world were filled homogeneously with matter this could not be at rest without any internal pressure or stress; for if it were, we would have the system A, with $g_{44} = 1$ for all values of the four coordinates (p. 20).

Thus, system B is *never* equivalent to system A, whatever the quantity or nature of matter (whether continuous or discrete) that is introduced in system

B. In this sense, it can be said that the world-matter is in effect necessary in system A, even if its sole purpose is to enable us to suppose that it does not exist; indeed, the ordinary matter of system B, as extended to the whole of space–time, can exist only under conditions alien to system A. Of the two ways by which one could conceive of the world-matter in A, neither is satisfactory: if it is a continuous fluid, the state of rest of this fluid implies a new material tensor; if it consists of discrete entities, these can only exist in motion. Of this alternative, De Sitter says: "Which way of treating it is chosen, is not essential for our purpose" (1917d, p. 21), the true aim being to disclose an irreducible antinomy within the system A.

De Sitter's discovery is that Einstein's universe can never be compared to an extended ordinary universe. Like he had claimed earlier about the reality of distant masses, De Sitter now claimed that the new world-matter "takes the place of the absolute space in Newton's theory, or of the 'inertial system'. It is nothing more than this inertial system materialized" (1917d, p. 9). This conclusion is derived from the application of what De Sitter called the "mathematical postulate of relativity" to the three-dimensional space–time. System B satisfies this postulate for the four-dimensional space–time as well. But, with respect to this mathematical postulate, system B is an extension of system A only if B is *completely* empty: if B entails world-matter, the material tensor has to be modified; if it entails ordinary matter, this matter cannot be at rest. De Sitter's scepticism acts at this level: he is willing to reject both A and B, return to the original field equations without a cosmological constant and without generally invariant $g_{\mu v}$ at infinity, and so leave unexplained the mystery of inertia:

we must prefer to leave it unexplained rather than explain it by the undetermined and undeterminable constant Λ. It cannot be denied that the introduction of this constant detracts from the symmetry and elegance of Einstein's original theory, one of whose chief attractions was that it explained so much without introducing any new hypothesis or empirical constant (1917a, pp. 1224–5).

With an undeniable feeling of premonition, De Sitter had already concluded his second paper for *Monthly Notices* in much the same way: the immense explanatory dimension of the theory of general relativity rested upon its ability to explain without reference to a new constant (1916, p. 184). True, the whole difficulty of conceiving system B as somehow derivable from an extension of system A comes from the status of the cosmological constant. An overall curvature is possible without matter, provided Λ is not set at zero. Thus, the fulfilment of the mathematical postulate "is brought about by the introduction of the term with Λ, and not by the world-matter, which, from this point of view, is not essential" (1917d, p. 6). It is only in system A that the field equations are not satisfied when the density is zero. The consequence is that "supposing it [the world-matter] not to exist thus appears to be a logical impossibility; in the system A, the world-matter *is* the three-dimensional space, or at least is inseparable from it" (1917a, p. 1222). This is the curious logic of Einstein's universe: the only basis for asserting that the model is logically consistent (as Einstein dearly wished) is that, if for a moment we give credence to the supposition that its matter does not exist (system B), there is no bridge allowing us to go back to the model in its original form.

6. Properties of light. Motion and the size of the universe

In a letter dated 22 June, Einstein expressed his delight that De Sitter and himself now shared an awareness of *common* problems in cosmology. He praised De Sitter for his ability to scrutinize what was "intellectually possible" in the construction of the field at large.

Via an impressive series of calculations, De Sitter had in fact systematically arrived at the actual, physical differences between the two models. It is important to follow his deductions, since Einstein was soon to realize that the work implied a contradiction within model B. In the same letter, he stated that his spherical universe should not be taken as any kind of near approximation to the facts but as a pure idealization, which had the aim of substantiating the *possibility* of a finite universe. Entire regions of space–time might well exist without matter in them: Einstein went on to compare the true universe to the surface of a potato — six months earlier he had spoken of the reference systems of general relativity as molluscs! The calculations provided by De Sitter, on the other hand, dealt with the real properties of light and motion of material bodies. From this, it was possible to draw inferences about the size of the universe (1917b, pp. 231–6; an exhaustive treatment of the question can be found in many valuable introductions to cosmology, such as J. Narlikar 1983 and H. Andrillat 1970).

De Sitter sought to establish the velocity of light in any radial direction for both systems A and B. In cosmology, photons are assumed to follow radial paths, this being a quite different assumption from that of the Schwarzschild exterior metric where they are conics. The equation $ds = 0$ represents the motion of photons, since this is the equation of geodesics whose length is zero in any space–time. What are the solutions which verify this equation in the two metrics for the universe?

Let us recall the two metrics in polar coordinates:

(A) $ds^2 = \dfrac{-dr^2}{1 - r^2/R^2} - r^2 d\theta^2 - r^2 \sin^2\theta d\psi^2 + dt^2$

(B) $ds^2 = \dfrac{-dr^2}{1 - r^2/R^2} - r^2 d\theta^2 - r^2 \sin^2\theta d\psi^2 + (1 - r^2/R^2) dt^2.$

In system A, it is easily seen that $\theta = cst$ and $\psi = cst$ verify $ds = 0$. Radial paths along the geodesics of the universe are thus always possible. The velocity of light along these paths is constant, since the coefficient of dt^2 in the coordinates r, θ, ψ, ct is just c^2. In the projection on to Euclidean space, straight lines (geodesics) are conserved, so that the ordinary formulae of spherical trigonometry can be used in order to compute the distances between heavenly bodies. The parallax vanishes for $r = (\frac{1}{2})\pi R$, that is, for the greatest distance. Theoretically, this means that we should be able to observe an image of the back of the Sun at the point of the sky 180° from the Sun. Since we obviously do not do so, certain very important conclusions can be drawn regarding the lower limit for the size of the universe.

Things look quite different in system B. The velocity of light in the radial direction is $v = c.\cos\chi$, where $\chi = r/R$. This velocity varies both with the distance between the source of light and the observer, and with the orientation. An appropriate change of variables allows the velocity of light to be constant: $\cos\chi = dr/dh$, where h is the radius-vector in the new coordinates. The metric (B) is now:

$$ds^2 = \frac{-dh^2 - R^2\sin^2 h/R(d\psi^2 + \sin^2\psi d\theta^2) + c^2 dt^2}{\cosh^2 h/R}.$$

The velocity of light is now constant in the radial direction, but the space is of the Lobachevskian type, i.e. of constant *negative* curvature. The rays of light are straight lines (geodesics) in the three-dimensional hyperbolic space h, ψ, θ. In other words, as far as the behaviour of light rays is concerned, this hyperbolic space in system B plays the role of the spherical space in system A, and of the Euclidean space in system C (see in particular 1917d, pp. 13-15). As a result, the parallax of remote heavenly bodies in system B can be derived from the trigonometric formulae of hyperbolic geometry: in this system, the parallax can *never* be zero.

The behaviour of material bodies in the two universes is also very different. In A, it is easily verified that a particle follows geodetic paths of the spherical space with a constant velocity: A particle at rest will thus remain at rest. In B, on the other hand, a particle follows a geodetic path only if it passes through the origin, however its velocity is never constant (1917d, pp. 16-17). In fact, the equations $r = cst$, $\theta = cst$, $\psi = cst$ do not all verify the equation of geodesics. Only the second and the third do (and this signifies that radial paths are always possible), but the first does not. Indeed, because the quantity dt^2 in the metric B is affected by the coefficient $1 - r^2/R^2$, the equation $r = cst$ verifies the equation of geodesics only for very special values of r, namely, $r = 0$, and $r = R$. Only when $r = 0$, i.e. when the particle coincides with the system of reference, is the state of rest possible in the same sense as in A. For any other value of r, however, the state of rest disappears together with the coincidence. Properties of motion are, of course, responsible for this situation.

They signify that no static system of reference is able to map completely the totality of the De Sitter universe, a fact which will gain significance only later but which, strangely enough, revives the very kind of problem Einstein wanted to overcome by postulating his cylindrical model. In his third paper for *Monthly Notices*, De Sitter computed a rough relation of proportionality between the velocity of any particle and the square of its distance from the origin, thinking not of a preference for one type of motion over another, but of some kind of equilibrium between receding and approaching motions.

How? Through the curious behaviour of particles at the periphery, when $r = R$. Consider the velocity of light in the radial direction. From the origin, the time taken by light to reach the periphery is equal to

$$T = \int_0^R \frac{dt}{dr} \cdot dr = \int_0^R \frac{1}{v} \, dr \cdot$$

Since $v = c.\cos\chi$, the result is $T = \infty$. The time taken by photons to travel round the world is infinite. *A fortiori*, this must also be true for the material particles. Thus, the metric B is affected by a *horizon of visibility*. For a clock at rest, χ, θ, ψ are constants, so that $ds = R\cos\chi dt$. The time elapsed between two beats of the clock is proportional to sec χ. The beats are more and more spaced as the distance from the origin increases. At the periphery, all physical phenomena cease to have duration.

The horizon is certainly not an effect of variability of the velocity of light, since in the system of reference in which c is a constant, the distance $r = R$ is infinite, and thus $T = \pm \infty$. The temporal periphery *exists*, but is *not observable*.* As De Sitter puts it:

All these results sound very strange and paradoxical. They are, of course, all due to the fact that g_{44} becomes zero at $r = (\frac{1}{2})\pi R$. We can say that on the polar line the four-dimensional time–space is reduced to the three-dimensional space: *there is no time*, and consequently no motion (1917d, p. 17).

In this way some sort of balance between receding and approaching motions can be made possible. A particle which leaves the origin, first acquires a certain velocity of recession; once it has covered half the distance to the polar line (the periphery), the velocity must decrease until no motion occurs on the periphery itself. In this latter part of its journey, the particle seems to be

---

*Synge (1960, pp. 256–65) has made an interesting development of the De Sitter universe with negative curvature. He finds that changing from positive to negative curvature is tantamount to interchanging the roles of the spacelike and the timelike geodesics. Thus, in the case of positive curvature, all the spacelike geodesics have finite length while the timelike geodesics never intersect again; in the negative case, spacelike geodesics are open but the timelike geodesics are closed curves. The latter case is very strange indeed: in Synge's words (p. 264) "this depicts what can only be described as a fantastic situation. We see a test particle repeating its history over and over again!"

'attracted' towards the observer at the origin. Equilibrium is brought about by these particles which leave the periphery and move towards the observer.

These results enable De Sitter to clarify the problem of how to compute the size of the universe in the two models.

Take system A first. Because the parallax is zero for the largest distance, its measure does not provide any information on the possible value of the radius of curvature. Other methods must be used. De Sitter shows a remarkable numerical agreement between contrasting methods: the value of the apparent angular diameter of some heavenly bodies, the density of matter in the centre of the Milky Way, and the density resulting from what De Sitter took to be the highly probable existence of other galactic systems (it being assumed that the whole of the world-matter is condensed in that form). In the latter calculations, the hypothesis of a true uniformity acquires a palpable significance: the spiral nebulae are supposed to be of the same dimension as the Milky Way, the distances between galactic systems are always supposed to be practically identical. An entirely different method is using the properties of light. Since no image of the back of the sun is seen, light must be absorbed during its journey towards us. The assumption that a reasonable degree of absorption must take place leads De Sitter to give a value for R which, once again, closely accords with that which the other methods yield.

Such a train of reasoning cannot be applied to system B, since light takes an infinite time to go around the universe. Apart from the parallax (which in this case does indicate a lower limit for R), De Sitter provided yet another method using light, this time one which was quite specific to system B. This method did not imply any hypothesis as to the physical 'reality' of the finite universe, partaking only of the remarkable properties of the metric. De Sitter explained that:

In the system A, g_{44} is constant, in B g_{44} diminishes with increasing r. Consequently, in B the lines in the spectra of very distant objects must appear displaced towards the red. This displacement by the inertial field is superimposed on the displacement produced by the gravitational field of the stars themselves (1917b, p. 235).

So here we hit on a third type of shift for the spectral lines, after the gravitational shift and the Doppler shift which De Sitter had so far encountered. The red shift is created by the inertial field: it is a property of the metric B, and proportional to distance.

By 1917, it was a well-established fact of observation that the majority of stars exhibit a slight shift towards the red. A part of this shift could be attributed to the inertial field, the rest being due to gravitational and Doppler effects. The whole problem, of course, was to sort out these different effects as they interacted in our measurements. Ever since the first use of spectroscopy in nineteenth-century astronomy, the myriad of shifts in spectral lines of stars or nebulae had been taken as indicating radial velocities. Now, the most spectacular phenomenon of all, as De Sitter reported it, was the recent measurement of significant radial velocities among three distinct nebulae. These are N.G.C. 1068, N.G.C. 4594, and the nebula in Andromeda, whose large radial velocities were measured by Slipher as well as such astronomers as Pease, Moore, and Wright. De Sitter suspected that independent tests of distance would reveal that these nebulae were in all probability extra-galactic. Among the three nebulae, only Andromeda showed a motion of approach; the two others receded, with much larger velocities than the oncoming velocity that distinguishes Andromeda. Interpreting these radial velocities in terms of a shift created by the *metric* was quite an alluring prospect because it would confirm De Sitter's prediction of g_{44} diminishing with distance. De Sitter, however, was filled with caution, and did not wish to pronounce in haste. It was clear that the reality of the effect depended upon the possibility of truly *systematic* displacements towards the red:

About a *systematic* displacement towards the red of the spectral lines of nebulae we can, however, as yet say nothing with certainty. If in the future it should be proved that very distant objects have systematically positive apparent radial velocities, this would be an indication that system B, and not A, would correspond to the truth. If such a systematic displacement of spectral lines should be shown not to exist, this might be interpreted either as pointing to the system A in preference to B, or as indicating a still larger value of R in the system B (1917b, p. 236).

Note that De Sitter is not directly concerned with the relation between the metrically predicted shift and the Doppler shift, which would be expected in the case of true motion. De Sitter reckoned on some sort of equivalence between the two, assuming one type of shift would be somehow translatable into the other. He took for granted, somewhat uncritically, that if the displacements were systematic, the velocities were therefore only apparent and need be interpreted only as properties of the inertial field. For De Sitter, only the homogeneity of the inertial field seemed suggestive of an explanation for this systematic feature; the directly visible, material universe still appeared to him to be essentially chaotic. At first it might seem that De Sitter did not really perceive the full implications of the relations between the theoretical and the observational. In fact, he was almost entirely preoccupied with a quite different contrast, the physical versus the geometrical. Properties of motion had to occur only because of a change in the *material* tensor, if one were to account for the existence of ordinary matter. The predicted behaviour of material particles (a balance between receding and approaching motions) had nothing to do with the systematic redshift required by the geometry.

7. Einstein's new reply

In the current state of records detailing the correspondence between Einstein and De Sitter, we have to wait until 10 April the following year, 1918, before there is trace of a new letter from De Sitter. This gap of almost a year makes it very difficult to evaluate how he gradually came to accept the existence of a higher form of connection between the physical and the geometrical in his own model. Two letters and three postcards from Einstein during this period are only guides here. Fortunately, in one of them, De Sitter has made notes in the margin. On 28 June, Einstein reacted to the conclusions De Sitter had drawn, the essentials of which were contained in the article "On the Curvature of Space" (1917b). The reaction was as short as it was lucid: Einstein realized that his earlier speculations about the occurrence of a singularity in model B were after all by no means completely wrong. There was now a new way of seeing things which, Einstein believed, made his own intuition an inescapable certainty.

On 9 August 1917, Einstein wrote to Elsa Einstein about his general state of mind: "For a change, there is. . .little new, except the correspondence with De Sitter and Levi-Civita, both highly interesting". Indeed, the discussion with De Sitter had reached a pitch of excitement, with Einstein writing to him on 31 July: "Your system is not a physical possibility". Take the expression for the energy of a material particle, $m \sqrt{g_{44}}$. In the metric B, dt^2 is affected by a coefficient $\cos^2(r/R)$. Therefore, at the distance $r = (½)\pi R$, this particle has simply lost its energy: *it can no longer exist*. It is therefore hard to maintain, as De Sitter had, that the periphery itself exists. A week later, on 8 August, Einstein managed to find a bridge between the models A and B. This argument represents his very first gropings towards a distinction between an event and a particle horizon. Einstein asked what the condition was for ascertaining the existence of the periphery. According to Einstein, the condition was precisely what made the De Sitter model nonphysical. The tendency of g_{44} to decrease to zero is, as he said, analogous to the phenomenon of clocks slowing down near large masses (like the Sun). In other words, all the matter of this universe has been concentrated on the equator of the spherical space. The non-physicality is due to the following fact: "The heterogeneity of the different points of space is not, in this conception, an 'autonomous' property of space". Einstein preferred an energy which was everywhere finite.

It is here that De Sitter added two comments in the margin. The first is a response to the supposition of concentrated matter at the equator: "That would be distant masses yet again". His position is quite understandable. As we have seen, the metric B seems to spell out the boundary conditions which Einstein tried to find before he invented the finite model, but De Sitter had no desire to attach a physical meaning to them. By so doing, Einstein simply reverted to his quest for distant masses. A general comment by De Sitter has pertinence here: "If g_{44} must become zero for $r = (½)\pi R$ by 'matter' present there, how large must the 'mass' of that matter then be? I suspect ∞! We then adopt a matter which is *not* ordinary matter." And he concludes with a pugnacious pun: "It is a *materia ex machina* to save the dogma of Mach". Just as De Sitter had given a version of the Einsteinian model in *his* own image, it seems that Einstein was now taking over De Sitter's model as part of his own imperial theme.

Alas, De Sitter's reply seems to be lost, but a new postcard from Einstein on 22nd August seems to widen the gap between them. Einstein reiterates his view, that all points of the metric B are spatially identical, but the frequency of a clock changes with the place, so that it reaches the inadmissible value of zero at the equator. As a result, the total energy of a particle would indeed disappear there. Einstein gave a clear-cut presentation of his objections in a paper which was published somewhat later by the Berlin Academy (see A. Einstein 1918b). There he says that a fundamental requirement of the equations of general relativity, even when they are supplemented by the cosmological constant, is an absence of discontinuity at any finite distance. The $g_{\mu\nu}$ and their dervatives should be everywhere continuous and differentiable. But this is not the case with the De Sitter metric. The determinant of the De Sitter metric is $g = R^4 \sin^4(r/R)\sin^2\psi\cos^2(r/R)$, and continuity requires that this determinant be everywhere different from zero. In fact, g already vanishes for $r = 0$ (as well as for $\psi = 0$). However, this is only an apparent violation of continuity, since the choice of a new reference system (a new centre) transforms it away. The fact is easily explained, since the choice of a centre must remain arbitrary. The determinant also vanishes for $r = (½)\pi R$, at the periphery. As De Sitter had demonstrated himself, the point located at the distance $(½)\pi R$ from the origin represents a finite distance in coordinate measure. What Einstein seems to be arguing is that in contrast with the centre, the periphery is *the same* for all observers. Consequently, no change of coordinates can prevent the periphery from being a singularity—it is a singularity of the field itself. If no inertial field is possible without matter, then all the matter must be concentrated at the singularity, since this may account in physical terms for why the clocks slow down.

It was bound to take a long time for De Sitter to draw up a consistent, i.e. *non*-Machian interpretation of the singularity. In his reply to Einstein's paper (10 April 1918), De Sitter wrote that he did not know whether the matter of his model was concentrated at the equator; that was something on which he could not commit himself. But the actual constitution of the periphery, and the relation of the periphery to the centre (that is, to observability) were two different things. He referred to a remark which he had included in his third paper for *Monthly Notices*, which he thought of as a decisive reply. This drives us on to the most speculative part of De Sitter's thought. He argues that there is no possible connection between the centre and the periphery:

> A particle which has not always been on the polar line can therefore only reach it after an infinite time, i.e., it can never reach it at all. We can thus say that all paradoxical phenomena (or rather negations of phenomena) which have been enumerated above can only happen after the end or before the beginning of eternity (1917d, pp. 17-18).

It is a stunning instance of an astronomer rejecting any physical meaning for the boundary conditions. However, De Sitter himself was not completely satisfied with this counter-argument, and he formulated a more detailed version of it in a new article. It would appear that Einstein wanted to show that the singularity in the metric B did not really differ from the better known, supposedly intrinsic singularity $r = 2m$, which occurs in the Schwarzschild exterior metric. Concerning the postulate of continuity enunciated by Einstein, De Sitter wrote:

> This postulate is not fulfilled by my solution B, as Einstein very correctly points out, and as is also shown very clearly in my communications. This postulate, however, in the form in which it is enounced by Einstein, is a *philosophical*, or metaphysical postulate. To make it a *physical* one, the words 'all points *at finite distance*' must be replaced by '*all physically accessible* points'. And if the postulate is thus formulated, my solution B does fulfil it (1917c, p. 1309).

The discontinuity is a point located at a finite distance in the space represented by coordinates, and no coordinates can change this unless they alter the space geometry altogether—for instance (as De Sitter had shown earlier in 1917d, p. 18), if the radius-vector is projected onto the Euclidean or the hyperbolic space, the paradoxical phenomena are also relegated to infinity in space. But by taking the spherical space of model B as its *actual* space, it is already clear that the time needed for light and material particles to reach the periphery is infinite in natural measure.

8. The scientific and philosophical outcomes of the Einstein–De Sitter controversies

(a) From a physical point of view

From the sequence of controversies between Einstein and De Sitter (from the first epistemological debate in 1916 to these cosmological speculations of 1918), it is clear that there is a seemingly irreconcilable conflict between what is regarded as 'physics' and what is regarded as 'philosophy'. Although this conflict does something to explain De Sitter's caution, scepticism, and humility, it has by no means hindered the practical development of ideas; what originated in the two protagonists' minds gained a quasi-autonomous status when it was taken up by other scientists, who called for an increasingly detailed interpretation of the available models as well as a radical solution to the problem. It took fifteen years for the problem of singularities, whether in the Schwarzschild metrics or in the cosmological metrics, to be properly understood. The key to its resolution was the recognition that changes of space-*time* coordinates (rather than of space alone) transformed what had been taken for intrinsic singularities into apparent ones. (Thus, apart from the cosmological problem, G. Lemaître 1933, pp. 62-4 was to show that the $r = 2m$ singularity in the Schwarzschild metric can be transformed away by an appropriate change of space-time coordinates. In order to get continuity between the interior and the exterior metric, suffice it to take the pressure as zero (or constant) and make c dependent on time in the interior case. The inferior limit for the radius of a given mass is now determined by the exterior field itself. Space-time coordinates create a truly dynamic picture of the universe, one that overcomes most of De Sitter's obsessions with such oppositions as motion and rest, discrete and continuous distributions of matter, and more generally his way of putting the tension between physics and geometry. Where De Sitter's feeling for dichotomies allowed him to see contradictions in Einstein's model as well as those between models A and B, it was Einstein's uncovering of a possible contradiction within model B which was decisive in determining a new strategy.

For Einstein the impact of the new cosmological synthesis was not to be belittled; it had an application that went far beyond the correct interpretation of continuity. On Max Planck's sixtieth birthday in 1918, Einstein delivered an address to the Berlin Physical Society in which he attempted to inculcate an understanding of the new world view derived from theoretical physics (1918a, pp. 3-4). The physicist is obliged to confine himself to describing very limited areas of the totality. And even then, in these limited areas, he can never expect a satisfactory description of anything but the most simple phenomena. This leads to tension between the method and the aim of the physical sciences because although he must rest content with "supreme purity, clarity, and certainty at the cost of completeness", it remains true that "the physicist's renunciation of completeness for his cosmos is. . .not a matter of fundamental principle". The aim, if not the realization, of a total and comprehensive vision was abiding; such vision would transcend all of the vicissitudes created by the tension between method and aim in physics. There is no doubt that Einstein thought that the problems endemic to global physics and cosmology, sprang from the peculiar status of what he called the "fundamental principle", which in his opinion vindicated the physicist's supreme task as a researcher. If it were ever possible to arrive at a logical understanding of it, which could mirror its state as a fact of nature is mirrored, then the physicist's desire would be consummated, he would find his resting place in "those universal elementary laws from which the cosmos can be built up by pure deduction". But there is no such thing: pure deduction in this sphere is a dreamer's hope because "there is no logical path to these laws; only intuition, resting on sympathetic understanding of experience, can reach them". We have no choice but to take refuge in the happiness of some "pre-established harmony" if our hope is really an agreement between our own intuition and things as they really are. It is primarily the quest for such a harmony, rather than any verifiable rationale, which in every period of the history of science accords superiority to one theory over another; the sense of superiority is incorrigible, the theoretical possibility of completeness is enough to justify and underpin the motive that finds faith in the chosen theory; it goes as close as anything could to *proving* that the theory, however local and however limited, actually matches the essence of reality.

Certain theoretical deductions and some experimental results have in quite a remarkable way lent support to Einstein's speculations about the large scale structure of space–time. As early as December 1917, physicist Hans Thirring published an article, which has remained famous, about the action of rotating distant masses in Einstein's theory of gravitation. His idea involved calculating the consequences of Einstein's applications of Mach's ideas to a concrete example. (H. Thirring 1918 and 1921. It should be noted that Thirring presupposed a flat background metric. It remains a matter of conjecture as to whether the results are also applicable to the spherical universe: see R. Torretti 1983, p. 200.) As we have already seen, Mach understood perfectly well that there was only one possible test which could validate his theory of relative rotation. This would be to compare Newton's bucket experiment (as it had been performed in practice) with a similar experiment whereby the universe would be made to rotate around an undisturbed bucket. Since such an experiment is impossible, Mach could see no reason why we should feel compelled to accept the Newtonian interpretation of the experimental result. Now De Sitter, while criticizing Einstein's first and more elaborate interpretation of Mach's idea, showed that Einstein's Machian interpretation could not really be proven unless measurable variations could be effected on the distribution and movement of the stars. It is, in fact, impossible to produce these variations, and the advent of a purely theoretical cosmology did not solve the practical problem. Thirring made a highly original contribution by showing that although performing an experiment on the whole universe remained impossible, the new theoretical concept of the universe provided a means for predicting effects of distant masses on certain local phenomena. Such effects were not reducible to a Newtonian interpretation. Because they are in principle measurable, they constitute, at the very least, an indirect proof of the soundness of Einsteinian cosmology. Thirring explained that a universe rotating around a stationary earth should produce a centrifugal force in order to offset terrestrial attraction. (It is this same force, for instance, which accounts for the Earth's equatorial bulge.) A rotating universe should also yield a Coriolis force which, for example, should turn the plane of a Foucault pendulum. By analogy, we should therefore expect the eventual manifestation of centrifugal and Coriolis forces, no matter how weak they may be, inside *any* rotating shell. Less than two months later, Thirring joined with physicist J. Lense and probed another kind of non-Newtonian prediction (see J. Lense and H. Thirring 1918). According to Newton's theory the fixity of the plane of Foucault's pendulum, relative to the stars when the Earth is rotating, should not be affected if the stars were to disappear. According to Mach, however, the plane in which the pendulum swings would remain fixed relative to the Earth if the stars were removed. Let us suppose that the stars were gradually brought back in, until their inertial effects prevail again. For reasons of continuity, these inertial effects would not any longer be capable of complete predomination. The Earth should therefore contribute ever so slightly to drag around the plane of the Foucault pendulum in the direction of its rotation. Thirring and Lense devoted themselves to a series of theoretical calculations with the purpose of predicting the extent of this effect. Obviously we are dealing with an effect so small that it can only be estimated from perturbations created by planets on their satellites; these perturbations are even more minute than the better known precession of Mercury's perihelion created by the Sun. Although it was quite impossible to measure these effects in 1918, more recent tests tend to confirm somewhat similar predictions to those conjectured by Thirring and Lense. (For a review of these questions, see C.W. Misner, K.S. Thorne and J.A. Wheeler 1973, §21.12.)

In his March 1918 article, Einstein for the first time drew a distinction between the relativity principle and Mach's principle. He also introduced important nuances as to how his own model was to be interpreted (1918c, particularly pp. 243-4). Einstein now explained that the static and spherical model which comprises Mach's principle cannot be constructed as a simple conceptual figure. In what way does it correspond to reality? The universe actually observed never exhibits the regular distribution of matter postulated in theory; only quite irregular matter may be perceived. It is only by taking the mean density of very large areas of space that the cosmological constant comes to seem inevitable, although the actual shape of the world is more ellipsoidal than spherical, and has a relation not unlike that of the shape of the Earth to a perfect sphere. Strikingly enough, Einstein's argument seems to take exactly the opposite line to that expressed in his 22 June 1917 letter to De Sitter, where the "potato-like" form of the universe (as Einstein called it) was seen as restricting cosmology to a mere intellectual game of possibilities. Einstein was not alone in hesitating over the difficulty of how to gauge the possible relevance of cosmological equations to local systems. The tension between local and global considerations kept cropping up in every discussion. For example, physicist Ludwik Silberstein could not fail to find interest in such an idealization, and in succeeding months he published a series of rather unprecedented speculations under the title "Bizarre Conclusion derived from Einstein's Gravitation Theory" (see L. Silberstein 1918 and Eddington's reaction 1918). The bizarre conclusion was that any homogeneous body has to be spherical if the cosmological equations are valid throughout. This result is obtained by keeping the cosmological form of the material tensor in Einstein's equations, i.e. by ignoring internal pressures and other forces interfering at the atomic scale. In fact, Eddington was able to show (without aiming at "depreciating Dr. Silberstein's beautiful result") that taking into account second-order effects (internal stress) would remove any ambiguity from this result.

From this point on, Einstein's concern for linking up the cosmological model with local physics dominated his research. The status of the cosmological constant therefore drew his attention in a systematic way. In fact, the cosmological constant had no bearing on Thirring's and Lense's calculations; the two scientists did not even mention it. In a 1919 contribution Einstein (1919, especially §§ 2 and 3) wondered if Λ could in fact be the link between the relativistic theory of gravitation and the electrical theory of subatomic particles. He believed that electricity-laden particles could produce a dynamic system when subjected to gravitational forces. If we were to write the most general form of space–time metrics of constant curvature, we would find that these space–times were homogeneous and that the metrical tensor could be written (see S. Hawking and G.F.R. Ellis 1973, p. 124): $R_{\mu\nu} - (\frac{1}{2})Rg_{\mu\nu} = -(\frac{1}{4})Rg_{\mu\nu}$. These spaces are therefore solutions to the field equations for an empty space with $\Lambda = (\frac{1}{4})R$. Starting from the field equations with the cosmological constant, and positing $T_{\mu\nu} = 0$ (that is to say, the $T_{\mu\nu}$ are mainly generated by the electromagnetic field), Einstein indeed found the relation $\Lambda = R_0/4$. (The meaning of R_0 is specified below.) In other words, there could be fundamental link between the cosmological constant and the possibility of unified gravitation and electromagnetism. This idea was never fully explored at the time, although it could be considered as the first step towards a unified field theory. Einstein did not come back to it until 1927, when he briefly commented on the way the mathematical aspect of such a model would confront the new quantum theory (see A. Einstein 1927). In fact, it is probable that Einstein did not linger over the issue because the whole idea of proceeding this way in order to establish a unified theory came from Hermann Weyl. Although Einstein admired Weyl's ingenuity, he could not help thinking, after a very short period of initial enthusiasm, that any eventual unified theory would have a completely different basis. (We will come back to this idea in the next chapter, when we look at Weyl's work.) More importantly, Einstein hit on the idea of considering Λ as a mere constant of integration rather than as a universal constant peculiar to a law of nature. This concept had the advantage of making its possible value entirely free of the problem of pressure. Einstein now identified the occurrence of negative pressure with what was responsible for equilibrium of electrical charge in the interior of particles. In so doing, he followed a hypothesis due to Poincaré according to which a pressure inside the particles balanced the electrostatic repulsion. The negative pressure was tantamount to $(R_0 - R)$, where R was the invariant of curvature and R_0 the value of R outside the particles. The quantity $R_0/4$, resulting from this negative pressure as a constant of integration, was itself identified with the cosmological constant. It is interesting to note that a comprehensive theory of negative pressure has been achieved only in recent years, with the investigation of the very early universe in terms of an inflationary model of the big bang (see Alan H. Guth and Paul J. Steinhardt 1984). This was made possible by a drastic revision in understanding of the physical nature of vacuum, which includes quantum effects. A vacuum can exhibit not only enormous changes in energy states but also in pressure; according to the model, the pressures of the peculiar vacuum prevailing during the very brief period just after the big bang were all negative and produced the cosmic repulsion.

It was discovered shortly after Einstein's early attempt that this theory of the electron, while meant to supersede the flaws of another theory (due to G. Mie), was far too general. Its conceptual background, however, is crucial in the context of the cosmological problem. In a 20 August 1918 letter to Besso, Einstein explained the true motivation which underlay the new conception of Λ: "Since the universe is *unique*, there is no essential difference between considering Λ as a constant which is peculiar to a law of nature or as a constant of integration" (A. Einstein and M. Besso 1972, p. 134). In other words, it is because the universe is one that we can think of a necessary harmony between the physical and the mathematical, over and above our limitations in understanding exactly why a particular type of harmony should prevail. From the mathematical point of view, all sorts of arbitrary properties of the universe can be built; this is the kind of argument in which De Sitter began to indulge when he criticized the cylindrical model. Einstein, however, in what was almost an ultimate appeal to what he called the all-encompassing "fundamental principle" of scientific research, argued that from a physical point of view, what appears *to us* arbitrary or merely possible cannot be so *in itself*. Because it is unique, the universe as a whole is a privileged place of research where our mental constructs, our experience, and the transcendent reality they finally uncover, loom up as transparent to one another.

This grave but private appeal was not strong enough to deflect the unrelenting course of ideas as it moved far away from Einstein's original ambitions. On 5 May 1920, Einstein returned to Leiden and held an extremely important conference. When the typescripts came out, De Sitter responded in two equally important letters, as well as in another paper which he delivered at the Amsterdam Academy on 27 November 1920. In his conference, Einstein introduced a small but noticeable subtlety to his original conception of his model of the universe: "By reason of the relativistic equations of gravitation. . .there must be a departure from Euclidean relations, with spaces of cosmic order of magnitude, if there exists a positive mean density, no matter how small, of the matter in the universe" (1920, p. 20). His central point was really a matter of emphasis: a departure from Euclidean space must necessarily take place, even in the case of the *smallest* density of matter. In actual fact, the whole purpose of his conference was to reintroduce the notion of ether in physics, or rather, to show the implications of the new meaning it could have in the general relativity theory. Without the notion of ether, Einstein declared, general relativity was simply "unthinkable" (p. 23). How was this possible? Einstein explained that: "To deny the ether is ultimately to assume that empty space has no physical qualities whatever" (p. 16). Such phenomena as light propagation or the very possibility for rigid rods and standard clocks to exist, hinged on the physical (that is, the non-empty) nature of space. For Einstein, the obsolete word ether signified something more subtle, the relativistic notion of field—a physical medium, likened to the space-time continuum "which is itself devoid of *all* mechanical and kinematical qualities, but helps to determine mechanical (and electromagnetic) events" (p. 19).

It is interesting that, within the context of this discussion, Einstein tackled the fundamental problem of rotation again. He began by saying that even when a system moves freely in empty space, rotation cannot be defined as an integral part of that system. Consequently, how are we capable of seeing rotation as something real? Instead of Newton's unobservable absolute space, Mach had introduced the notion of mean acceleration with regard to all the masses comprised in the universe. What Einstein recognized now was that this implied some kind of action at a distance. He proceeded to reject this idea as strongly as he opposed the unobservability of absolute space. On this particular point, we can therefore claim that Einstein tacitly deferred to De Sitter's criticisms: 'Action at a distance' refers to the strange behaviour of distant masses which De Sitter had encountered, namely, that these masses are always pushed beyond a point where further observation ceases to be effective. The newly discovered ether, Einstein declared, conditions the inert masses but is also conditioned by their behaviour. Clearly, Einstein's fresh interpretation of Mach's principle involved the idea of interaction between metric and matter, rather than complete subordination of the former to the latter. Similarly, as a consequence of rejecting action at a distance, Einstein also rejected the idea of a lower limit for the density of matter in the universe. From now on, the smallest possible density could induce a constant positive curvature for space.

THE SCIENTIFIC AND PHILOSOPHICAL OUTCOMES

This development led Einstein to herald that: "the metrical quantities of the continuum of space-time differ in the environment of different points of space-time, and are partly conditioned by the matter existing outside of the territory under consideration" (p. 18). Three years earlier, at the time of "Cosmological Considerations," Einstein would have said something rather different—that the metrical quantities were *totally* conditioned by matter existing outside the territory under consideration. In line with this, the notion of a necessarily non-empty space tended to substitute itself for the quite strict form of Mach's criterion. It is precisely because space has a structure, and even a non-isomorphic one, that it is not empty. Within these general terms, Einstein probably now viewed as more firmly grounded both his early thought experiment on rotation and the cosmological model. Because, as he said, " 'empty space' in its physical relation is neither homogeneous nor isotropic", the differential effects of rotation would be liable to occur even in a supposedly free system, and the whole of space as postulated in cosmology should not be isomorphic. To quote Eddington's later and humorous metaphor, the real figure we are talking about is the "pimply Einstein world" (1933, p. 52).

De Sitter should obviously have been satisfied with Einstein's new idea. From the epistemological point of view, Einstein had considerably modified his previous position and seemed to reject, as De Sitter wanted him to do, the last residue of mechanical interpretation for gravitational forces. And from the point of view of cosmology, Einstein forsook the idea of a law which would inflexibly determine the quantity of matter in the universe. De Sitter, however, chose to dispute Einstein's brief statement about the density of matter, first in his 4 November 1920 letter, and later in an article. He argued that in order to relate consistently the curvature of space to the smallest quantity of matter, there was another hypothesis which had to be introduced and which Einstein had not mentioned—the notion of matter in static equilibrium.

De Sitter proceeded in the following way (1922, pp. 866–7). Three types of relation between the cosmological constant and the mean density of the universe can be recognized: model A, in which $\rho = 2/R^2$ and $\Lambda = 1/R^2$; model B, in which $\rho = 0$ and $\Lambda = 3/R^2$; and model C, in which $\rho = 0$ and $\Lambda = 0$. In order to derive a matter density different from zero, model A has to be chosen; also it implies a material tensor representing matter without motion, pressure, or internal stress. If pressure or internal stress were incorporated in the material tensor, it would follow that all three solutions A, B and C would no longer be exact, to the effect that even models B and C can contain matter. In short, De Sitter wanted to prove that the fundamental objections are by-passed when the density of matter in the cylindrical universe is lowered without limit. The difference between the two models lies not in the interpretation of the cosmological constant, but in the form of the material tensor. Therefore the whole question cannot be settled in terms of presence or absence of matter; the conditions formally imposed on the material tensor should allow us to decide in the last resort.

So, in a direct response to Einstein's conference, De Sitter now compared the world-matter at rest in the cylindrical model with a state of statistical equilibrium. He did not explicitly develop what he meant by this, but instead immediately applied himself to criticizing it:

Now the possibility of statistical equilibrium of large portions of the universe is, to my mind at least, by no means self-evident, or even probable. The idea of evolution in a determined sense appears to me to be rather opposed to the actual existence, if not to the possibility, of equilibrium (1922, p. 867).

The notion of statistical equilibrium is enmeshed with evolutionary processes in a determined direction of time. Such a laconic comment makes it difficult to understand what De Sitter meant by 'evolution'. His letter to Einstein disclosed valuable clues. He explained that one of the least attractive features of Einstein's model was the possibility for an observer to perceive several images of one and the same star. These images depict the star at epochs separated by intervals of time during which light travels once entirely round the world. Thus, spherical space would be filled with a multitude of 'ghost-images' images whch would be visible and yet would not represent anything material. Of course, as mathematician Felix Klein had already pointed out some two years earlier, this kind of difficulty (discussed in Chapter 4) is easily overcome with the aid of purely topological considerations. Thus, it can be argued that two value-systems (x_1, x_2, x_3, x_4) and (x_1, x_2, x_3, x_4) of spherical space correspond to the same point, thereby transforming this spherical space into an elliptical one. Accordingly, all physical quantities of the universe, like its total mass, should be divided by two. There is no doubt that both De Sitter and Einstein knew very well of this topological argument; in fact, they rapidly joined with Klein's views. But what was afoot here was the attempt to understand the really physical predictions of model A. This is why Einstein did not object to this aspect of De Sitter's speculations; the topology only modifies the magnitude of the relevant quantities. De Sitter calculated that a new image of the sun should probably appear approximately every 500 million years, a very short period as far as astronomy and geology are concerned. The universe should present a greater number of young stars than old ones, but observation proves the contrary. De Sitter considered two possible explanations for this: either the process of star creation is practically over, or the first phases of star formation develop more quickly than subsequent ones. As an astronomer, De Sitter believed that the second explanation was the more likely, that is, the universe is endlessly involved in the evolutionary process of its fundamental constituents. This obviously refers to the stars' internal processes, and in his article De Sitter tended to identify the existence and the activity of these processes with the reality of motion contending with statistical equilibrium. In Einstein's metric there is no provision for taking into account the pressures and internal forces at work any more than there is any allowance for large-scale motions. As a result, all mechanisms of evolution which initiate motion have to be thought of as neutralized. Statistical equilibrium is precisely a state which borders a final one. If we compare stars of the universe with molecules of a gas, here also the disturbances are at their lowest and particles with small velocities are more numerous than those with large ones when the whole system reaches the penultimate stage of its evolution. Such a prediction is a direct consequence of applying Maxwell's law of distribution for particles of a gas. Neither De Sitter nor Einstein mentioned these laws, but both could have had in mind Hermann Weyl's reflections, in which such analogies had already been developed. (We shall examine this in the next Chapter.)

At this stage it seems more opportune to follow the progression of De Sitter's explicit train of reasoning. Beyond all theoretical distinctions, De Sitter advocated a final resort to *observation* as the only discriminatory criterion between the various models:

The systems...differ...in their physical consequences at large distances, and an experimental discrimination between them may be possible in the future. The decision...may be brought about by the study of systematic radial motions of spiral nebulae (1922, p. 868).

De Sitter's doubts about the theoretical foundation of relativistic cosmology remained so deeply rooted that even then he kept counting three models, the third one being nothing other than Minkowski's pseudo-Euclidean space. The criterion based on observation appeared tenuous back in 1917. Only three important radial velocities were then known; among them, one was negative. Three years later, De Sitter reported 25 measurements carried out at Mount Wilson, and only three of them were negative. Although the existence of motion made the limitations of Einstein's model immediately apparent, De Sitter did not yet undertake a detailed examination of the relation between the redshift specified in his theory (the metrical property) and the redshift as it is measured by astronomers (the Doppler effect). He simply declared, as if it could be taken for granted, that his system required "a (spurious) positive radial velocity for distant objects".

Again, De Sitter's new conjectures give the same impression of an insuperable partition between the two models. In his analysis of Einstein's model, De Sitter tended to favour the existence of an evolutionary process which he regarded as instigating motion that countermanded statistical equilibrium. In his analysis of model B, however, he kept ignoring that the apparently systematic measurement of redshifts could indicate a reality of movement. In fact, if observation could indeed be used to *discriminate* between the two models, he seems also to have thought that a more fundamental debate on the nature of concepts overshadowed any positive contribution to the *relationship* between the two models.

In the article I have just commented on, and more obviously in his letter of 4 November 1920, De Sitter had plenty of reason for feeling pleased with Einstein's rejection of the idea of action at a distance. Placing himself at the level of concepts, he wanted to dismiss not only action at a distance, but the whole mechanical interpretation of inertia as well. And the existence of evolution was quite a minor issue in this respect. As De Sitter wrote in his letter: "The *ether* carries the inertia". In other words: "The field itself is the real". As he wrote this, De Sitter was only foreshadowing Einstein's later attitude towards a unified field theory. It becames obvious that the remaining duality of metric and matter had to be overcome by deriving the properties of matter from those of space–time, rather than the other way round.

A response from Einstein (24 November 1920), followed by De Sitter's reply (on the 29th), represents the final episode of controversy between the two scientists. It would not be an overstatement to say that the progress they achieved in relation to the early form of the problem tempered the ways of dealing with the whole cosmological issue for the next ten years. As we shall see, it is undeniable that the form any future solution to the problem might take was in large part set by De Sitter, because it would necessarily be linked with the way he had perceived the difficulties.

Einstein rejected De Sitter's arguments concerning the possibility of the sun's ghost-images in model A: the matter distribution, he repeated with some insistence, was not homogeneous, so such images could never take shape. De Sitter replied with a remark that was to have important consequences. "The world", he said, "is incredibly empty". It was precisely the recently proved existence of extra-galactic nebulae that De Sitter took as confirmation of his earlier conjectures. Because the universe was practically empty, absorption of light by matter was certainly not sufficient to prevent the formation of images in model A. Einstein's criticism of De Sitter's statistical argument was even more shattering. It proceeded quite simply. Consider an isolated system. According to a well-known theorem of Newtonian mechanics, the virial theorem, the average kinetic energy of each particle in this system is half the total potential energy. Consequently, if the system is to be stationary, it cannot be so in terms of the kinetic theory of gases, which is statistical. Only gravitational forces are required to prevent the whole system from dispersing. Einstein went on to say that if it could be demonstrated by observation that stellar velocities were compatible with an equilibrium of this type, then the value of the cosmological constant could be estimated: "This would be a great result". Dealing then with model B,

Einstein for the first time stated quite unambiguously that it predicted a repulsion which increased with distance. This was the effect of upholding Λ in a space–time whose structure was conceived as independent of the presence of matter. In these conditions, Einstein suggested, "would it not be more satisfactory to take $\Lambda = 0$", since no value for the mean density was available? Einstein thought that in this model B, $\Lambda = 0$ would be the only way to avoid cosmic repulsion, and thus also the only way *to preserve inertia as an interaction between bodies*.

Einstein came to the conclusion that the idea of cosmic repulsion contradicted the concept of inertia as interaction. He was provoked to this view by his new attempt to find some consistency within De Sitter's argument on the two models. In his 29 November reply, De Sitter stressed the fact that observation of star velocities in the Milky Way was certainly in Einstein's favour. It seems that gravitational pull alone was enough to account for the stability. But the problem remained unsolved: "In order to know how the Milky Way holds together, Λ is of no use". Indeed, as De Sitter explained with the aid of figures, if Λ was derived from the mass of the Milky Way, space becomes too small to contain it; if, on the contrary, it was derived from the density of matter around the sun, the cosmos becomes so large that the Milky Way could no more hold together than if Λ were left out. In view of the uselessness of Λ, De Sitter once more underlined his preference for the theory minus the cosmological constant. "But. . .the existence of an apparent force of repulsion seems to be truly confirmed!" he exclaimed immediately. However, he specified, very carefully now, that the apparent force of repulsion was nothing but an effect of $g_{44} = \cos^2 r/R$. He also added that the extra-galactic nebulae, whose velocities were radial, seem to be scattered at random throughout the universe. For De Sitter, the existence of extra-galactic nebulae meant that it was ordinary matter which showed its true distribution, not the world-matter which would start making itself visible.

The disagreement between Einstein and De Sitter was thus never more extreme than at this time, when each one of them started paying off old scores. In his interpretation of the redshift predicted by model B, Einstein argued that in Λ there was a real force of repulsion, where De Sitter saw only a metric effect. Broadly speaking, De Sitter refused to concede that observed irregularities in the matter distribution could ultimately conform to the regularity postulated by Einstein. There was no compromise possible, because Einstein's account of the observed irregularities had nothing in common with the reasons De Sitter put forward. In any case, Einstein's reasons were *not* in conflict with the postulated homogeneity. Subsequent evolution of observational cosmology has repeatedly proved, in a virtually systematic way, that in this respect Einstein, as theoretician, was right and De Sitter, as astronomer, was wrong: the universe is indeed homogeneous in very large portions of space–time (see the discussion in P.J.E. Peebles 1980,

Ch.1). At the same time, however, if Einstein was right, it was clearly not for the reasons he gave on this occasion. He confined his argument specifically to an analysis of the Milky Way, where the configuration of the stars presents a conspicuous lack of uniformity. In fact he did not even bother to extrapolate from observations; he left this up to De Sitter, who had always been reluctant to do so. The truth of the matter is that Einstein tended to idealize, as any good theoretician would, while always feeling uneasy about not knowing enough astronomy (see for instance his confession to A. Sommerfeld, in A. Herman 1968, p. 39).

It is probably in his famous paper "Geometry and Experience", addressed to the Prussian Academy of Sciences, Berlin, on 27 January 1921, that Einstein outlined for the first time what he held to be the impact of extra-galactic discoveries. Therein he denied any possibility of finding the mean density of the whole universe from observations in the observable part of it, because "however great the space examined may be, we could not feel convinced that there were no more stars beyond that space" (1921a, p. 43). The only experimental technique he thought suitable for proving that the world was finite depended not on the idea of homogeneity, but on the static nature of the universe. Once the statistical distribution and the masses of the stars were known, it would be possible to calculate, with the aid of the available Newtonian laws, the intensity of the field of gravitation as well as the stellar velocities necessary to maintain the Milky Way in a stable equilibrium. Einstein deduced from this: "Now if the actual velocities of the stars, which can, of course, be measured, were smaller than the calculated velocities, we should have proof that the actual attractions at great distances are smaller than by Newton's law" (pp. 44–5). So Einstein could see nothing decisive in De Sitter's reports on the reality of extra-galactic systems; even the possibility of systematic movements of recession did not discompose him. Certainly, De Sitter himself did not interpret them as movements, and in any case, Einstein believed that cosmic repulsion as a consequence of Λ occurred *only* in model B. The upshot was that Einstein never again mentioned the cosmological constant as a 'real' feature of *his* cylindrical model, neither in this address nor in the four Princeton Lectures of May 1921. Instead, he talked about theoretical proofs lending credence to a spatial finiteness, using such terms as "mass-density of negative sign" or "hypothetical pressure" which cannot vanish (p. 44, and also A. Einstein 1921b, pp. 117–18). Furthermore, he inferred a relation between pressure and negative density, thereby reverting to the first premises of the problem he tackled in 1917, when he postulated a cosmological constant to prevent the effect of a negative pressure. After his 1919 paper, he was in a position to extend the hypothesis of negative pressure to the material tensor used in cosmological considerations. His argument rested, as he said, on the requirement of "phenomenological presentation". This was his way of coming to terms with De Sitter's repeated

THE SCIENTIFIC AND PHILOSOPHICAL OUTCOMES 221

comments on the presence of internal forces in the material tensor. De Sitter, however, was ready to believe in the existence of these forces by virtue of quasi-logical qualities of consistency whereas Einstein, by contrast, acknowledged them in order to take into account yet another aspect of physics, electromagnetism. In reality, Einstein argued, the new pressure had nothing to do with the hydrodynamic equilibrium of macroscopic systems: "it serves only for the energetic presentation of the dynamical relations inside matter" (1921b, p. 118).

These 1921 addresses point to Einstein's new awareness of the problems, an awareness that defined his attitude towards cosmology for the next ten years. What is noticeable is that he gradually came to seek a unification of various areas of physics. In this project, the achievements which had been forged so far in cosmology formed the background knowledge which Einstein had no wish to modify in depth. Thus, he hoped to wrest from a better theoretical knowledge of the electromagnetic field the complete vindication of the pressure term, the physical nature of which remained unclear (1921b, p. 117). In "Geometry and Experience", Einstein talked about the so-called "practical" foundations of geometry (stressing in the process the limitations of Poincaré's conventionalism), and then examined the possible application to domains which in principle had no necessary relation to these foundations. As far as the submolecular domain was concerned, "success alone can decide as to the justification of such an attempt, which postulates physical reality for the fundamental principles of Riemann's geometry outside the domain of their physical definitions" (1921a, p. 40). It is well known how fiercely Einstein fought that least classical position in the emerging theory of submolecular physics, quantum mechanics. Conversely, he maintained that things were "less problematical" when one was dealing with the other extreme of the scale, cosmology. It was at this stage that he presented the theoretical and experimental arguments I have touched on. In the last part of "Four Lectures", Einstein was also eager to elucidate the cosmological problem, "for without this, the considerations regarding the general theory of relativity would, in a certain sense, remain unsatisfactory" (1921b, p. 108).

In the years that followed, Einstein's contributions to the cosmological problem were discussed by such scientists as Klein, Lanczos, Friedmann, Eddington, and Weyl. Einstein himself contributed little more. As early as 1922, Friedmann gave the first satisfactory formulation of non-static solutions for isotropic and homogeneous distributions of matter. Einstein then wrote to Weyl on 23 May 1923: "If there is no quasi-static world, then away with the cosmological term". Doing away with Λ also implied (but Einstein talked about this only in the early thirties after Friedmann had been rediscovered) that the finiteness of space was no longer a prerequisite for constructing cosmological models. And it was during 1922 and 1923 that

Einstein had his debate with physicist Franz Selety, who thought that he could re-establish a Newtonian cosmology. (See Chapter 5 for an analysis.)

Let us return once more to Einstein and De Sitter. If Einstein, as theoretician, was right about the observational knowledge that could be derived from the universe, De Sitter, as astronomer, was right when he placed the cosmological debate at the level of examining the nature of concepts implicated in the new physics. As Einstein explained in his address at Leiden, things like the concept of ether in the theory of general relativity are extremely hard to understand. The only thing which can be said about it is what it is not: "a medium which is itself devoid of *all* mechanical and kinematical qualities". Or else: "this ether may not be thought of as endowed with the quality characteristic of ponderable mass, as consisting of parts which may be tracked through time" (1920, pp. 23-4). Here we pass beyond the sphere of logical deductions drawn from known premises. Einstein was quite aware that this had to have an effect on data that related to the cosmological problem. In a remarkable letter to Arnold Sommerfeld (28 November 1926), he criticized Eddington's concept which presented the theory of relativity in such a way as to imply that it was a logical necessity, saying: "God might also have decided, instead of the relativistic ether, to create one absolutely at rest" (in A. Hermann 1968, pp. 109-11). An ether at rest would be no other than the one Lorentz had imagined before the advent of the theory of relativity; with an ether at rest the space functions are replaced by constants (i.e. the law of action and reaction does not apply to inertia). All of which would have been inevitable, Einstein went on to argue, if God "had set up an ether according to De Sitter's model, which is essentially independent of matter". Einstein closed his letter with the following exclamation: "It is amazing that most people's heads lack an organ capable of judging these matters". We can see how, with the passage of the years, Einstein lost much of his confidence. Once everything was reduced to divine choice, our own understanding of the logical consistency of general relativity, Einstein's dream when he created cosmology in 1917, seemed forever out of reach.

The problem of finding a correct interpretation of ether was nevertheless one of the fundamental tasks which preoccupied Albert Einstein until the end of his life. Ever since his controversy with De Sitter, he was convinced that the notion of field (discovered in the nineteenth century) was insufficiently radical, because it could never succeed in eradicating from the image of the world an implicit reference to neo-Newtonian mechanical properties. This reference was revealed in the duality of source and field that Einstein wished to abolish by describing the particles as singularity-free solutions of continuous fields. That is, Einstein struggled to advocate a world view very different from that of quantum theory, in which ultimately matter would appear as a particular property of continuous space. There is no empty space in the sense that space is always filled by field. Returning very late to his popular exposition of relativity, Einstein made clear in a new opening note that the purpose of his attempts to unify all forces of nature under the guidance of continuous fields was to show that "physical objects are not *in space*, but...are *spatially extended*" (1954, p. vi). This specifically created the problem of how to interpret the solutions of the field equations apart from the equations themselves. On the issue of singularities Einstein was sharper than ever when he came to re-polish the ultimate version of *The Meaning of Relativity*. It is simply not "reasonable", as he said, "to introduce into a continuum theory points (or lines, etc.) for which the field equations do not hold" (1956, p. 156). As for boundary conditions, the absurdity of having infinite solutions at infinity prompted him to believe that "the postulation of boundary conditions is indispensable"; but the turn of the argument showed how the earlier decision to abolish the infinite had lost its impact for, in his terms, this absolute necessity was determined only "in case the space is an 'open' one" (p. 157).

As Abraham Pais has pointed out (1982, pp. 289–90), there was a brief period in the development of Einstein's thinking when this belief in singularity-free solutions came under question. Within the framework of his research on the problem of motion (the possibility of conceiving the equation of motion of a source as a direct consequence of the equations of the gravitational field), he wrote with Jakob Grommer (who had been Einstein's associate in the early search for boundary conditions) that so far, it was the natural way of thinking to suppose "that there are no field variables other than the gravitational and the electromagnetic field (with the possible exception of the 'cosmological term')." However, the series of failures in constructing physics in terms of continuous fields had resulted in something quite different, so that it was now imperative "to conceive of elementary particles as singular points or singular world lines" (A. Einstein and J. Grommer 1927, p. 4); that is, the field theory could include discontinuities. Given this, it is worth emphasizing that Einstein came to perceive the cosmological constant, in terms of continuity, as an impediment to any progress in the theory of fields. As he suggested in Princeton in 1921, scientists had to look for a physical interpretation of the cosmological constant which did not reduce it to a mere force of repulsion. If it were a cosmic repulsion and nothing else, then De Sitter's model would be acceptable because it had been constructed on the single presupposition that the constant could be maintained, without taking into account any of the wider implications of general relativity. As a result, it is perhaps hardly surprising that Einstein reacted in such an ambiguous way to the type of solution proposed by Friedmann. The form of his reaction to Friedmann's article was disconcerting: first he found an error in the calculations, but later admitted that his objection was incorrect. All this was expressed in two short notes written only a few months apart. It is true that Friedmann's solution was highly provocative, since it did away with Λ as a necessary part of the cosmological solution. But, by the same token, it was equally clear that it could be a highly significant step towards the search for a unified field theory. The profound difficulty with the expansion of space is the way it seemed to represent the qualities of ponderable mass entirely in terms of geometry. For this involved a new time function, which certainly proved very hard for Einstein to take up: contrary to his own view, the field (or ether) which became constitutive of the expansion of space could be 'tracked' through time. The remaining parts of this book are devoted to showing the highly tortuous advent of an entirely new conceptual background necessary to acceptance of that very peculiar time function. We must immediately note how Einstein's early reaction to Friedmann reveals the extraordinary difficulty of thinking of time on a large scale in line with space (as form of things existing independently of us): for some reason time always sends us back to an irreducible dimension of human experience.

We may reflect on the fact that De Sitter, Einstein's early sparring partner, chose to remain silent throughout the 1920s. He made no public statement on cosmology until 1930, when other scientists had found the supposed solution to his problem. The silence has its own significance because it testifies to the growing relevance of purely conceptual considerations that have no immediate place within the workings of the formal apparatus.

(b) The wider context of philosophical implications

Significant philosophical issues are discernible in several physics articles that immediately followed Einstein's "Cosmological Considerations". In the wake of Kretschmann, who had shown that the Newtonian law of inertia allows expression in a form independent of coordinate system, such people as Mie (1917, pp. 599–602), and Holst (1920) questioned the possible difference between a general-relativistic and a Newtonian determination of the inertial system for the entire universe. Such questions send us back to De Sitter's original critique and its implications in terms of philosophy.

In 1916, when Ehrenfest suggested application of four-dimensional relativity to the whole of the space–time continuum, he was trying to unify physics and geometry at the level of large-scale space–time. When De Sitter gave substance to the suggestion in the spring of 1917, he had behind him something quite unknown to Ehrenfest, Einstein's model of the universe. And that model offered another persuasive argument for harmonizing physics and geometry: time had to be distinct from space, for otherwise boundary conditions could not be eliminated entirely. Strikingly enough, De Sitter's metamorphic reformulation of Ehrenfest's original idea had the opposite effect; it disintegrated most of Einstein's ambitions and much of his hope to achieve an eventual harmony. This baffling sequence of ideas calls

THE SCIENTIFIC AND PHILOSOPHICAL OUTCOMES 225

for clarification on philosophical grounds; the actual status of boundary conditions needs to be understood in terms of their implications at the highest level of generality.

As early as April 1917, De Sitter was to lay the foundations for such a clarification when he explained what he meant by the *new* sense boundary conditions had acquired with the invention of cosmology by Einstein. By boundary conditions, one was from now on to understand the replacement of Mach's principle by a *purely ideal* invariance of the $g_{\mu\nu}$ at infinity. This is the mathematical postulate of relativity De Sitter was to illustrate in his 1 April letter by means of a startling example: "The world as a whole can perform arbitrary motions without us (in the world) being able to ever observe them...These motions...are a purely mathematical concept". De Sitter went on to explain that model B is in better agreement than model A with this mathematical requirement of relativity.

The purpose of such a comparison is far from self-evident. In the article on the relativity of inertia, De Sitter is somewhat more explicit:

The three-dimensional world must, in order to be able to perform 'motions', i.e., in order that its position can be a variable function of the time, be thought movable in an 'absolute' space of three or more dimensions (*not* the time-space x, y, z, ct). The four-dimensional world requires for its 'motion' a four- (or more-) dimensional space, and moreover an extra-mundane 'time' which serves as an independent variable for its motion. All this shows that the postulate of the invariance of the $g_{\mu\nu}$ at infinity has no real physical meaning. It is purely mathematical" (1917a, pp. 1222-3).

De Sitter is speaking quite confusingly here about the difference between models A and B: he speaks of B, apparently in contrast with A, as requiring an extra-mundane time. Does he want to suggest that the cosmic time in A, if identified with t, might make its motions physical—an impossibility by virtue of the mathematical postulate of relativity?

However obscure it may be, an example like this carries many implicit references to traditional discussions on the possible movement of the whole universe. Recalling the early modern origins of this argument will help clarify its significance in the context of the Einstein–De Sitter dispute. In his controversy with Leibniz, Clarke had tackled a similar issue in the framework of Newtonian physics (*Clarke's Fourth Reply*, § 12 and *Clarke's Fifth Reply*, §§ 26–32 in H.G. Alexander 1956, pp. 48 and 65). Were it to happen that God's will modified the state of the universe, an acceleration would result which could be perfectly well felt and measured by an observer inside the universe. Clarke's contention in essence is that absolute space acts as the appropriate reference system in relation to which the 'true magnitude' of any acceleration can be derived, whether the subject of motion lies within the universe or is itself the whole universe. The case of the rotating bucket was of course the inverse of this situation: absolute accelerations continue to occur even if the whole universe is supposed to be amorphous, or even if it is supposed not to exist at all—in the latter case, the bucket is a universe unto itself. The whole universe is quite comparable to such a bucket, except that the supposition of non-existence of the universe no longer makes sense in the case of an accelerating universe. In this respect, Leibniz's objections have more metaphysical point than physical application. He argues that the displacement of the whole universe does not make sense from the logical point of view, since no sufficient reason could ever motivate God's will to impart such a motion (*Leibniz's Fifth Letter*, § 52, p. 56). Because the universe is all that is, any two of its different states should remain indiscernible. Translated into physical terms, this means that nothing enables us to ascertain that the whole accelerated universe would create effects similar to those of a bucket within the world, just as the effects of the rotating bucket in an otherwise empty universe cannot be predicted to remain identical to what is actually observed within the universe.

The Leibniz–Clarke controversy is, of course, not directly related to the kind of cosmological question involved in the discussions between Einstein and De Sitter. It had a far greater impact in the discussions about the foundations of the new space–time theory of general relativity (e.g., whether space was an absolute entity or an order of things coexisting), where cosmology played no direct role. However, among the classical philosophies that were undoubtedly very influential on how cosmology was perceived before Einstein, that of Kant was crucial (I. Kant 1781). In the first cosmological antinomy of pure reason, Kant claimed to have discovered an irrefutably rational proof that any assertion of either the finiteness or the infinity of the universe was a false statement. In fact, Kant discredited on a rational basis any attempt at a scientific model of the universe. The impact of his argument was such that Kant has often been hailed as the philosopher who did most to damage the dignity of cosmology as a science, to the effect that he hampered for more than a century the articulation of a genuine science of the universe. (This is an almost universal feature of the literature on this topic. A thorough historical analysis has been given by J. Merleau-Ponty 1983, pp. 255–73.) Kant's importance in the emergence of relativistic cosmology is twofold. First, Einstein felt it desirable to explain the radical breakthrough achieved by his model when he was led up to see its relation to the inevitable conflict envisaged by Kant. And second, Kant himself, in his "Observation on the First Antinomy", had argued in relation to the Leibniz–Clarke controversy and motion for the whole universe. We will see how De Sitter's objections are an extremely original contribution to this implicit but crucial background of cosmological debates.

Clearly enough from Einstein's point of view, the new space of his cosmological theory embodied a way out of the Kantian antinomy. In his conversa-

THE SCIENTIFIC AND PHILOSOPHICAL OUTCOMES   227

tions with Moszkowski, shortly after his 1921 address to the Berlin Academy, Einstein emphasized the ultimate aspect of his cosmology: there was a relation between the limits of the world and the bounds of understanding (A. Moszkowski 1921, p. 120). As Moszkowski went on to comment, Einstein's universe seemed to offer

> a source of consolation to tormented spirits which have sickened of Kant's antinomies. For in this still almost immeasurable world the fateful conception 'infinite' has been made bearable for the first time. . .[It] forms a bridge between the thesis 'finite' and the antithesis infinite. We are brought to a common stream, in which both conceptions peacefully flow together (p. 131).

Moszkowski's statement certainly represents the general impression Einstein's cosmology made on philosophers. Of course, the advent of non-Euclidean geometries in the second half of the nineteenth century had already produced something similar, but what was now decisive was the apparently successful application of non-Kantian views on geometry to the entire physical world. Moszkowski himself discussed the physical sense of this application, taking the by then well-known example of an observer who moves along a "straight line", returning to his initial point of departure without ever contacting a boundary (p. 122). But then that is precisely the problem: Einstein builds what he claims to be the physical counterpart of space as a 're-entrant' form, thereby depriving space of all but its allegedly physical characteristics. The question arises as to what actually supports the conjecture that the physical sense of space is all that is required by the theory to be complete.

The Kantian critique was already concerned with this kind of attempt, and this analysis remains pertinent in our context even if it ignores non-Euclidean geometries. Kant addresses himself to the arguments that tend to get around the question of the world's infinity without falling into antinomy. While Kant discussed both space and time in this respect, we will here limit ourselves to space; including time would lead us too far away from the present context. The supposed way out of the antinomy is that "a limit of the world" does not necessarily involve "an absolute space extending beyond the real world" (Kant, p. 399). But it is precisely in the nature of the antinomy, Kant argues, that this non-entity (*Unding*), empty space outside the world, has "to be assumed if we are to assume a limit to the world" (p. 400). The reason is that space, whether full or empty, is not *only* something that can be determined or even something that can be limited by the physical phenomena (the appearances); it is *also* the "form of possible objects" that may be intuited and for that reason "it cannot be regarded as something absolute in itself that determines the existence of things". In fact, the antinomy arises because, while infinite space as a non-entity has to be maintained even in the positing of a finite world, nevertheless the positing of an infinite world unduly transforms this non-entity into a reality. Kant went as far as to assert that the substitution of "boundaries" for "the limits of extension" is simply a way of getting rid of space (p. 401).

One of Kant's examples had been none other than the motion of the whole universe:

> If we attempt to set one of these factors outside the other, space outside all appearances, there arise all sorts of empty determinations of outer intuition, which yet are not possible perceptions. For example, a determination of the relation of the motion (or rest) of the world to infinite empty space is a determination which can never be perceived, and is therefore the predicate of a mere thought-entity (pp. 398-9).

Kant does not specify here the kind of motion of the world that he has in mind, whether it is meant to be a uniform translation or an acceleration. Clarke had shown that only an acceleration could be measured, but Kant has no need of any such distinction because his problem is not that of observable *effects* of matter in space, but of the true *relation* of matter to space. While Clarke established a physical expression of those conditions that do not distinguish the physics of the whole universe from the physics of bodies within the universe, Kant demonstrates that any such expression is a mere figment of the mind, that is, it lends itself to inevitable contradiction when applied to a physical system.

Of course, the application of differential equations to physics since the time of Kant has altered the actual terms of Kant's problem. In his own reflections on the problems raised by Newtonian cosmology, however, Einstein began by disentangling a conflict of that classical kind between matter and space. He did this in the first place by showing that the actual structure of the Newtonian universe can never be accommodated to the requirements of a physical representation of boundary conditions. As he went on to show, reverting to an ideal, purely mathematical representation of boundary conditions was simply a return to the sense of local physics, and had no application to cosmology. In fact, Einstein provides us with an interpretation of the insuperable problem of physically consistent boundary conditions which reflects in its own way the classical form of Kantian antinomy. The more sophisticated concept of boundary conditions may be substituted for the rather loose, classical form of the relation between matter and space, but the basic problem of the Newtonian universe is still there: space remains a peculiar entity which can never be completely fitted to the body of our purely physical knowledge of the external world. Now, because of the parallel between Newtonian and general relativity theories as to the nature of boundary conditions, Einstein was able to demonstrate that dropping the boundary conditions altogether led to a physically acceptable cosmology attuned to relativity. However, a question remains. After all, Einstein did not revert to Newtonian cosmology *after* he had disposed of boundary conditions by means of Λ, so in what sense can Einstein's relativistic cosmology offer a way out of the classical antinomy?

The rejection of all ideal, pre-physical space is tantamount to a new geometry, and the complete rendering of this geometry in physical terms is achieved by the cosmological constant. At first, Einstein seems to have firmly believed that the new geometry supplemented by Λ expressed fully the physical importance of his cosmology. Thus, in a 12 March 1917 letter to De Sitter, he wrote about how one could possibly know whether the space of the universe was infinite or finite. "Heine has given an answer in a poem, *'and an idiot expects an answer'* ". To Einstein, the answer to the question of dimension is also an answer to all antinomical questions of matter and space.

De Sitter's reply testifies to the incompleteness of the answer. His first move was to characterize the logic of Einstein's cosmology as being based on a nonsense premise: "if the world were not there". Interestingly enough, in Clarke's Newtonian arguments, the significance of this premise is taken for granted, so that, while Newton's science remains impervious to cosmology, Einstein's is wide open. However, it is essentially the far broader distinction between mathematical and physical postulates of relativity that inherits, and so also re-activates, the perennial problem that dominated the discussions of Leibniz and Clarke or of Kant's antinomy. (No doubt a good many of their disciples influenced Einstein and De Sitter more directly, but this need not be discussed here.) De Sitter's most important point is that one can still consistently speak of boundary conditions, very much after the classical manner of conceiving them, in a naturally finite universe. Just as Kant had realized that the self-contained universe did not do away with non-entities, De Sitter realized that the physical meaning of the new cosmology did not inhere only in the non-Euclidean conception of its size, not even when Λ was supposed to bridge the gap with the physical. With the example of the motion of the whole universe, Kant had demonstrated how the extension of Newtonian physics to the whole universe was in contradiction to that variety of physics. Likewise, the space used by De Sitter to perform a devastating projection of Einstein's universe is strikingly similar to the 'non-entity', that yet may exist, formulated by Kant. Thus, the 'absolute space' that coincides with the mathematical postulate of invariance at infinity is a sort of pure form in the Kantian sense, i.e. it has no metrical structure defined by phenomena. In his very first paper on the relativity of inertia, De Sitter came to this remarkable conclusion:

The constant R only serves to satisfy a philosophical need felt by many, but it has no real physical meaning, though it can be mathematically interpreted as a curvature of space (1917a, p. 1224).

Not only the $g_{\mu\nu}$ at infinity are purely mathematical, but all relativistic universes have to be. Only in the wider philosophical context does this remark make sense; otherwise, it can all too quickly be put on the list of flat mistakes committed in the early days of relativity, when even local curvature was taken to be a mere manner of speaking.

However, because Einstein still believed that he had completely exhausted the matter/space antinomy *even after* De Sitter's critique, he went on to radicalize his Machian commitment in a broad philosophical sense precisely when Mach's principle in its original, technical form began to lose strength. In his 1921 conversations with Moszkowski, this remarkable idea was reported:

We spoke of the 'Properties of Things', and of the degree to which these properties could be investigated. As an extreme thought, the following question was proposed: Supposing it were possible to discover *all* the properties of a *grain of sand*, would we then have gained a complete knowledge of the *whole universe*? Would there then remain no unsolved component of our comprehension of the universe? Einstein declared that this question was to be answered with an unconditional affirmative. 'For if we had completely and in a scientific sense learned the processes in the grain of sand, this would have been possible only on the basis of an exact knowledge of the laws of mechanical events in time and space. These laws, differential equations, would be the most general laws of the universe, from which the quintessence of all other events would have to be deducible' (A. Moszkowski 1921, p. 202).

The statement may sound rather rhetorical. It could have been said by Einstein at any time in his career, or even by someone who had no particular concern with over-arching issues of relativistic cosmology. But the analogy with the grain of sand may be carried further. Of course Einstein knew Newton's celebrated words: "I seem to have been only like a boy playing on the sea shore. . .and then finding a smoother pebble or a prettier shell than ordinary, whilst the great ocean of truth lay all undiscovered before me". Where Newton sees no way of grasping this infinity, Einstein believes he can master it when he knows one particular grain. What he has in mind is an exaltation and extension of Mach's principle, a formula that any form of physical science would be compelled to take over. Such a general truth would go far beyond any kind of antinomy created by the reduction of gravitation to inertia, or by the inverse reduction of inertia to gravitation.

References

Alexander, H.G. (1956). *The Leibniz-Clarke correspondence*. Manchester University Press.
Andrillat, H. (1970). *Introduction à l'étude des cosmologies*. A. Colin, Paris.

REFERENCES

De Sitter, W. (1916). On Einstein's theory of gravitation and its astronomical consequences. Second paper, *Monthly Not. Roy. Astr. Soc.* **77**, 155-183.

De Sitter, W. (1917a). On the relativity of inertia. Remarks concerning Einstein's latest hypothesis. *Proc. Kon. Akad. Wet. Amst.* **19**, 1217-25.

De Sitter, W. (1917b). On the curvature of space. *Proc. Kon. Akad. Wet. Amst.* **20** 229-42.

De Sitter, W. (1917c). Further remarks of the solutions of the field equations of Einstein's theory of gravitation. *Proc. Kon. Akad. Wet. Amst.* **20**, 1309-12.

De Sitter, W. (1917d). On Einstein's theory of gravitation and its astronomical consequences. Third paper. *Monthly Not. Roy. Astr. Soc.* **78**, 3-28.

De Sitter, W. (1922). On the possibility of a statistical equilibrium for the universe. *Proc. Kon. Akad. Wet. Amst.* **23**, 866-8.

Eddington, A.S. (1918). Silberstein's paradox and Einstein's theory. *The Observatory.* **41**, 350-2.

Eddington, A.S. (1933). *The expanding universe.* Cambridge University Press.

Einstein, A. (1918a). Principles of research. (trans. A. Harris) In *A. Einstein: essays in science.* 1934, pp. 1-5. Philosophy Library, New York.

Einstein, A. (1918b). Kritisches zu einer von Herrn De Sitter gegebenen Lösung der Gravitationsgleichungen. *Sitz. Ber. Preuss. Akad. Wiss. Berlin*, 270-272.

Einstein, A. (1918c). Prinzipielles zur allgemeinen Relativitätstheorie. *Ann. Phys.* **55**, 241-4.

Einstein, A. (1919). Spielen Gravitationsfelder im Aufbau der materiellen Elementarteilchen eine wesentliche Rolle? *Sitz. Ber. Preuss. Akad. Wiss. Berlin*, 349-56.

Einstein, A. (1920). Äther und Relativitätstheorie: Rede gehalten am 5.Mai 1920 an der Reichs-Universität zu Leiden, J. Springer, Berlin. In *A. Einstein: sidelights on relativity*, (trans., G.B. Jeffery and W. Perrett), 1922, pp. 3-24. Methuen, London.

Einstein, A. (1921a). Geometrie und Erfahrung, erweiterte Fassung des Festvortrages gehalten an der Preussischen Akademie. Springer, Berlin. In *A. Einstein: sidelights on relativity* (trans. G.B. Jeffery and W. Perrett), 1922. pp. 27-56. Methuen, London.

Einstein, A. (1921b). *The meaning of relativity.* Four lectures delivered at Princeton University, (trans. E.P. Adams), 1922. Methuen, London.

Einstein, A. (1927). Allgemeine Relativitätstheorie und Bewegungsgesetz. *Sitz. Ber. Preuss. Akad. Wiss.*, 235-45.

Einstein, A. (1945). *The meaning of relativity.* Princeton University Press. Second edition of (1921b).

Einstein, A. (1954). *Relativity. The special and the general theory* (transl. R.W. Lawson), Methuen, London. 15th edition.

Einstein, A. (1956). *The meaning of relativity.* Sixth edition of (1921b).

Einstein, A. and Besso, M. (1972). *Correspondance 1903-1955*, ed., P. Speziali, Hermann, Paris.

Einstein, A. and Grommer, J. (1927). Allgemeine Relativitätstheorie und Bewegungsgesetz, *Sitz. Ber. Preuss. Akad. Wiss.*, 2-13.

Guth, A. and Steinhardt, P. (1984). The inflationary universe. *Scientific American*, May, 90-102.

Hawking, S. and Ellis, G.F.R. (1973). *The large-scale structure of space-time.* Cambridge University Press.

Hermann, A., ed. (1968). *A. Einstein—A. Sommerfeld, Briefwechsel*. Schwabe, Basle.

Holst, H. (1920). Wirft die Relativitätstheorie den Ursachsbegriff über Bord? *Zeitschr. für Physik*, **1**, 32–9.

Kant, I. (1781). *Critique of pure reason*. (trans. N.K. Smith) 1929. MacMillan, London.

Klein, F. (1928). *Voslesungen über Nicht-Euklidische Geometrie*. J. Springer, Berlin.

Lemaître, G. (1933). L'univers en expansion. *Ann. Soc. Sci. Brux.* **53**(A), 51–85.

Lense, J. and Thirring, H. (1918). Ueber den Einfluss der Eigenrotation der Zentralkörper auf die Bewegung der Planeten und Monde nach der Einsteinschen Gravitationstheorie. *Phys. Zeitschr.* **19**, 156–63.

Merleau-Ponty, J. (1965). *Cosmologie du XXème siécle*, Gallimard, Paris.

Merleau-Ponty, J. (1983). *La science de l'univers à l'âge du positivisme*, J. Vrin, Paris.

Mie, G. (1917). Die Einsteinsche Gravitationstheorie und das Problem des Materie. *Phys. Zeitschr.*, **18**, 596–602.

Misner, C., Thorne, K., and Wheeler, J.A. (1973). *Gravitation*. Freeman, San Francisco.

Moszkowski, A. (1921). *Einstein the searcher: his work explained from dialogues* (transl. H.L. Brose), Methuen, London.

Narlikar, J. (1983). *Introduction to cosmology*, Jones and Bartlett, Boston.

Pais, A. (1982). *Subtle is the Lord*. Oxford University Press.

Peebles, P.J.E. (1980). *The large-scale structure of the universe*. Princeton University Press.

Rindler, W. (1956). Visual horizons in world models. *Monthly Not. Roy. Astr. Soc.* **116**, 662–77.

Silberstein, L. (1918). Bizarre conclusion derived from Einstein's gravitation theory. *Monthly Not. Roy. Astr. Soc.* **78** pp. 465–7.

Synge, J.L. (1960). *Relativity: the general theory*. North-Holland, Amsterdam.

Thirring, H. (1918). Ueber die Wirkung rotierender ferner Massen in der Einsteinschen Gravitationstheorie. *Phys. Zeitschr.*, **19**, 33–9.

Thirring, H. (1921). Berichtung zu meiner Arbeit: 'Ueber die Wirkung. . .'. *Phys. Zeitschr.* **22**, 29–30.

Tipler, F., Clarke, C., and Ellis, G.F.R. (1980). Singularities and horizons, a review article. In *General relativity and gravitation* (ed., A. Held) Vol.2, pp. 97–206. Plenum Press, New York.

Torretti, R. (1983). *Relativity and geometry*, Pergamon Press, Oxford.

4

Matter without motion or motion without matter?

1. Questions of dynamic cosmology

In the last part of the preceding Chapter, I have deliberately omitted any discussion of the role of time in the controversy over Einstein's arguments in his alleged solution to Kant's first antinomy. Such a discussion would have led us too far afield since it would have involved a much more systematic analysis of the problems of creation—for instance, of what we mean by talking about something 'before' creation. We shall now turn to an almost exclusive focus on the problem of time. Here we will find that physical and philosophical considerations will ultimately coalesce into one vast, new schema of understanding, in which early relativistic cosmology will consort with classical philosophy and its perennial speculations about creation, time, and the cosmos. More precisely, the as yet imperfectly comprehended but clearly crucial requirement of a new time function (resulting from the need of dynamic cosmology) and De Sitter's obscure reference to an 'extra-mundane' time in his model *do* exhibit an essential kinship.

In his 1927 paper (p. 50), Lemaître gave this description of the problem arising from the two different models: "De Sitter's universe is empty; that of Einstein has been described as containing as much matter as it can contain; it is remarkable that the theory can provide no mean between these two extremes." Such a compromise leads to no less than a theory of the expanding universe. How is this possible? We have seen that De Sitter was literally obsessed with making every kind of opposition abide his question: motion versus rest, continuous versus discrete distribution of matter, model A versus model B, not to mention the contradiction within model A. Einstein was to discover, in turn, a contradiction that seemed to affect model B itself. Now, the key to non-statical cosmology lies in the explicit rejection of antinomies. Thus, it was Lemaître's idea that not only motion and rest, but the two models at issue, were not in opposition to each other: they may both exist, *but at different times*; the two universes may be limiting cases that are separated by a long evolution that extends between them. Lemaître's work came amidst growing scepticism with regard to the achievements of cosmology as a science. By the end of the 1920s, philosophers like Gaston Bachelard deplored the fact that the models of Einstein and De Sitter on the nature of time (linear versus curved) lacked any sharp distinction, for we could not be quite sure that the two hypotheses covered the field of all possible hypotheses. What was missing, in Bachelard's words (1929, p. 180), was some higher standpoint from which to judge that one large scale structure of time was the dialectic opposite of the other. In this chapter, we shall be mainly concerned with how the immediate successors of Einstein and De Sitter, between 1918 and 1923, struggled with the limits imposed by the static representation of cosmic time. Something very significant about the problematic nature of the future non-static solution lies deeply embedded in the way these limits were integrated and overcome.

On 15 April 1918, Einstein sent De Sitter a postcard, in which he expressed dissatisfaction with De Sitter's attempt to save his space-time metric by using an allegedly more physical postulate of continuity. Certainly there was nothing wrong with the explanation that a motion until the equator of this metric takes an infinite time. But the point was that this was merely formal. Einstein referred to a new work just published by Hermann Weyl, in which the right physical and realistic requirements made it appear that the De Sitter continuum was nothing but the limiting case of matter extended on the equator of an otherwise empty continuum. Weyl's *Space-Time-Matter* was a landmark in the arduous journey towards the breaking of antinomies, because it offered the possibility of a connection between the two models in terms of the necessary existence of matter in what Weyl took to be *any* cosmological solution. Weyl's analysis proved to be the first and the most comprehensive elucidation of any conceivable cosmic-scale application of Einstein's theory. He began with an epistemological investigation of the foundations of general relativity, which was aimed at showing how two requirements seemingly independent of each other could be combined: namely the very idea of cosmology as being an almost straightforward corrollary of general relativity *and* the De Sitter discovery that no special values were to hold for the constants of integration in that theory.

But Weyl's search for continuity came under intense challenge from others, most notably Eddington, whose reasons for attacking the compromise were based on a quite different kind of epistemological analysis. Rather than simply undermining the objective validity of a compromise, his rebuttal contributed to a basic change in the general desire for a compromise: one where the emphasis would be on the occurrence of motion and not of matter.

In fact, Eddington's arguments were so crucial in refining Weyl's investigations that they initiated a fundamental form of cosmological controversy. In order to find our bearings in this area, it may be helpful to introduce at this stage two definitions of the universe that come from Hermann Bondi (1960, p. 10). According to the first, the universe is "the largest set of events that can be considered to be physically linked to us". The second defines it as "the largest set to which our physical laws (extrapolated in some manner or other) can be applied". Surprisingly enough, these two definitions are actually different and do not overlap completely, because, at the most general level, what is understood by a physical link is included as only one part of what is covered by physical laws. As it will appear in the course of this and the next chapter, the relevance of the debate between Eddington and Weyl is that it may be taken as the first explicit articulation of the general problem which underlies the hiatus between the two definitions. In the first definition, we will have Weyl's conception of the universe, which is admittedly the standard definition today. The second definition, however, is more sophisticated; something like it was seen by Eddington (in the static frame of reference) as restoring a higher unity to the universe. In the context of Lemaître's problem and with the advantage of hindsight, the controversy took on an aspect which is traditionally described as the need for an *intermediate* solution: one that would lie between Einstein's "matter without motion" and De Sitter's "motion without matter" (Bondi 1960, p. 99). Eddington himself later summed up the essence of the early approaches to the problem with a droll rhetorical question: "Shall we put a little motion into Einstein's world of inert matter, or shall we put a little matter into De Sitter's Primum Mobile?" (1933, p. 46).

Early in 1918, it was already clear that the De Sitter universe was 'static' only at its centre ($r = 0$) and periphery ($r = R$). In the words of De Sitter himself, the relation of the centre to the periphery could not be thought of as something real, for it was the prerogative of only an extra-mundane time to connect them. If, however, such a universe tallies with the only possible physical meaning the postulate of continuity can have, it is all the more natural to require a thoroughly physical account of the centre/periphery relation as well. It was this task that Weyl saw as his first objective. But it was left to Eddington to provide the clues for putting together this theoretical reconsideration and the observational data. The possibility of systematic displacements towards the red of the spectral lines emitted by distant nebulae was based, in De Sitter's conjectures, on some sort of postulated equivalence between the Doppler effect and the metrical properties of model B; the problem was how to elaborate the consequences of these systematic redshifts in a model which, as Eddington was later to say, was taken to be static "by a fortunate piece of gate-crashing" (1933, p. 46).

2. New ideas of cosmic time

The years from 1918 (which saw the reception of general relativity) to 1923 (when a cosmological principle was formulated) constitute a period of extraordinarily intense intellectual activity for Weyl. During that period, he was not only trying to nail down and lay bare the essence of general relativity, he was also searching for a more coherent view of physical science, a view whereby general relativity would be only the starting point for a comprehensive integration of all the forces of nature. This led him to lay down some principles of unification of gravitation and electromagnetism. The gradual development of the theory can be seen from the successive editions of Weyl's major work, *Space-Time-Matter* (*Raum, Zeit, Materie*, first edition finished at Easter 1918, second edition in early 1919, third edition in August 1919, fourth edition completed in November 1920 and published in 1921, and fifth edition finished in the Autumn of 1922 and published in 1923). The changes from the first edition to the third and from that to those subsequent are quite numerous, and Weyl often seems to modify his views, though on the whole the changes are mainly additions borrowed from articles published in various journals. This is especially true of the paragraph devoted to the cosmological problem, which Weyl entitled, "Concerning the Inter-connection of the World as a Whole".

In the fourth edition, Weyl went on to a radical extension of what he had done; somehow he managed to perceive the true direction of his earlier grapplings with the cosmological problem. What he came to realize was that the whole debate, in which the opposition between the two known models had been accentuated, ought to be redirected to an analysis of the very concept of 'totality' as it had hitherto been used. Weyl had certainly a very strong underlying and independent motivation for doing that, for he then saw convergence looming up between a way out of the Einstein–De Sitter antinomy and what had preoccupied him right from 1917, namely, an entirely original theory of unified field based on a development of Einstein's general relativity. The unified theory really attempted to push to its limits Minkowski's programme for complete harmony between mathematics and physics. In the preface to the first edition of *Raum, Zeit, Materie*, Weyl spoke of the effect of the general theory that it was "as if a wall which separated us from Truth has collapsed", the strength of the theory lying less in experience than in its logical texture (1918b, p. 198); the preface to the third edition went on to describe the further developments as making us "realize the 'harmonia mundi' ". Weyl's formidable task involved the reunion of the infinitely small and the infinitely great: it amounted to bringing out the affinity between the global view of the universe and the intellectual movement that had determined progress in the nineteenth century, that is, the transition from finite (Euclidean) to differential (Riemannian) geometry paralleled by the evolution in physics from action at a distance to action of contact (1922, p. 91).

As a consequence, his discussion from the early 1920s was remarkable for the way it was then conducted with such an emphasis on a priori considerations. I think we have to begin our investigations of Weyl's contributions with this form of 'solution', i.e. the more or less complete clarification of the technical difficulties with cosmic time in the De Sitter model of a static universe. The point of starting with this later clarification is that it is the only way we can make sense of why and how cosmic time began to be seen as a central problem in the first place. Beyond this there is the more speculative aspect of the theme of cosmic time in Weyl's thought. In fact Weyl derived much of the clarity he brought to bear on the subject from his correspondence with Einstein and with the mathematician Felix Klein (from 1918 to 1921). Understanding both technical and speculative backgrounds will require close attention to the specificity of Weyl's arguments about the limits of both the Newtonian and the Einsteinian approaches to physical science. (The English translation of Weyl's book, *Space-Time-Matter*, is based on the fourth edition.)

(a) Difficulties with cosmic time in De Sitter's static universe

The clarification of the problem created by the existence of two conflicting metrics for the universe goes like this: it should be possible to deduce the two metrics in terms of something they have in common. Hence, an expression of the cosmological metric must be possible at the highest level of generality. This is a metrically homogeneous form which should satisfy two criteria:

(a) a *formal* conformity to the Minkowskian requirement on space and time, which De Sitter had already postulated in order to do justice to the mathematical spirit of the theory of relativity; and
(b) some constraint on this form, based on the *material* necessity of having a physically meaningful distribution of matter; as Einstein's analysis had revealed, the homogeneous distribution of matter seemed to follow from the positing of a cosmic time.

Clearly, De Sitter's discovery of internal contradictions in cosmology arose from his belief that it was impossible to reconcile (a) and (b). Weyl goes rather more thoroughly into the question by developing the full consequences of what he claims is the common denominator of the two models. As he puts it,

A metrically homogeneous world is obtained most simply if, in a five-dimensional space with the metrical groundform $ds^2 = -\Omega(dx)$, ($-\Omega$ denotes a non-degenerate quadratic form with constant coefficients), we examine the four-dimensional 'conic-section' defined by the equation $\Omega(x) = 6/\Lambda$. Thus this basis gives us a solution of the Einstein equations of gravitation, modified by the Λ term, for the case of no mass"(Weyl 1922, p. 281).

(As Lanczos remarked (1922, p. 539 n. 2), the radius of curvature of the empty metric is usually taken to be $\sqrt{3/\Lambda}$ instead of $\sqrt{6/\Lambda}$. This is due to a change of units which will be introduced later.) Five-dimensional space permits the discerning of the global properties of the four-dimensional world. The form can be written:

$$\Omega(x) = x_1^2 + x_2^2 + x_3^2 + x_4^2 - x_5^2. \tag{4.1}$$

In this homogeneous form, Weyl explores the following transformation of coordinates:

$$x_4 = z \cosh t, \quad x_5 = \sinh t \tag{4.2}$$

A first property of this transformation is that x_4 is always greater than or equal to x_5. Indeed, $x_4/x_5 = \cosh t/\sinh t$, and for all t the quantity $\cosh t/\sinh t$ is always greater than or equal to 1. Furthermore, two other properties of the transformation can be singled out. For any fixed t, we have $x_4 \geqslant x_5$, and for any fixed z, we have $x_4^2 - x_5^2 = z^2$. In other words, all straight lines $t = cst$ and all hyperbolas $z = cst$ meet in a series of points that define a hyperboloid (see Fig. 4.1).

By inserting the change of variables (4.2) into Eqn (4.1), we get

$$x_1^2 + x_2^2 + x_3^2 + z^2 = R^2, \tag{4.3}$$

so that an elementary interval of the hyperboloid is defined by

$$-ds^2 = (dx_1^2 + dx_2^2 + dx_3^2 + dz^2) - z^2 dt^2. \tag{4.4}$$

In these new coordinates (z, t), the variable t does represent a cosmic time in the De Sitter static metric. Indeed, the coefficients of dz^2 and dt^2 are independent of t. The proper time along the world line of a particle is given by $z dt$, z being constant throughout the trajectory. The geodesics of null length are defined by $z^2 dt^2 - dz^2 = 0$. As a result, when a light ray is sent from one particle to the other, the change of variable t depends only on the two values of z, not on the time when the light was emitted. Because t is a cosmic time, the transformation (4.2) defines new 'contemporary' spaces in the De Sitter universe, and thus new geodesic paths in this universe. Let us see the results with the help of a geometrical, rather intuitive representation of the De Sitter universe.

Such a model of the De Sitter universe can be built by following a technique developed by E. Schrödinger (1956, pp. 1–40). Take x, y, z, u, v to be Cartesian coordinates of a five-dimensional pseudo-Euclidean space. The fifth, auxilary dimension helps the four-dimensional world to be visualized in terms of its global properties. The equation of the four-dimensional hyperboloid embedded in this five-dimensional pseudo-Euclidean space is:

$$x^2 + y^2 + z^2 + u^2 - v^2 = R^2. \tag{4.5}$$

The intrinsic geometry on the surface of this hyperboloid must be obtained from the $g_{\mu\nu}$ defined by the square of the distance between any two points:

$$(x_1 - x_2)^2 + (y_1 - y_2)^2 + (z_1 - z_2)^2 + (u_1 - u_2)^2 - (v_1 - v_2)^2. \tag{4.6}$$

NEW IDEAS OF COSMIC TIME

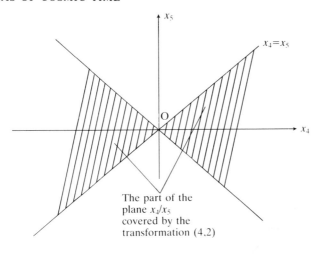

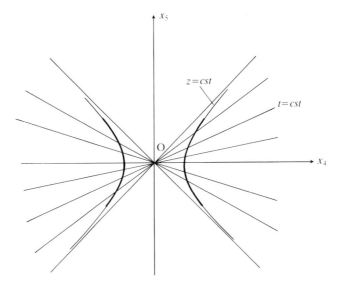

FIG. 4.1. Weyl's reconstruction of the De Sitter metric with the use of cosmic time (adapted from Weyl 1923b, p. 293).

The hyperboloid (4.5) results from the equation of the hypersphere,

$$x^2 + y^2 + z^2 + u^2 + v^2 = R^2, \qquad (4.7)$$

in which iv is substituted for v, where i is an imaginary number. All points of this sphere are equivalent and, at each point, all directions are equivalent.

Since the change from (4.7) to (4.5) is linear and homogeneous, there is no reason why the hyperboloid should not be everywhere homogeneous. The geodesics on the surface of (4.5) must correspond to those of (4.7), since they result from the intersections of the hyperboloid with all planes through the origin of the five-dimensional pseudo-Euclidean space. To the great circles of the sphere correspond the conic sections of the hyperboloid. An intuitive representation of this situation on the hyperboloid, obtained by suppressing two spatial coordinates ($y = 0$ and $z = 0$), is drawn on Fig. 4.2.

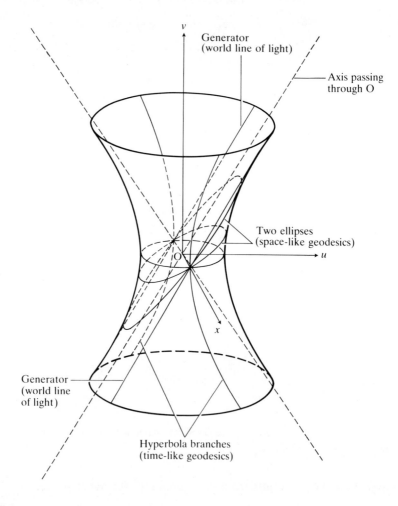

FIG. 4.2. World-lines in De Sitter's universe (adapted from E. Schrödinger 1956).

Let us write the metric of the three-dimensional, pseudo-Euclidean space in which the De Sitter universe is embedded after the suppression of these two spatial coordinates:

$$ds^2 = dx^2 + du^2 - dv^2. \tag{4.8}$$

The universe (4.5) is now reduced to a one sheet, equilateral hyperboloid:

$$x^2 + u^2 - v^2 = R^2. \tag{4.9}$$

Suppose v plays the role of 'time' in the metric (4.8). Then, clearly, the transformations that leave this metric unaltered by carrying any 'slice' of the hyperboloid into the 'neck' are Lorentz transformations. With v taken as time, all parallel circles on (4.9) represent space at different times (see Fig. 4.3). Contemporary spaces contract up to the epoch $v = 0$, and then expand. It seems that the epoch $v = 0$ (all those events that occur in the neck of the bottle-neck) is really distinguishable from all other epochs and events. In fact, all parallel circles are not equivalent, since $v = 0$ defines a space-like geodesic while all other circles do not define geodesic paths (they do not intersect the origin). Because all points of the hyperboloid have been defined as equivalent, one is left with the task of finding new contemporary spaces that pass through the origin.

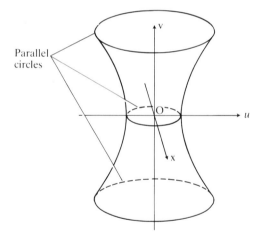

FIG. 4.3. The circle passing through O seems to be 'naturally' privileged.

This is precisely the problem solved by the coordinate transformation (4.2) introduced by Weyl. Contemporary spaces are the ellipses passing through the origin and containing the x-axis. All ellipses, which result from the intersection of planes whose slopes are less then 45°, define space-like geodesics. Because Weyl selects those ellipses that contain the x-axis as contemporary spaces, the new time must be a function of the ratio v/u, that is to say, of the parameter of the Lorentz transformation in the plane (v, u) that carries all ellipses into $v = 0$. Let us verify that $\tanh t = v/u$ is the appropriate function. Using the classical form of the Lorentz transformation $x' = (x - st)/(1 - s^2/c^2)^{1/2}$, $y' = y$, $z' = z$, $t' = (t - sx/c^2)/(1 - s^2/c^2)^{1/2}$ (where s is the velocity), the coordinate transformation $x' = x\cosh\theta - ct\sinh\theta$, $y' = y$, $z' = z$, $ct' = -x\sinh\theta + ct\cosh\theta$ yields the following identity: $ct'^2 - x'^2 - y'^2 - z'^2 = ct^2 - x^2 - y^2 - z^2$. This defines the invariant of the pseudo-Euclidean metric, provided that $\tanh\theta = s/c$. Consequently, Weyl's coordinate transformation is tantamount to the 'velocity' in the Lorentz transformation, since $\tanh\theta = \tanh t$.

Weyl immediately indicated that his transformation had yet another remarkable property:

These 'new' z, t coordinates, however, enable only the 'wedge-shaped' section $x_4^2 - x_5^2 > 0$ to be represented. At the 'edge' of the wedge (at which $x_4 = 0$ simultaneously with $x_5 = 0$), t becomes indeterminate (1922, pp. 281–2).

Indeed, for $z = 0$, the corresponding hyperbola branch is simply identical to the edge of this wedge-shaped section; namely, the two straight lines at right angles on Fig. 4.4. The contemporary space is now this wedge. The edge of the wedge is the origin $x_4 = x_5 = 0$, and it is clear that the intersection of the wedge with *any* straight line t = constant occurs at this one single point. Thus, all straight lines t pass through the origin O for the particular value $z = 0$. That the intersection of z with t occurs for any t, means that cosmic time becomes indeterminate at this edge of the wedge. Weyl started from a homogeneous form, and he discovered that the introduction of cosmic time destroyed the postulated homogeneity at this particular point.

Having performed the coordinate transformation (4.2), the unavoidable result is that, as Weyl says, "we have a static world that cannot exist without a mass-horizon" (p. 282). In other words, Weyl seems to subscribe to Einstein's interpretation of the De Sitter model. But he arrives at this interpretation in a most novel manner, since he *derives* the discontinuity from the fact that cosmic time breaks down at some point of an homogeneous metric form. Weyl then goes on to raise the question "whether it is the first or the second coordinate system that serves to represent the whole world in a regular manner". The first system of coordinates is the unaltered homogeneous form from which we started. In that case, Weyl adds, "the world would not be static as a whole, and the absence of matter in it would be in agreement with physical laws". The very term "non-static" makes its entrance here, apparently for the first time, into cosmological literature. Yet, Weyl is very far from perceiving its implications, since the statement remains a by-remark and there is no trace of an analytic treatment of the idea. The reasons for this

NEW IDEAS OF COSMIC TIME

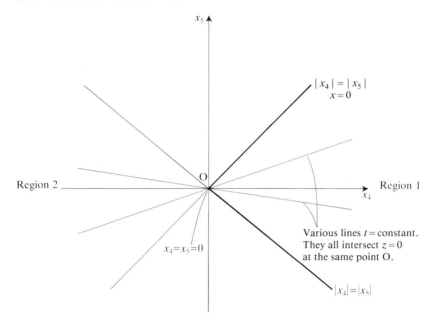

FIG. 4.4. O as the edge of the wedge-shaped section $x_4^2 - x_5^2 > 0$ (based on Weyl 1923b, p. 293).

are not all too clear. There is the seemingly obvious paradox (at least in the Machian spirit of Einstein's conceptions) of having no matter and an agreement with physical laws. But it does not seem that Weyl holds the identification as a paradox in itself. Rather, there is an aspect of the non-static world that brings it into conflict with the other system of coordinates: that is the fact that the non-static character refers to a metric form in which cosmic time has been ruled out.

These considerations end the paragraph of *Space-Time-Matter* devoted to the inter-connection of the world as a whole. Weyl began his investigations with the remarks about the topological implications of the large-scale structure of space-time, which I have commented on at the end of Section 7 in Chapter 1. As a result of the strictly local aspect of the metric used in general relativity, it would always be possible to have, at some point in the universe, a situation where the cone of the active future overlaps with that of the passive past. This would require considerable fluctuations of the $g_{\mu\nu}$ over large portions of space-time, and only the *facts* directly presented to experience protect us against such paradoxes. But Weyl finds "a certain amount of interest in speculating on these possibilities inasmuch as they shed light on the philosophical problem of cosmic and phenomenal time" (1918b, p. 220;

1922, p. 274). In fact, it is certainly not an overstatement to assert that all of Weyl's ensuing analysis is directed to a clarification of this problem. Theoretical speculation on cosmological models is a form of philosophical approach to the relation of phenomenal to cosmic time. But the world as a whole is only one end of the fundamental problem of time; the other end starts from the human self, as we see in the next section.

(b) The philosophy of cosmic time

The book *Space-Time-Matter* begins with a general introduction that puts forward an extremely subtle philosophy of time. In fact, the introduction endeavours to justify the 'natural' character of cosmic time in relativity (1922, pp. 1ff). Weyl argues that the very advent of modern science was conditional on the rejection of all sensible and individual perception from the properties belonging to real things in themselves. The realization of the subjectivity of sense qualities appeared together with the fundamental thesis of the mathematico-constructive method. In accordance with the allegedly universal scope of this method, the essence of the real was declared to be deprived of these qualities. Weyl addresses himself to the question of the ultimate validity of this kind of claim. The question arises as to what really remains of the 'I' in modern and contemporary science, of the always individual and particular acts of knowing leading to the constitution of an independent, transcendent realm of things. His answer is that all the determinations of the real world "are, and can only be given as, intentional objects of acts of consciousness" (p. 4). In this sense, independent things *are* such objects. This is the core of Weyl's debt to the then very new phenomenological philosophy created by Edmund Husserl. Weyl refers himself to Husserl's major work where the thesis of 'intentionality' was developed for the first time quite systematically (E. Husserl 1913). (On Husserl's influence on Weyl's thought, particularly in the context of unified field theory which I will discuss later, see V.P. Vizgin 1984. And on Weyl's own phenomenological philosophy of physical and mathematical science, see J. Kockelmans and T. Kisiel 1970, pp. 91–118.)

For all his approbation of the foundations of phenomenology, Weyl's concern with the actual *limits* of Husserl's thematization of the *Wesenschau* (intuition of essences) dominated his thought throughout the decisive years of his life. Thus, in a late account of his philosophical motivations, an address on "The Unity of Knowledge" delivered at the Bicentennial Conference of Columbia University, he spoke of intuition as mind's originary act, "limited in science to the *Aufweisbare* [that to which we can point in concrete], but in fact extending far beyond these boundaries". Weyl then remarked, "how far one should go in including here the *Wesenschau* of Husserl's phenomenology, I prefer to leave in the dark" (1954a, p. 629). In a sort of report of his activities and thoughts, Weyl was also quite eager to recall this influence of Husserl on his early philosophical interests when he was a student in Göttingen, along with Fichte's idealism and Meister Eckhart's mysticism (Weyl 1954b, p. 637). Philosophy, as he said, occupied him mostly in the years 1913–1922. The 1922 presentation of Husserl's *Ideas* was very important, but most decisive was, during the winter 1922/1923, the reading of Meister Eckhart: this made the strongest impact on his thought (p. 647). And indeed, perhaps nothing less than such a speculative revolution was required to pave the way to a very peculiar state of mind which coincided with Weyl laying down in 1923 his famous cosmological principle of origin. Let us not anticipate, however, and let us go back to Weyl's arguments in the introduction of his book on relativity.

According to the phenomenological thesis, so Weyl argues, the foundation of all possible meaning lies in what is given immediately and unconditionally in the very nexus of conscious experiences. The real thing is an idea forged from within the data of consciousness and the flow of perceptions. But because it belongs to the essence of a real thing to be inexhaustible as far as its content is concerned, the empirical character of all of our knowledge arises and with it the thesis of independent reality as indissoluble with the merely phenomenal existence of objects. Now, Weyl asks himself, if the original form of material reality may indisputably be identified with space, what is the original form of the stream of consciousness? This is none other than *time*. What is presented in intuition is not simply a set of atemporal things such as numbers. Rather, what is presented is always being-now, filling the form of the enduring now with a continually changing content. By exercising pure reflection on this raw material, we pass from the series of 'nows' to the actual relation earlier and later. This comes about when the original stream of consciousness is itself taken as an object over against us.

With the help of this background, Weyl is in a position to dramatically reassess the views on the relations of the world of consciousness to transcendent reality. The bridge between the two is not merely perception. More fundamental than perception is that which he calls the pair, action and passion, what there is in us as experience of strife and resistance. First and foremost, it is this primitive form of '*causality*' which is immediately responsible for the indissoluble coupling of space (as form of material reality) with time (as form of consciousness). Furthermore, in the experience of action and passion, "I become a single individual with a psychical reality attached to a body which has its place in space among the material things of the external world" (1922, p. 6). At the same time that communication between individuals arises, the ego "becomes a piece of reality". The pair action/passion is thus more fundamental than perception, because it allows the self-identity of myself to myself and things to themselves to be constituted. That is why and how conscience legitimately spreads its form (i.e. time) over external reality.

Change, motion, and becoming exist in time itself and are thus real. Because this reality is correlated, in the last resort, to my will which "acts on the external world through and beyond my body as a motive power", the external world as reality must be said to be active. Active means: it is working, as Weyl indicates the etymology of the German noun *Wirklichkeit*, reality, in which *wirken* stands for the verb to act, to work. Likewise, all appearances of the world are subordinated to one another in a universal causal connection. Weyl draws out from this analysis that the theory of relativity ushered in a most remarkable return to this original way of conceiving reality. The theory, as he says, "brings with it a deeper insight into the harmony of action in the world", for it shows that the objectivized stream of consciousness (what Weyl calls cosmic time) and space (what he calls physical form) "cannot be dissociated from one another".

Weyl conceives of this cosmic time as the only actually existing entity bearing the name time. But the need to apply mathematical conceptions to questions of time, as he goes on to say, has been instrumental in modifying much of its original form as continuous flow. Mathematics captures and conceptualizes the idea of "order-relation". The now becomes a point of cosmic time, while the relation earlier and later marks off a length of cosmic time. Point and length fix what is in itself alien to fixation. Thus, the amalgamation of space and time in a four-dimensional world (Minkowski's space–time) certainly goes in the sense of what is required by the return to original conceptions, but it also abolishes the qualitative differences separating the intuitive nature of space from that of time. As a result, only consciousness experiencing a detached piece of this world experiences "*history*, that is,. . .a process that is going forward in time and takes place in space" (p. 217). What is gained in the emphasis on the centrality of consciousness is lost in the formalization which cannot thematize history. By contrast, what consciousness lost was in a sense regained by the mathematics of classical physics: Weyl notes that the particular role ascribed to time in the Galilean principle of relativity arose from its remaining invariant (that is, 'cosmic') under any transformation of the Galilean group (pp. 151–4). That is why Weyl is eager to further highlight the fact that we are never allowed to by-pass the speculative nature of the problems scientific theory is dealing with. He closes his Introduction by pointing out that the mathematization of nature provides the scientist with methods that debar him from questioning the greater depths of the origin of things. As a matter of fact, "all beginnings are obscure" (p. 10), yet they determine a genuine way of *comprehending* that precedes and constrains all processes of formalization that take place in natural science.

The Introduction to the first edition of *Raum, Zeit, Materie* allegedly contains all that is required for the understanding of time. In the rest of his book, Weyl says, attention is confined at much greater length to space, and then to matter. But consideration of time also concludes the fourth edition of the book. Weyl draws attention there to the clash between, on the one hand, the idea of causation in which a unique temporal distinction past → future is prescribed and borne out by the most elementary perception of the world, and, on the other hand, field physics in which this uniqueness is not a priori embodied (1922, p. 310). The recapture of this sense can be made by imposing certain (limiting) conditions on the field. But in following these conditions so well illustrated by Einstein's cosmology, Weyl will gradually perceive the limitations of field physics altogether, arguing in the fourth edition that the very concept of field (even when appropriately 'dynamized' in accordance with the law of action and reaction—see below, Section 3a) remains one of transmitter of effects, incapable of mastering the primary freedom of natural action (p. 311). In the first three editions (see 1918b, p. 227 and 1919b, p. 263) Weyl was keen on concluding with a striking analogy: physical science is to reality what formal logic is to truth, that is, we can do generalizations but we are unable to really deduce the existence of the particular entities that happen to exist. (It may be recalled in this context that, as a mathematician and philosopher of mathematics, Weyl espoused a view on the nature of mathematics which developed that of the intuitionists, i.e., the realm of mathematics cannot exceed the bounds of human intelligence.) In the fourth edition, these considerations disappeared in the light of a proposed unification of gravitation and electromagnetism (an enlarged field physics) which was intended to yield some insight into the existence of the electron. But then, he expected quantum physics to overcome the shortcomings of any possible field physics and penetrate "a deeper stratum of reality" (p. 311) in which, perhaps, the concept of causality would at last be founded on rigorously exact laws.

This is how the short and introductory philosophical account of time dictates some of the fundamental motives of Weyl's thought. The core of what follows is the analysis of the foundations of general relativity in relation to the cosmic dimension of time; we find there the culmination of this philosophical motivation.

(c) The search for a new expression for boundary conditions and the idea of cosmic time

Weyl's model is G. Mie's electrodynamic theory. By means of this theory, which was worked out in the years 1912–1913, it became possible to conceive of an entity like the electron as deriving from the electrodynamic field, while the more classical theory of Maxwell–Lorentz viewed the electrons as added to the pre-existing field (Weyl 1922, pp. 206ff). In fact, because it conceives of the electrons as singularities of the field, the classical theory cannot hold for the interior of the electron. On the other hand, it can be said that Mie's theory satisfies the *principle of causality*, in the sense that it captures the idea of continuity. The field is constructed in this theory with a system of equations in which the derivatives of the physical quantities with respect to time are expressed in relation to themselves and their spatial derivatives. The resulting electrodynamic laws express something intrinsic to nature, because different possible states of a physical system are connected to one another in a continuous way. Weyl assumes that this should remain true if the number of physical quantities is increased so as to include the gravitation potentials, the $g_{\mu\nu}$ (pp. 274–5). However, apart from causality, the theory of gravitation is based on yet another requirement, that of general invariance for the physical laws. This implies two things: "we must formulate our statements so that, from the values of the phase-quantities for one moment, all those assertions concerning them, *which have an invariant character*, follow as a consequence of physical laws; moreover, it must be noted that this statement does not refer to the world as a whole but only to a portion which can be represented by four coordinates" (p. 275). Applied to electrodynamic laws, the idea of continuity leads to the requirements stated in the principle of causality. But with the principle of general invariance, which arises from gravitation laws, the idea of continuity means something different from the stretching out of a physical system in time: it requires all possible viewpoints on some given quantities to be connected to one another. What, then, does it really mean to say that general invariance and causality are compatible with one another? In his following analysis, Weyl investigates the consequences of these new requirements of invariance set by the relativistic theory of gravitation on the view of causality derived from electrodynamic laws.

The geodesics of the general-relativistic space–time are invariant in the sense that they are defined irrespective of any particular frame of reference. It is Weyl's contention that the superimposition of causality leads to a specific type of separation of space from time. In the neighbourhood of a point O, four coordinates are introduced such that, at O itself, the ds^2 is that of the special theory of relativity:

$$ds^2 = dx_4^2 + d\sigma^2 = -c^2 dt^2 + (dx_1^2 + dx_2^2 + dx_3^2) \qquad (4.10)$$

The neighbourhood is thus represented by pseudo-Euclidean coordinates. The three-dimensional space $x_4 = 0$ surrounding O marks off a region R in which ds^2 is always positive. Through every point of this region, draw the geodetic world-line orthogonal to the region. This orthogonality defines all geodesics with a time-like direction. Suppose these lines cover completely a given four-dimensional neighbourhood of O. Weyl now introduces new coordinates

which will coincide with the previous ones in the three-dimensional space R, for we shall now assign the coordinates x_1, x_2, x_3, x_4 to the point P at which we arrive, if we go from the point $P_0 = (x_1, x_2, x_3)$ in R along the orthogonal geodetic line passing through it, so far that the proper-time of the arc traversed, P_0P, is equal to x_4.

The new system of coordinates carries P_0 to P only because some amount of time has elapsed. On all time-like geodetic lines we have $ds_2 = dx_4^2$; that is to say, for all four coordinates:

$$g_{44} = 1. \tag{4.11}$$

Further, orthogonality leads to (for $x_4 = 0$)

$$g_{41} = g_{42} = g_{43} = 0. \tag{4.12}$$

By construction, this is also valid when the other three coordinates are kept constant, so that we have for all four coordinates:

$$g_{4i} = 0 \; (i = 1, 2, 3). \tag{4.13}$$

As a result,

The following picture presents itself to us: a family of geodetic lines with time-like direction which covers a certain world-region singly and completely (without gaps); also, a similar uni-parametric family of three-dimensional spaces $x_4 = 0$. According to (4.13) these two families are everywhere orthogonal to one another, and all portions of arc cut off from the geodetic lines by two of the 'parallel' spaces $x_4 = $ const. have the same proper time (p. 276).

Weyl's construction amounts to the following thing. The pseudo-Euclidean manifold of space–time is 'physical' only in the surrounding of some given point-event. Weyl defines a time-like field of vectors on this manifold, as well as a family of hypersurfaces whose normals coincide with this field of vectors. Because the pseudo-Riemannian metric must keep the same Minkowskian signature at each point, the time coordinate can be intrinsically distinguished from that of space in the neighbourhood of *any* such point. In fact, the distinction is similar everywhere, *irrespective of how the particular distribution of matter or its density may vary from point to point*.

This a priori universal distinction of space and time is Weyl's way of conceiving of a causal connection implicated in the requirement of general invariance for physical laws. The definition of the time-like vector field now yields two distinct senses of speaking of *global* properties for the space–time manifold: one is related to the form of physical laws, the other to the distribution of matter in the universe. Let us develop these two aspects in turn.

In the chosen system of representation, calculations allow the equations of gravitation to be expressed without ambiguity in terms of their $g_{\mu\nu}$, their first and second-order derivatives, as well as of the Christoffel symbols involved in the equations of motion. There result twelve quantities, the ten $g_{\mu\nu}$ and the two Christoffel symbols, which lend themselves to the following representation:

The cone of the passive past starting from the point O' with a positive x_4 coordinate will cut a certain portion R' out of R, which, with the sheet of the cone, will mark off a finite region of the world G (namely, a conical cap with its vertex at O'). If our assertion that the geodetic null-lines denote the initial points of all action is rigorously true, then the values of the above twelve quantities as well as the electromagnetic potentials ϕ_i: and the field-quantities F_{ik} in the three-dimensional region of space R' determine fully the values of the two latter quantities in the world-region G (Weyl 1922, p. 276).

(The quantities F_{ik} refer to the electromagnetic field law.) In other words, Weyl believes that, if his construction of naturally distinguishable time is correct, he thereby proves how and why gravitation constrains electromagnetism. The full demonstration of this point will occupy him later, when he will turn to his original insight into the unity between the two domains (see Section 3b). For the time being, he is in possession of an important result as to the status of global assumptions in general relativity. The hypothesis which is used as starting-point is that any possible general-relativistic space-time is *locally* of the same qualitative causal structure as in special relativity. But globally very significant differences can occur because of non-trivial topology, as Weyl had emphasized in the opening part of his cosmological arguments. Consider the regions defined in the following way. A three-dimensional space V_1 (with x_4 = const.), encloses a set of time-like geodesics which are, as we would say in today's language, time orientable; that is, a continuous designation of past and future can be made at each event of the space-time bounded by V_1. These geodesics intersect another three-dimensional space $(x_4)'$ = const. > (x_4) = const., in another volume V_2 of R (see Fig. 4.5). The tube-like region T is marked off by V_1, V_2 and the above-defined geodesics. The region is 'causally well behaved' in the sense that no closed time-like curve is allowed to exist. Only the relation 'later than' ensures this behaviour: it rules out topological pathologies such as the identification of any two hypersurfaces x_4 and x_4' (which would induce closed time-like curves). In that way, the equations of motion (characterized by the Christoffel symbols) can be entirely derived from the law of gravitation, without making use of boundary conditions. This is the first, crucial conclusion drawn out by Weyl: "*In any case, we see that the differential equations of the field contain the physical laws of nature in their complete form*, and that there cannot be a further limitation due to boundary conditions at spatial infinity, for example" (p. 276). The word "complete" is the formal counterpart of 'global' in the concrete sense of the entire universe. The nuance between the two is of great importance. "Complete form" expresses the idea that the field equations are already by themselves something uni-

NEW IDEAS OF COSMIC TIME

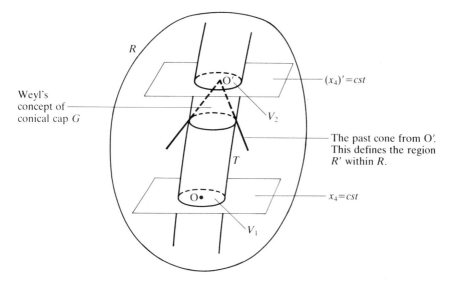

FIG. 4.5. The time-orientability of space–time according to Weyl.

versal even though they are local, just because they apply everywhere in the same way; Weyl's contribution in this respect is to show that somehow we need a special time-function in order to conceptualize this situation. The time-function is not the immediately global one of Einstein's model. Einstein did place time on the same footing as truly universal and global properties like constant spatial curvature and homogeneity, but Weyl's restriction of the time-function to something universal but local embodies the first theoretical justification of the irrelevance of boundary conditions to the relativistic field equations. Only the everywhere-identical distinction of space from time on a priori grounds leads to the view that the differential equations express completely by themselves the laws of Nature.

The argument has indeed some interesting consequences in the context of Einstein's cosmology, since the universal distinction of space from time is meant to 'naturalize' the abolishment of boundary conditions. That is, this abolishment no longer appears as just a mathematical makeshift due to some putative incomplete overlap between the equations and their solutions. Nor can it simply result from the independent decision to incorporate finite space at the global scale. All of which leads to the attempt to actually deduce and naturalize along the same lines the other essential feature of Einstein's cosmology, the closedness of space. Global and universal properties are likely to be also (in general) global and particular, but Riemannian geometry in itself is certainly not one that lends itself to global considerations without some qualifications. This geometry only says how bits of spaces may be 'sewn' together, that is, it is local and particular. The specification of the connection between these bits is a matter of restrictions and limitations which do not come exclusively from the global side. For it should be noted that not just 'artificial' topological identifications can be blamed for the occurrence of closed time-like curves. There is also the possibility of light cones that would be sufficiently 'twisted', i.e. they would 'tip over' sufficiently to permit the existence of such curves, for instance under the influence of large gravitational fields. Here Weyl encounters the possibility of restructuring yet another of Einstein's original motivations, the critique of Newtonian cosmology. If we look at the ds^2 (4.10), we see that the time component represents the Newtonian potential of gravitation. This potential must vary as the time along the geodetic lines that cover the region R. In other words, Weyl is led to think of the physical conditions underlying the acceptability of well-behaved time-like curves throughout space-time in terms of the possibility of a Newtonian cosmology. Einstein's original question was: How come the stellar system persists and has not dispersed itself? An answer is now vital to Weyl's fundamental claim that his geodetic null-lines denote the initial points of all possible action.

3. Beyond the limits of both the Newtonian and the Riemannian world views

(a) Weyl's statistical interpretation of Newtonian cosmology

We have seen in Chapter 2 how Einstein's critique of Newtonian cosmology was tacitly imbued with relativistic presuppositions. In particular, the rejection of all disjunction between matter and space seemed to be resulting from the interpretation of Mach's principle as the key to the relativity of inertia. But the further distinction of curved space from linear time was allegedly commanded by the facts themselves. Weyl's procedure is very different, since he begins with a purely theoretical justification of the distinction between space and time, and then turns to the consequences on the matter distribution in terms of this justification. Hence, his fundamental critique of all relation between cosmic time and Mach's principle.

The critique is rooted in an alternative approach to the essence and the foundations of the theory of general relativity. The geometrization of the forces of gravitation was based by Einstein on the epistemologically incomplete character of Newtonian theory; in this theory, there is no connection of essence between a field of gravitation and a field of acceleration, and nothing empirical justifies the distinction between inertial and non-inertial frames of reference. Weyl finds the extension of the relativistic way of thinking to gravitational forces in a different kind of epistemological criterion. He starts from the Newtonian problem of the origin of inertia, and concludes that the Newtonian law of action and reaction *already* postulates that if a physical entity (such as absolute space) has observable effects (the inertial forces), then this entity must also be affected by the reaction of the objects which it acts upon (1927, p. 105). The only way of fulfilling this requirement is to substitute the metrical field (in the Riemannian sense of the term) for absolute space. In Weyl's language, that absolute space is rejected by general relativity means that its effects are now thought of as exhibiting the reality of the generalized metrical field (1922, pp. 220-1). For the "forcible guidance" (*zwangweiser Führung*), inherent in the inertial tendencies of bodies in classical relativity, is substituted the "guiding field" (*Führungsfeld*) impressed upon all types of motion—those that are classically distinguished as inertial and non-inertial. In classical physics, the non-inertial reactions do not assert themselves as part of the guiding field, while general relativity assumes all local dynamic effects to result from this field. Of course, the new metrical field is not really more observable than the old absolute space is; but its merging with gravitation bestows upon it an observable *origin*, that is, the distribution of matter and energy in the universe (p. 228).

Weyl's principle of universal reciprocity works as a critical development of the Machian critique of Newton's theory. For it is not simply the existence of distant matter, but the rejection of any separate form of 'unconditional' in physical science that implies the rejection of both Newton's absolute space and absolute rest. Only this double abolishment allows general relativity to overcome the classical dualism between the purely geometrical field of inertia and the mechanical field of impressed forces. But an identity of essence such as that between the inertial and the gravitational mass applies to local portions of the universe. Clearly, the question arises whether the classical dualism between matter and space-time metric can not only be overcome but be completely eliminated, and this amounts to asking whether relativity can be applied without contradiction to the entire universe. The problem is one of contrast between what is understood by 'world' in pre-relativistic and in relativistic physics. In the context of the Galilean principle of relativity, 'world' is the homogeneous, immutable space-time as the *form* of phenomena, distinct from the *accidental*, non-homogeneous material content (p. 155). According to Weyl this distinctness finds implicit resolution in the way the Galileian principle of relativity uses only one type of coordinate systems—the Cartesian coordinate systems. For it is precisely this limitation which makes all systems ultimately equivalent, thereby constituting a 'world'. By contrast, it is primarily the interaction of the metrical field and the material content which defines the essence of general relativity (p. 220). As Weyl says, gravitation becomes "a mode of expression of the metrical field" (p. 226). Now, the resulting arbitrariness of the four world coordinates creates a distinction between their being assigned a priori and the possibility of actually measuring the coincidences (physical phase-quantities) of matter in motion. On this account, the ideas of Riemann and Einstein are "the first to give due importance to the circumstance that space and time, in contrast with the material content of the world, are *forms* of phenomena" (p. 227). That is, they remain separate forms for reasons exactly opposite to those which allowed a Galilean 'world' to be structured: in relativity, it is no longer possible "to represent the phase-quantities extending throughout the world by means of mathematical functions (of four independent variables)". How, then, are we to represent them? This is where it must be acknowledged that the identification of an observable origin for the generalized metrical field does not dispense the relevance of the question as to the possible pre-existence of the metrical field over matter on the large scale. It is the case of relative rotation that remains the paradigm in this context. Weyl explains: "In so far as the state of the guiding field does not persist, and the present one has emerged from the past ones under the influence of the masses existing in the world, namely, the fixed stars, [centrifugal effects] are partly an effect of the fixed stars, *relative to which* the rotation takes place" (p. 221). This passage requires careful explanation.

The essence of this statement lies in the only partial influence of matter on the field. This is clarified in a footnote: "We say 'partly' because the distribution of matter in the world does not define the 'guiding field' uniquely, for both are *at one moment* independent of one another and accidental." Thus, Weyl identifies the existence of a guiding field *that does not persist* (i.e. that is opposed to the classical conception of the immutable field) with the *global influence* of the masses existing in the world. Local effects are dependent on the material totality of the universe, only in the sense that the totality of *past* states uniquely determines the emergence of the present one. There is even no other sense of 'global' than that which results from the entire course of the history of the universe. But any possible physics of fields starts from the nature of this present: "Physical laws tell us merely how, when such an initial state is given, all other states (past and future) necessarily arise from them". Observation shows the existence of a guiding field which is most remarkably stationary (the world of stars still assumed by Weyl to be at rest) but this by itself can have no theoretical significance. Deriving the global time function from this fact would obscure the role of the guiding field in explaining the only partial relation of distant masses to local phenomena. If the world seems to be stationary, this must be explained as the result of a series of accidental conditions: "The statement that the world in the form we perceive it taken as a whole is stationary (i.e. at rest) can be interpreted, if it is to have a meaning at all, as signifying that it is in statistical equilibrium".

Statistical principles are indeed those that imply something universal from the accidental. Weyl raises the question of an analogy between the stellar system and a gas in statistical equilibrium (p. 227). According to Maxwell's law of distribution, the molecules of an isolated volume are distributed more or less uniformly. At least if sufficient time is allowed, small velocities are also much more frequent than large ones. The problem is thus to show that, if the ideal state of statistical equilibrium may rightly account for the fairly uniform and stationary distribution of stars, this identification does not contradict the relativistic law of gravitation. Using the already constructed cosmic time and assuming this kind of stationary world, Weyl obtains a series of simplifications analogous to (4.11), (4.12), and (4.13). Not surprisingly, he deduces from these approximations inserted into the field equations an equation quite similar to that of Poisson. That is, $\Delta c = (\frac{1}{2})\rho$, which is obtained from the 'Newtonian' approximations, $R_{44} = \Delta c$ and $T_{44} = T = \rho$. As a result, "the ideal state of equilibrium under consideration *is incompatible* with the laws of gravitation, as hitherto assumed" (p. 278). For, exactly as in the case of Poisson's equation, the only possible uniform and stationary distribution is that for which there is no matter at all (the density is made to vanish everywhere so long as the potentials remain constant). Following Einstein, Weyl goes on to introduce the cosmological constant. He indicates that R is not the most general invariant that can be linearly dependent on the $g_{\mu\nu}$, their first and second differential coefficients. The most general one is $\alpha R + \beta$, where α and β are numerical constants. Accordingly, it is possible to replace R by $R + \Lambda$. A constant density of matter is now given by

$$\rho = \Lambda. \tag{4.14}$$

When the masses of the universe are distributed with the density Λ, the ideal state of equilibrium is realized. (It was not until 1922 that a rigorous demonstration could be given of the fact that the most general metrical tensor corresponding to the requirements of the energy–momentum tensor must indeed incorporate a constant of the type of Λ; this was provided by Elie Cartan 1922.)

The Newtonian and the relativistic cosmology lead to the same equation but differ in its interpretation. In the relativistic case, a metrically homogeneous space is obtained most simply by assuming that the intrinsic metric tensor $g_{\mu\nu}$ and the curvature tensor $R_{\mu\nu}$ must be proportional to one another:

$$R_{\mu\nu} = \Lambda g_{\mu\nu}, \tag{4.15}$$

where Λ is the constant of proportionality. These equations are satisfied for the ds^2 of a three-dimensional, spherical surface of revolution. In other words, the closedness of space is just the relativistic way of formulating the classical fact that you get statistical equilibrium in a closed volume.

Weyl's strategy is a most interesting case, because it is simply a reversal of Einstein's arguments. First by using statistical considerations, Weyl assumes uniformity of matter distribution to be a true prediction of both the Newtonian and Einsteinian theories. The next step is that where Einstein claims that Λ has the sole purpose of complying with static equilibrium, Weyl proceeds to show that it closes space, the equilibrium being the necessary premise. What this boils down to is that the difference between a Newtonian and a relativistic cosmology does not lurk in the latter's closure of space, but rather in the hierarchy, in the mere order of what is to count as premise and what as a conclusion. This leads Weyl to refute the cogency of Einstein's approach to boundary conditions. He emphasizes the statement that was already obtained above without cosmological considerations: "The differential equations in themselves, without boundary conditions, contain the physical laws of nature in an unabbreviated form excluding every ambiguity" (p. 277). This results from the requirement of causality and cosmic time being inserted in the form of the laws of general relativity. By contrast, in Einstein's deduction, the closedness of space reflects a constraint in the solution of the field equations: it is the relativistic language for the classical fact that the gravitational potentials are made to vanish at infinity. Hence, the whole idea of identifying closed space with the large gravitational fields required by the world matter of Mach's principle cannot express the essence of the laws themselves. According to (4.14), it is already clear that a null density at infinity cannot receive a coherent physical interpretation since *the laws themselves*, at least if the cosmological constant is an integral part of them, *would not exist at all*. In this sense, equation (4.14) would express a sort of natural agreement between physics and mathematics in the finite universe.

However, in the fourth edition of *Space-Time-Matter*, Weyl finally found that such an upshot caused the gravest doubts. He added a few comments on his own deduction, in which he said that if Λ should define the radius of curvature of the whole universe, this would be "great demands on our credulity" (p. 279). The suspicion is not so much related to the cosmological constant as to the direct ratio of this constant and the total mass of the universe. Several reasons could motivate Weyl. A first reason might be astronomical. In order for the stellar system to be justifiably compared with the molecules of a closed volume of gas, the stars should exhibit the statistical pattern of an almost final state of rest. Now, astronomical research by the end of the second decade of the twentieth century tended to reveal that the stars of the Milky Way were distributed as if they were rather near a common origin. A conclusion of that kind was already apparent in Eddington's first book of 1914, a book quoted by Weyl only in the fifth edition of *Raum, Zeit, Materie* (1923).

There is another, probably more plausible reason that might justify doubts in the foundations of Einstein's model. During the years 1918-23, Weyl was engaged in the attempt to unify gravitation and electromagnetism. This proved to entail the elements of a complete restructuring of Einstein's world view, not just his cosmological model. The problem was how to conceive of an extension of Einstein's field equations (supplemented by Λ) if they are to really provide a sound basis for unification of gravitation with other fundamental forces. I now turn to an analysis of this crucial project, since it is only in the course of this research that Weyl spelled out his reasons for being incredulous at the relativistic derivation of (4.14).

(b) Weyl and the project of unification of physics

Einstein had concluded his 1920 lecture at Leiden with a rather negative note: physics is still concerned with two very different types of reality, gravitation and electromagnetism (Einstein 1920, p. 22). He was quite critical of the new remarkable theory of the relationship of the electromagnetic and gravitational fields, given by Weyl two years earlier in a paper that Einstein himself had communicated to the Berlin Academy (H. Weyl 1918a, 1918c; on Weyl's theory, see Whittaker 1954, pp. 189–90). In the last two sections of the third edition of *Raum, Zeit, Materie* (1919b), Weyl began to elaborate a systematic presentation of his ideas.

Riemann's geometry is not the most general geometry conceivable. In fact, it involves a certain limitation which Weyl found illogical. The essence of Weyl's ideas is that, when this limitation is removed, we find that the resulting generalized space is characterized, not only by the property which is interpreted as the gravitational field, but also by something else which turns out to have all the well-known properties of the electromagnetic field. Thus Weyl's theory is intended to extend to electromagnetic forces what Einstein has accomplished for gravitation; both are reduced to a natural geometry of space and time. In Weyl's words, what from now on emerged was that "Descartes' dream of a purely geometrical physics seems to be attaining fulfilment in a manner of which he could certainly have had no presentiment" (1922, p. 284). True, the general theory of relativity was already on the way of sealing "the doom of the idea that a geometry may exist independently of physics in the traditional sense" (p. 220).

The basis of Einstein's theory is that, corresponding to two neighbouring events, there is a quantity ds^2 (the interval between them), which can be measured in an absolute way. Absolute means here that it is not necessary to specify the motion of the observer who is measuring the interval. The change from Euclidean to Riemannian geometry in the theory of general relativity was described by Weyl as laying in the change from a finite geometry to an infinitesimal one, in which a metric could only be applied directly to infinitesimally small regions (pp. 92–3). Now, if two points are *not* near together, we have to connect them by a series of intermediate points. The measurement of the interval then involves an integration, and, in general, the result depends on the path chosen. Another way of seeing the manner in which pre-relativistic physics valued an independent homogeneous space background (over and above the material content of the universe) lies in the fact that whereas in the flat Euclidean geometry the directions of vectors at different points could be directly compared, their relative direction in the curved Riemannian geometry is dependent upon the choice of path joining the points. In Einstein's theory, the parameters defining this choice are identified with those defining the gravitational field. The new argument put forward by Weyl is that both Riemannian geometry and Einstein's inclusion of this geometry into physics still contain an element of finite geometry. This is the whole idea of ideally rigid bodies: while the relative *direction* of two vectors acting at distant points cannot be directly compared, their relative *length* is declared to be directly comparable. Thus, when the ds^2 is expressed in units of length, we are virtually comparing it with a standard interval defined once and for all at some remote place and time. What seems illogical with this comparison at a distance is that in connecting distant points both Riemann and Einstein proceeded by the step-by-step method. Clearly, a more consistent interpretation is that the ds^2 ought to be compared with the standard interval by transferring the interval through a series of connecting steps. It is in admitting the possibility that the result of the comparison may well happen to depend on the path followed that Weyl's theory differs from Einstein's. The consequence is quite far-reaching, for it amounts to arguing that although we can make a relative survey of the neighbourhood of a point, the absolute scale of our map remains arbitrary, because there is no unique way of comparing it with our distant standard. The arbitrary multiplier, denoted 'gauge', will clearly involve a line-integral taken along the path by which we travel to reach the standard interval. The remarkable thing found by Weyl was that the four functions which appear as the coefficients of dx, dy, dz, and dt in the line-integral can be interpreted as the four potentials of the electromagnetic field (one scalar-potential and the three components of the vector-potential of Maxwell's theory). When the electric and magnetic forces vanish, general relativity is recovered in its original form, in the sense that the gauge is independent of the path of integration, and the distinctive feature of Weyl's theory disappears. In short, Einstein's theory is valid only in the absence of an electromagnetic field. In order to make this generalized geometry the true foundation of a physical theory, Weyl introduced the requirements of gauge invariance, just as Einstein had introduced general covariance in order to conceive of Riemannian geometry as the foundation of a geometrical theory of gravitation. This gauge invariance was to account for the conservation of electricity.

Weyl's new theory shows that the idea of a priori homogeneity of lengths throughout the universe is but a remnant of the old notion of action at a distance. Two intervals, far apart from one another, can be directly compared without specifying the mode of comparison (the system of gauges) only when the electromagnetic field between them vanishes. In the third edition of his book (1919b, p. 263), Weyl emphasized that this generalization of Einstein's theory leads to a further metamorphosis of physical reality into mere geometrical form. This statement was replaced in the fourth edition (1922, p. 311) by the idea that no field whatsoever can be the objective reality, since a field remains a transmitter; it is a concept that still falls short of what material things are in and for themselves. But even if the avowed purpose of Weyl's whole theory were claimed to be further progress towards the deduction of the existence of the electron and of the quantum behaviour of the atom, it cannot be denied that this theory interacted with, and perhaps indeed motivated, some essential aspects of Weyl's critique of relativistic models of the universe.

Thus, the upholding of the geometrical properties of congruence in Riemannian space defines the integral properties of space altogether. There is little doubt that, from the standpoint of the proposed generalization of Einstein's theory, this type of property refers to what Weyl called above the independent homogeneous *form* of phenomena which was given explicit (and thus also problematic) pre-eminence in the general-relativistic way of thinking—in particular, the success of Einstein's theory in subsuming the matter distribution to such form. But in Weyl's theory, if it can be said that the combination of the gravitational and the electromagnetic fields determines the influence of the whole universe on our measurements of space, this influence makes itself felt through a continuous, and thus imperceptible transformation of the length vector in its displacement with a given direction. As a result, the idea according to which the differential equations of physics already contain the laws of Nature in their complete form is now buttressed up by the fact that boundary conditions are of no use if we want to calibrate the system of gauges. Thus, Weyl's incredulity at the consequences of Eqn (4.14) reflects doubts on the ability of general relativity *alone* to be cosmologically consistent. A higher degree of consistency indeed arises from the fact that the new type of curvature which characterizes the non-transferability of length implies, in turn, a new constant. Following Weyl's assertion that in Einstein's theory Λ is substituted for boundary conditions, there is no real surprise to find out that Weyl was able to discern a direct relationship between this new constant and Einstein's cosmological constant. It is just because of the very existence of such a relationship that, as Weyl said, "our theory necessarily gives us Einstein's cosmological term" (p. 296). In fact, the new generalization is just "what makes Einstein's cosmology physically possible". Given (4.14), there is a proportionality between Λ and the total mass of the universe, and by (4.15) Λ establishes a further proportionality between the metric tensor and the curvature of the world. So, Weyl says, on this account there would be a sort of "pre-established harmony" between the constant and the total mass which is always "fortuitously present in the universe" (1921a, p. 801), in the sense that the constant and the total mass cannot be distinguished as causal factors determining the curvature. In the generalized theory, however, the cosmological term is dependent on both the gravitational and the electromagnetic potentials, so that the constant and the total mass can be distinguished. When the density occurring in (4.14) includes electromagnetic interaction, so Weyl argues, this equation can no longer express mere substitution of one term for the other: the total mass present in the world now truly-*determines* the curvature, while Λ, in Weyl's language, simply *denotes* it.

Not surprisingly, the unified theory was to stir up vivid interest but also severe criticism. In the first years of the theory, both Weyl and his critics realized that its main commendation was its appeal to logical instincts, rather than experimental tests which remained highly speculative. Although the treatment of the electromagnetic equations was one of the most elegant features of Einstein's theory, the fact remained that the electromagnetic vector appeared as something extraneous. No reason is given by Einstein's theory for expecting such a vector to exist, nor is there any reason why there should be only one such vector. It was therefore particularly attractive to obtain an explanation of the electromagnetic field, not by *introducing* an artificial assumption, but by *removing* one. In fact, the difficulty of deriving successful experimental tests of the theory was mainly in connection with its application to the concept of *time* (see Section 6c). Here, the assumption of a static world was to play the key role since, according to Weyl, it was this particular assumption that provided natural calibration (1922, p. 298).

Einstein's reaction was at first very enthusiastic. As he wrote to Weyl on 8 March 1918 upon reception of the book: Your *Space-Time-Matter* "is a masterful symphony". But just a month later, he began to have some doubts. He insisted on the physical significance of ds by arguing that the trajectories of particles cannot reasonably be seen as depending on the past history of these particles. As he said in a postcard of 15 April, the invariance of the ds is the empirical basis of relativity. Take two points P_1 and P_2 connected by a time-like curve. In fact, the two infinitesimal intervals of these points, ds_1 and ds_2, can be connected by more than one curve of that kind, but this freedom in picking out the curve does not affect the ratio ds_1/ds_2 which remains independent of the choice. Then Einstein immediately moved on to other considerations; on the very same day (15 April) he wrote a letter in which he pointed out a mistake Weyl had made in the last part of the last section of his book (the cosmological question), and wrote again the day after in order to correct his own criticism.

The concluding development in the first edition of *Raum, Zeit, Materie* is already one in which Weyl sought to derive the consequences of Einstein's modified field equations on the *local*, radially symmetrical solutions of general relativity. Clearly, this development came as an implication of Weyl's commitment to his tentative unification. For what Weyl had in mind was that the supposedly cosmological constant is already a necessary ingredient of local physics, provided local physics can be unified along the proposed line of argument. Via the correspondence with Einstein and Felix Klein during this period, the triple context of the whole of Weyl's argument gradually emerged: the unification of gravitation and electromagnetism in terms of generalized metric of the world, the formalization of cosmological models, and the attempt to view the unification as a foundation for the ultimate form of cosmology.

4. Global implications of extended local physics

(a) The impossibility of a completely empty world

What does the cosmological constant bring out in terms of local physics? Weyl asks the reader to indulge in "the allurement of an imaginary flight into the region of masslessness', for otherwise it would be forbidden "to make clear what the new view of space and time bring within the realm of *possibility*" (1918b, p. 226; also 1922, p. 281). The empty space–time deprives the form of phenomena of its observable origin, thereby opening up the field of pure possibility; and this possibility should throw light on what actually characterizes the existence of matter.

The massless, radially symmetrical and homogeneous solution is Schwarzschild's exterior metric. Its well-known form is given by Eqn (1.10). By introducing the cosmological constant in the solution, the coefficient

$$\gamma = 1 - 2m/r \tag{4.16}$$

becomes (taking into account the change of units mentioned in Section 2a)

$$\gamma = 1 - 2m/r - (1/3)\Lambda r^2. \tag{4.17}$$

A completely empty and homogeneous world would be one in which $m = 0$. This means that γ now becomes

$$\gamma = 1 - (1/3)\Lambda r^2. \tag{4.18}$$

With $R^2 = 1/\Lambda$ and $c = x_4/R$, the spatial part of the metric is congruent to the three-dimensional sphere

$$x_1^2 + x_2^2 + x_3^2 + x_4^2 = 3R^2 \tag{4.19}$$

embedded in the four-dimensional Euclidean space with Cartesian coordinates x_1, x_2, x_3, and x_4 (Λ being positive). Taking the time to be linear, the world is a cylinder erected in the direction of the fifth dimension, the t-axis. This is the geometry of the Einstein world from which all matter has been removed.

Rather, this is only apparently so. Take a three-dimensional Euclidean space x_1, x_2, x_3 in which x_3 is the axis of rotation of a surface of revolution

(1922, pp. 252ff). In a radially symmetrical statical gravitational field, this surface is a sphere, in which $x_3 = 0$ defines a plane passing through the centre. This is the greatest possible plane. Likewise in the four-dimensional space, $x_4 = 0$ is the greatest sphere: this is the equator or space-horizon for that centre. Because the velocity of light is defined by $c = x_4/R$, it decreases from the pole (where it takes the value 1) to the equator (where it takes the value 0). Thus, $x_4 = 0$ leads to $c = 0$, that is, the velocity of light vanishes on the space-horizon of the centre. This bears a special relation to the spatial part of the metric. Taking $ds^2 = c^2 dx_4^2 - d\sigma^2$ (in which $d\sigma^2$ is a definite positive quadratic form in the three space variables x_1, x_2, x_3), radial symmetry allows the spatial coordinates to be chosen such that both c and $d\sigma^2$ are invariant with respect to linear orthogonal transformations of these coordinates. The distance from the centre can thus be defined as $r = \sqrt{x_1^2 + x_2^2 + x_3^2}$. The velocity of light is only a function of that distance, and $d\sigma^2$ must have the form, $(dx_1^2 + dx_2^2 + dx_3^2) + \ell(x_1 dx_1 + x_2 dx_2 + x_3 dx_3)^2$, in which ℓ is likewise a function of r alone. Setting $h^2 = 1 + \ell r^2$, we have, by taking into consideration that $ds^2 = 0$ for light, $1/c^2 = h^2$. When c vanishes, which is the case of the space-horizon of the Einstein empty world, h becomes infinite; the spatial part of the metric is singular.

Weyl infers from this result that the space-horizon is also a physical horizon. For from what has been found, "we see that the possibility of a stationary empty world is contrary to the physical laws that are here regarded as valid". Indeed "there must be least be masses at the horizon" (pp. 279–80). This conclusion is derived from the sole incorporation of Λ in Schwarzschild's exterior metric. In fact, the physical nature of the inescapable horizon can be ascertained from the general solution of (4.17) in which the mass is not taken to vanish. This equation defines an equation of the third degree in r. We can find the positive roots of this equation by using reasonable approximation. For a sufficiently small cosmological constant, it appears that $r_0 = 2m$ is an obvious solution; this is the already well-known singularity occurring in Schwarzschild's metric. But there is also another solution, far more remote from the centre, which betokens yet a new singularity, and thus also a new mass-horizon: this is $r_1 = \sqrt{3/\Lambda}$. It is nothing other than the radius of curvature of De Sitter's metric. Thus, any distance r falls between two values r_0 and r_1; within this interval $1/h^2$ is positive, and $1/h^2 = 0$ at the limiting points. Because of the cosmological constant, however small it may be, empty space around a mass falls between these two horizons; in the case of no mass, the more remote horizon cannot be transformed away.

(b) On the physical nature of the mass-horizon

Also in the first edition of *Raum, Zeit, Materie*, Weyl developed a series of investigations on the physical conditions prevailing inside the mass-horizon (1918b, p. 226). These conjectures proved to be most controversial as

Einstein took exception to them in his letters of 15 and 16 April 1918. Weyl immediately acknowledged the relevance of a mistake pointed out by Einstein in his calculations (27 April) and made some changes on the proofs of his book. But the main point of his argument he was not ready to abandon very quickly: this was the conclusion that the symmetric structure of the mass-horizon would tend to prove that the topology of the elliptic kind for the whole world should be preferred to the spherical, that is, that the identification of antipodal points was "much more natural" (Weyl to Einstein, 19 April). Einstein did not like this reasoning and replied that "it is impossible to settle the question in a speculative way" (31 June). As he went on to say, he was willing to favour the spherical topology on the basis of the "obscure feeling" that "any closed curve should be allowed to contract in a continuous way until it shrinks to a point"; the elliptic case does not permit this, in that it seems to be physically reminiscent of the finite Euclidean space with its abrupt edge. (In a similar vein, Einstein wrote to Felix Klein on 16 March 1919 about a non-relativistic way of obtaining finite space: take a Euclidean line element and identify periodically the points separated by some fixed distance.)

In the third edition, these considerations were completely left out and replaced by a conclusion that Weyl had reached in an article published in the meantime (his 1919c paper). This conclusion was now quite independent of topology and simply focused on the development of the original idea, which was a derivation of the relationship between the quantity of matter and the size of the horizon (this material reappears with little change in Weyl 1922, p. 280). In this context, Einstein's early objection was that not all spherically symmetrical solutions of the field equations with Λ entail a region of matter symmetrical around the equator. In the more general analysis, Weyl was cautious enough to start with symmetry just as a hypothesis, and noted that, as this only yields the simplest case, the following calculations merely help to orient ourselves in the question. However, as we shall see, the conclusions were extremely important.

Weyl assumes the horizon to consist of an incompressible fluid, with constant density μ, situated between two meridians and filling in a region symmetrical around the equator. As the horizon can be regarded as a finite amount of matter embedded in empty space, the prevailing metric must be that of the interior type. Weyl begins with the equation of the interior metric supplemented by the cosmological constant. The term $1/h^2$ becomes

$$1/h^2 = 1 + (2M/r) - [(2 + \Lambda)/6]r^2 \quad (4.20)$$

in which M is a constant of integration. (See the complete deduction of the interior and exterior metrics, 1922, pp. 252-68.) Let $r = r_0$ be the value of the radius of the two meridians $x_4 =$ const. in empty space. Because we pass over in a continuous manner from $c = 1 - 2m/r$ to (4.20), it is possible to derive the following value for the constant: $M = \mu r_0^3/6$. This is nothing other than the expression of the constant, the 'relativistic mass', which defines the gravitational radius of any sphere of fluid (pp. 264–5; see Chapter 1, where it was mentioned that the relativistic mass has the dimensions of a length). Hence, the horizon is a singularity, retiring from the visible world for an observer from the outside. As before in the case of the exterior metric, the space of the whole world may still be represented as congruent to the sphere (4.19). However, this is not valid for the zone occupied by fluid. Therefore, the whole space is best described as being a *spheroid*. This is obtained by continuous 'warping' of the original sphere: the spheroid remains symmetrical with respect to the axis $x_4 = 0$, and its radius $r = \alpha$ is composed of the distance between r_0 and R. If the velocity of light vanishes on the equatorial sphere, the radius of the spheroid should be determined by the solution of equation (4.20) for $1/h^2 = 0$. At this point, Weyl's calculations become very complex. Let us more modestly try to follow his mode of reasoning rather than the actual calculations.

He sets $r/\alpha = \rho$, and $r_0/\alpha = \rho_0$, which enables him to take α as the new unit. He also sets a quantity p, determined by a non-parametric relation between r_0, α, Λ, and μ; this complicated relation is $(r_0/\alpha)^3/(1 - \Lambda/2\mu) = \rho$. All this helps to find a solution of equation (4.20) in terms of a relationship between the radius and the quantity of matter. With the boundary condition of vanishing pressure at $r = r_0$, it can be verified that, for any value of α between r_0 and R, the quantity $1/h^2$ is indeed equal to zero. Hence, "*c* remains positive throughout the zone of fluid, including the equator" (Weyl 1919c, p. 32). The result is a transcendental (non-parametric) relation between μ/Λ and r_0/α. This yields a uniquely determined solution, $\rho = \rho_0(p)$, for any allowable value of p, a solution which satisfies the physical conditions of positive density and pressure. On the sphere, the equations of the statical gravitational field thus have this unique solution in which $\rho_0 = r_0/\alpha$. Now, the value of the cosmological constant Λ here plays the role of the boundary condition for the metric prevailing within the mass-horizon, for if $\Lambda = 0$, $r_0 = \infty$ and space becomes Euclidean. For $\rho_0 = 1$, the zone of fluid is infinitely thin and calculation leads to $\Lambda/2\mu = 0$, that is, the density of matter becomes infinite (for $\Lambda \neq 0$). For ρ_0 almost equal to 1, the zone is very narrow and $\Lambda/2\mu$ is a very small quantity ϵ. The smaller this quantity is, the greater is μ. When ϵ tends to zero, it becomes possible to evaluate the corresponding quantity of matter in the horizon, for in the case of an infinitely narrow zone, the corresponding total mass defines the lowest limit of the quantity of matter.

This derivation of the physical conditions governing the mass-horizon has some far-reaching consequences. That, in equation (4.20), the total mass does not become less than a certain positive limit, means that the radius of those meridians which enclose the mass-horizon can always fall within the limits of the gravitational radius. The determination of the density corresponding to the radius results in a statical solution of the cosmological equations which is everywhere continuous. Most important in that derivation is that Λ does *not* fix a priori the quantity of matter in the universe. On the other hand, if the zone of fluid is gradually enlarged, the universe is ultimately filled in with a pressureless, homogeneous matter of density Λ. This is some sort of extension of the mass-horizon to the whole of spherical space, in which Λ recovers the meaning it had in the original form of Einstein's model. The fundamental result is expressed by Weyl in the following terms: "As the zone is smaller. . .its mass is greater; for an infinitely small zone the mass is $4\sqrt{6/\pi}$ greater than in the case of homogeneous fulfilment of the whole world" (p. 33). Thus, the mass of an infinitely small mass-horizon is greater than that of the homogeneous world-matter. As the singularity of an apparently empty world is swept aside towards the periphery, the quantity of matter in the singularity increases so as to exceed the quantity which initially corresponded to the 'full' universe.

This plasticity or malleability of the mass-horizon, as it may appropriately be called, drastically changes the whole approach and the whole status of cosmological considerations in the context of general relativity. Two consequences could be derived from this new perspective on the mass-horizon brought about by Weyl's "imaginary flight" into the realm of possibilities. In the first place, it appears that the relationship between Λ and the total quantity of matter in the universe is no longer to be fixed uniquely in terms of cosmological theory. Rather, this relationship is now regaining (at least in principle) its connection with empirical determination, of just the kind that Einstein refuted in his 1921 lecture, 'Geometry and Experience', when he declared that "however great the space examined may be, we could not feel convinced that there were no more stars beyond. . .it seems impossible to estimate the mean density [of the universe]" (1921, p. 43). In other words, following Weyl, the motivation for a cosmological model is *not* that it provides us with a theoretical law which is somehow substituted for the impossible empirical knowledge of the total mass of the universe. And it seems to be this critique that, in turn, motivated Weyl's incredulity at the definite relation of the total mass to the cosmological constant. A second implication of the malleability of the mass-horizon is that the existence of the horizon, in the apparently empty universe, allows a link to be found between the two known models; so that this malleability, precisely because it offers the only possible bridge, supports the view that the interpretation of the singularity in terms of matter is correct.

It was in the fourth edition of *Raum, Zeit, Materie* that Weyl finally discussed the implications of the two solutions in the most general terms. But this discussion involved more than just the new perspective on the mass-horizon, since it also included the fundamental aspects of cosmic time. Weyl was driven to this generalization because he had to tackle an alternative, more geometrical interpretation of the mass-horizon in which time played a central role. The interpretation was mounted by Felix Klein, the celebrated mathematician of the Erlangen Programme.

5. Weyl and his critics: the geometrical versus the physical approach

(a) Felix Klein

On 18 April 1918, Klein wrote to De Sitter about the geometrical implications of cosmology. As De Sitter observed, the major geometrical concepts used in the different stages of the development of relativity are contained in Klein's 1872 programme which synthesized the various senses and possible uses of non-Euclidean geometries. (See F. Klein 1872 and the comments of R. Torretti 1978, pp. 137–42. Klein's letter to De Sitter is quoted by De Sitter in his 1917 paper, p. 1310.) With this programme, Klein was the first to delineate with full clarity the links between the new geometries on the one hand and group structure and the theory of invariants on the other. In 1918, he gave three last contributions to the development of this programme, in which the theory of relativity was connected to his old ideas. (On Klein's contributions to general relativity, see J. Mehra 1973, pp. 138–40 and A. Pais 1982, pp. 274–8. In fact, Klein belongs to the category of those mathematical writers who, by the end of the nineteenth century, had entertained the possibility of non-Euclidean geometry for the whole universe.) In his letter to De Sitter, Klein gave a purely geometrical explanation of the fact that the De Sitter hyperboloid is characterized by a barrier of time. If we travel along a straight line which intersects the polar line of our starting point, we return to the starting point with the original direction reversed. De Sitter even went on to outbid Klein when he took this argument as a proof that the motion in question, though mathematically thinkable, is physically impossible. In order to have what he called a physically possible circuit, a closed curve should not intersect the polar line of any of its points; on returning to the starting point, we find the initial direction unaltered if that condition is fulfilled.

Klein, however, did not draw any conclusion as to the reality or the non-reality of this 'motion'. What he wanted to point out was that the De Sitter model provides geometrical reasons for rejecting Einstein's scheme of linear flow of time from the past to the future. In his first note on the problem (1918a, p. 615), he described as "amusing" the consequences of the complete relativity of time. Events that, for one observer, lie in eternity and remain imaginary, are quite real and accessible for another observer. Thus, because the singularity is itself subject to the relativity of time, it would seem difficult

THE GEOMETRICAL VERSUS THE PHYSICAL APPROACH 267

to agree with the opinion that this singularity is something absolute, physically identical for all observers. In his first popular exposition of the theory of general relativity, Eddington echoed this view when, speaking about the greatest possible distance between two points in the De Sitter universe, he explained with remarkable lucidity and simplicity:

> Time in the two places is proceeding in directions at right angles, so that the progress of time at one point has no relation to the perception of time at the other point. The reader will easily see that a being confined to the surface of a sphere and not cognizant of a third dimension, will, so to speak, lose one of his dimensions altogether when he watches things occurring at a point 90° away. He regains it if he visits the spot and so adapts himself to the two dimensions which prevail there (Eddington 1920a, p. 160).

The idea is presented on Fig. 4.6. The singularity, at which time stands still, is created by the chosen *coordinates*, not by the *intrinsic* nature of spacetime. This geometrical reduction raises a problem as to the compatibility between De Sitter's postulate of continuity and the allegedly 'absolute' barrier of time.

Klein wrote two papers in 1918, in which he endeavoured to investigate the connection between the laws for the conservation of momentum and energy and Einstein's theory of gravitation. Already by the end of 1915 Klein had come very close to a correct derivation of the field equations from a variational principle. That is, he sought a way of looking at the conservation of energy–momentum in such a way that it would result from general covariance. (It was Weyl who did this successfully in August 1917, after others such as Lorentz and Hilbert had tried by using various formalisms.) Klein's motivation in his paper which was read before the Göttingen Society of Sciences on 6 December 1918 was to spell out the conservation laws in their integral form, after he had examined them some five months earlier (19

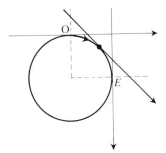

FIG. 4.6. The singularity of the De Sitter universe in terms of geometry. For a one-dimensional observer at O, point *E* is a 'singularity', as it lies in the 'second' dimension. But the singularity vanishes as O moves towards *E*.

July) in their differential form (Klein 1918b and 1918c). The last part of these contributions was devoted to the cosmological problem, in particular the connection between the conservation laws and the closed universe. Klein's point was one of bewilderment: different formulations of the conservation laws do not predict the same total gravitational energy for a finite universe (1918c, pp. 413-14). (Hilbert had even envisaged that energy conservation was perhaps not at all part of general relativity.) As Klein said in the December paper: "My goal is to state the *mathematical* connections as clearly as possible, while touching upon the physical question only briefly" (1918c, pp. 394-95). He found satisfaction in the relation of this analysis to his "old ideas" of the Erlangen Programme (1872), namely, the distinction of geometries in terms of the invariants under transformations permitted by the relevant groups of transformation. In fact, "there are as many different kinds of theories of relativity as there are [mathematical] groups" (1918c, p. 399). Klein went on to discuss the symmetry groups related to the generalizations of Einstein's closed universe, only to discover that the irreducible diversity of theories remains in force in the case of the application of relativity to cosmology. Klein made use in this cosmological context of his earlier description of the entities preserved when the groups are associated with manifolds of constant curvature. In particular, in his celebrated 1873 dissertation, "On the So-Called Non-Euclidean Geometry" (p. 124), Klein had defined the transformation group "which preserves the metric on a manifold of constant curvature". This consists, "for a suitable choice of coordinates, in the group of linear transformations that preserve a given quadratic equation". So, in short, his problem was now to identify the nature of the linear transformations that characterize the two cosmological metrics.

From this geometrical vantage point, the disjunction brought about by De Sitter's distinction between mathematical and physical principles of relativistic cosmology gains new impetus. Klein begins by noting that Einstein's model is already a simulation (*fingieren*) of the mean state of matter in the universe, and that cosmic time is an amazing reinjection of the classical standpoint into relativity (1918c, p. 409). This means that Einstein's model implies a restriction of the group of allowable transformations of coordinates, such that the group in which the ds^2 of the cylindrical universe remains invariant is enlarged to the Lorentz group when the radius of curvature is taken to be infinite (pp. 407-9). But making the radius of curvature infinite is not the only way of restoring a more general group. Quite generally, the ds^2 can be written:

$$ds^2 = \Sigma g_{\mu\nu} dw^\mu dw^\nu, \qquad (4.21)$$

in which w is an arbitrary parameter. By imposing constant curvature, the components $R_{\mu\nu}$ become $R_{\mu\nu} = (3c^2/R^2)g_{\mu\nu}$. If the field equations are taken so as to include the cosmological constant, $R_{\mu v} - \Lambda g_{\mu v} - \kappa T_{\mu v} = 0$, these equations will be satisfied if $\Lambda = 3c^2/R^2$ and all $T_{\mu v} = 0$. This is exactly De Sitter's result, which proves that the elimination of matter leaves the original generality of the ds^2 unaltered. Thus, in Klein's mind, the finiteness of the radius of curvature does not jeopardize the original combination of space and time, provided that matter is made to vanish. In this respect, Λ has nothing to do with the existence of matter, even less with its distribution. True, one of the purposes of Klein's articles was to show that the adjunction of Λ in the field equations does not change anything in the consequences which can be drawn from the conservation law in Einstein's theory (Klein 1918b, pp. 187-8). Furthermore, as Klein went on to show, the elimination of matter seems to imply the equivalence of all points in the universe. This results from equation (4.15), in which it is sufficient to assume that either one of the tensors has the same value everywhere. One is thus prompted to ask how it is at all possible to conceive of the occurrence of a singularity in the De Sitter model.

In his 1918 review article (p. 243) on the different principles involved in the theory of relativity, Einstein mentioned that, as far as he knew, the modified field equations do not admit a singularity-free empty solution. This article was received for publication in *Annalen der Physik* on 6 March 1918. Shortly after, however, in June of the same year, Einstein sent a postcard to Klein in which he reacted to the very different interpretation Klein had mounted in the meantime: "You are perfectly right. De Sitter's world is in itself singularity-free and all its world-points are equivalent. . .My critical remark with regard to De Sitter's solution requires correction". The cosmological part of Klein's December paper develops a series of arguments and counter-arguments that he and Einstein discussed from March to July; it was this correspondence that led Einstein to revise his earlier position about the nature of the De Sitter universe. As Klein acknowledged that Einstein had privately taken exception to his earlier position, he wrote that the very existence of the ds^2 in which w is an *arbitrary* parameter (it can be replaced by the coordinates of either the Einstein or the De Sitter world) proves that the allegedly physical claim as to the non-existence of a singularity-free empty solution is in itself nothing more than mathematical (1918c, p. 419).

It all began with Klein's concern about topology. Klein was, in fact, the first author to pay attention to the topological issues (as he says explicitly in 1918c, p. 409 n. 1). His starting point was the observation that topologically different space–times for the same Einstein model of the universe exist (namely, the elliptical and the spherical) and that their difference manifests itself in the time variable: because the elliptic plane is one-sided, it becomes impossible to distinguish past and future in this representation. This, of course, runs against the very idea of cosmic time (Klein's postcard to Einstein, 25 April). Then Klein went on to show in an analytic way how, in a world of constant curvature, elliptical and spherical space are not equivalent (letter of 31 May); his discussion also involved an attempt to geometrically define the time variable in such a way that the distinction between past and future would be always possible in both the Einstein and the De Sitter worlds.

Einstein's reply (letter of 2 June) focuses on the physical interpretation of g_{44}. He recalls that a non-constant g_{44} requires enormous star velocities, which are not observed, and moreover, that (as Weyl had just shown in his *Space-Time-Matter*) the De Sitter world is simply the limiting case of the more general Schwarzschild type of solution (see Einstein's drawings on Fig. 4.7). Klein was not convinced by the correctness of Einstein's communication. On 16 June he stated the difficulties and showed how it is possible to transform away the singularity pinpointed by Einstein. The argument goes as follows.

In order to elaborate an interpretation of the singularity which can be understood consistently without matter, Klein introduces the elements of projective, non-Euclidean (hyperbolic) geometry (1918c, pp. 415-8). (For an introduction to the concepts used here, see R. Torretti 1978, pp. 125-42. A more complete development of De Sitter's universe in terms of projective geometry was offered by P. du Val 1924 and F. Gonseth 1926, pp. 75-97 and 141-5; see the survey in H.S.M. Coxeter 1943.) The De Sitter world is a manifold of constant curvature, which can be represented as a five-dimensional pseudo-sphere on which Euclidean measures hold:

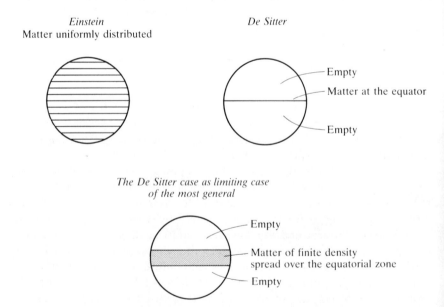

FIG. 4.7. Einstein's drawings in his letter to Felix Klein, 2 June 1918.

THE GEOMETRICAL VERSUS THE PHYSICAL APPROACH

$$\xi^2 + \eta^2 + \zeta^2 - v^2 + \omega^2 = R^2/c^2 \tag{4.22}$$

(the pseudo-sphere results from the change in the sign of one of the variables in the equation of the sphere). The corresponding line element is

$$-ds^2 = d\xi^2 + d\eta^2 + d\zeta^2 - dv^2 + d\omega^2. \tag{4.23}$$

Klein builds a pseudo-elliptic world, in which two antipodal points are identified as one and the same point, by introducing the following parametric representation:

$$x = \frac{R}{c}\frac{\xi}{\omega}, \quad y = \frac{R}{c}\frac{\eta}{\omega}, \quad z = \frac{R}{c}\frac{\zeta}{\omega}, \quad u = \frac{R}{c}\frac{v}{\omega}.$$

The equation of the pseudo-sphere becomes $x^2 + y^2 + z^2 - u^2 + (R^2/c^2) = R^2\omega^2c^4$, in which the right-hand side is always positive if the original coordinates ξ. . . . ω are limited to real values. As a result, this parametrization furnishes a two-sheet hyperboloid, and any point of the real world is now situated *between* the two sheets (Fig. 4.8). In the projective representation, the fundamental figure is the two-sheet hyperboloid, $x^2 + y^2 + z^2 - u^2 + (R^2/c^2) = 0$. In homogeneous coordinates (that is, the coordinates that allow the points at infinity to be included in the general expression of the coordinates), we have $\xi^2 + \eta^2 + \zeta^2 - v^2 + \omega^2 = 0$. This is the equation of the asymptotic cone. The two-sheet hyperboloid is limited by the plane which cuts off the asymptotic cone at infinity and defines the circle at infinity. The hyperboloid can thus be said to be the *absolute* (the set of points at infinity) of the hyperbolic plane, and the real world is the interior of this absolute.

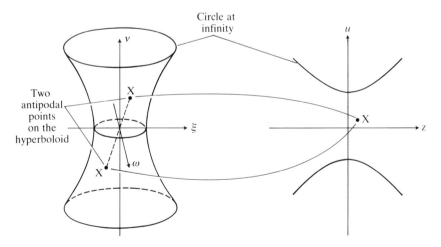

FIG. 4.8. Klein's parametric representation of the De Sitter universe.

Klein goes on to provide us with a definition of spaces: spaces are the new figures represented by a unique linear equation in x, y, z, u (or by a corresponding homogeneous equation in $\xi, \eta, \ldots$). When these spaces intersect the fundamental hyperboloid in imaginary points (for instance, $y = 0$), they correspond to the ellipses of our intuitive model (Fig. 4.1) on which distances are always finite. Spaces such as

$$u = \pm (R/c) \text{ or } v \mp \omega = 0 \tag{4.24}$$

intersect the hyperboloid in one point: they can be called tangential spaces. Any two tangential spaces bound a connected slice of the world inside which the hyperboloid does not penetrate. Klein calls this region of the world a double wedge; the tangential spaces define the double edge of the double wedge, and correspond to what Weyl denoted by equator or space horizon. The real universe comprises what is between all possible pairs of tangential spaces. In the example (4.24), the points of the double wedge are defined by (Fig. 4.9)

$$- (R/c) < u < + (R/c) \text{ or } -1 < (v/\omega) < +1.$$

The points -1 and $+1$ are excluded from the real world, for $u = \pm (R/c)$ leads to $v \mp \omega = 0$. Thus, the only way of getting $v = \mp \omega$ (which represents the double edge of the double wedge) is to have $v = \omega = 0$; u becomes indeterminate at the edge of the wedge.

In the projective conception, a distance is represented by an angle. Here, any two elliptic spaces (situated between the two edges) determine a pseudo-angle. The two 'flanks' of the double wedge are tangent to the absolute, from which it follows that the cross-ratio of the two spaces under consideration defines the pseudo-angle:

$$\phi = \frac{R}{2c} \log \frac{\omega + v}{\omega - v} = \frac{R}{2c} \log \frac{R/c + v}{R/c - v} \tag{4.25}$$

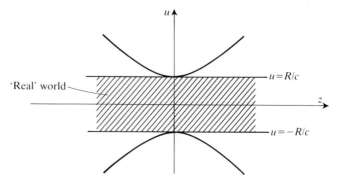

FIG. 4.9. The 'real' part of the De Sitter universe between a pair of tangential spaces.

in which ϕ is counted from $u = 0$. The distance between the two spaces is thus defined by a time-like geodesic on the hyperboloid. When u varies from $-(R/c)$ to $+(R/c)$, ϕ increases from $-\infty$ to $+\infty$, that is, the pseudo-angle wanders through the whole of the double wedge. However, at the points of the edge itself, ϕ becomes indeterminate. In other words, it is impossible to find a time-like geodesic which would connect two points of two generators, forming the double wedge of the hyperboloid.

On substituting x, y, z, u for the arbitrary parameter w in the ds^2 (4.21), this ds^2 becomes infinite along the hyperboloid. Klein takes it that this result is "tantamount to the nonexistence of matter at the non-singular points of the world" (1918c, p. 420). But what does this indicate about the singularity itself, which has been derived from purely geometrical considerations? Klein discovers here the nature of the general requirement of cosmic time, for he identifies its role not with the *interpretation* of the singularity but with its very *occurrence*. From a geometrical point of view, cosmic time in (4.21) amounts to limiting the allowable transformations in accordance with the group: $\bar{w}^{I} = w^{I}$, $\bar{w}^{II} = w^{II}$, $\bar{w}^{III} = w^{III}$, $\bar{w}^{IV} = w^{IV} + C$, which C is an arbitrary constant. Such a group of transformations represents a continuous rotation of the pseudo-elliptic world around some fixed axis. Thus, the time $t = w^{IV}$ must correspond (up to one constant) to the pseudo-angle of the double wedge defined by (4.25). Ignoring the value of this constant, and keeping to the same choice of tangential spaces as above, we have to take

$$t = \frac{R}{c} \log \frac{\omega + v}{\omega - v}. \quad (4.26)$$

The quantity t is real only in those parts of the four-dimensional world for which $(\omega + v)/(\omega - v)$ is positive; it is imaginary everywhere else. Now, in developing his Erlangen Programme, Klein had been able to show that any quadratic surface is left unaltered under ∞^6 linear transformations. In the present context, this means that ∞^6 pairs of tangential spaces satisfy Eqn. (4.26). The result is expressed by Klein: "*Accordingly we have ∞^6 ways of introducing t*" (p. 420).

As Klein goes on to show, the quantity t is quite consistent with the De Sitter metric. The ds^2 of this metric can easily be obtained from (4.23) when appropriate hyperbolic functions are applied (p. 421). Now, the following relation is found to hold between these functions:

$$\tanh \frac{ct}{R} = \frac{v}{\omega}, \quad (4.27)$$

which is in agreement with (4.26). Indeed, the quantity (v/ω) takes the values -1 to $+1$ when t goes from $-\infty$ to $+\infty$. The two flanks $v \mp \omega = 0$ represent the infinite past and the infinite future for any given t in the De Sitter world. Consequently, the 'world' which Klein has erected on purely geometrical grounds is the same as the De Sitter world, not only as far as space is concerned, but also time.

This formal equivalence brings out the essence of Klein's critique, for it reveals that the Einstein–Weyl interpretation of the De Sitter world is as yet not sufficiently constraining so as to rule out an alternative in which this world remains truly empty (pp. 421–3). Equation (4.27) yields the indeterminate value 0/0 for t when both ν and ω are zero, that is, at the singular edge of the double wedge. The distinctive feature of this quantity t is that it is not unique: *each* equivalent observer in the De Sitter universe is equipped with a *different* cosmic time. Take any two observers belonging to the same De Sitter universe, selecting a different 'universal' time t. The pairs of tangential spaces which, for each of them, intersects the fundamental hyperboloid, have always only some portion of the universe in common. As a result, it can be expected that the very same events will not be uniformly characterized as real or imaginary *already within* the fields of observability. In regard to reality or imaginarity, therefore, the singularity beyond the fields of observation is definitely not distinctive. In fact, the pseudo-elliptic world is tangent to the fundamental hyperboloid only in two *points*, namely, $\xi = 0$, $\eta = 0$, $\zeta = 0$, $v \mp \omega = 0$, while the coordinates used by Einstein and Weyl, so Klein argues, allow for the singularity to be represented as an *extended region*. It is essentially this difference, as he surmises, that prompted these two authors to fill in the singularity with matter. For it can be verified that the expression (4.26) for time is consistent with an entirely empty world: what was immediately established in the derivation following (4.21) can be reproduced along these lines without much difficulty.

Given the equivalence between a purely geometrical derivation of the properties of the De Sitter model and allegedly physical principles, the question arises as to what is exactly the cause of the singularity. Quite correctly, Klein identifies it with the choice of the coordinate t, not with a property of the original ds^2. He also notes that the universe is completely empty *only* if we adopt the transformation whch represents *static* coordinates (p. 420). It seems that Klein's point is to contradict the Einstein–Weyl rejection of a static and entirely empty metric, by arguing that a distinction must be made between the geometry and the physical requirements of matter distribution. Einstein did acknowledge this point in his June 1918 postcard to Klein: the De Sitter world, he now said, is "originally" empty and homogeneous, in the sense that "a singularity arises through the substitution that furnishes the static form of the line element". But Einstein fell back on the need of a physical interpretation in terms of cosmic time. What deprives this world of physical possibility is no longer the absence of matter, but the fact that "it is not possible to establish in it a time t such that the three-dimensional sections $t = $ const. do not intersect each other and that these sections are equal (metrically)". Thus, Einstein came to realize that his argument for cosmic time is just the requirement of universal time partitioning the world into isometric space-like slices. In a non-static world, the ds^2 would change all the time, and that is what he found to be physically unacceptable.

(b) Towards the most general cosmological form

The property of the quantity t of becoming indeterminate at the periphery makes it quite similar to the cosmic time function introduced by Weyl (fourth edition of *Raum, Zeit, Materie*) in his analysis of the apparently empty world. Also, quite striking are the emphases in both Klein and Einstein on the term 'static' and the complete absence of development on the meaning of 'non-static'. A really empty world is not necessarily non-static: it was Klein's contribution to *oppose* the empty and the non-static, but the considerations on the non-static world *as such* did not go any further. In a postcard of 22 January 1919 that he sent to Weyl, Klein reacted to Weyl's description of the singularity in the De Sitter model in terms of matter, pointing out the coordinate character of this singularity. Then on 2 March, he compared the horizon with the kind of singularity you get by using polar coordinates in the description of a sphere; he also begged Weyl to think seriously about the limits of Einstein's static conception.

So, as we may now understand, Weyl's analysis of the most general cosmological form (in the concluding part of the section on cosmology in the fourth edition of *Raum, Zeit, Materie*) was directed to a twofold reconsideration: an explicit reference to the non-static nature of the De Sitter world, and a reconstruction of what it means to speak a priori (independently of *any* cosmological model) of 'one' cosmic time for the whole world. Let us go back to this analysis with which we started (Section 2a): we are now in a position to develop its implications.

The unaltered metrical groundform (4.1) is called by Weyl non-static; this is the only one that represents an actually empty world, for no singularity occurs. The introduction of cosmic time in this form changes it into a static metric, with an unavoidable horizon: this is the place at which cosmic time becomes indeterminate. Now, Weyl did not obtain anything else than this latter form when he started from a very different metric, i.e. Schwarzschild's exterior metric with $m = 0$, $\Lambda \neq 0$ and no cosmic time. This yields a most remarkable agreement between a local metric extended to the whole universe (by means of Λ) and a metrical form immediately applied to the whole universe (by means of cosmic time).

In all, three ways of combining time and metric at the cosmic level have been contemplated by Weyl. It is worth the effort to summarize them in the following table:

Time:	non-cosmic	non-cosmic	cosmic
Metric:	static	non-static	static
	↓	↓	↓
	horizon	no horizon	horizon

To conceive of the singularity in an empty world in terms of a mass-horizon is, according to Weyl, the only way of making the Einstein and the De Sitter models comparable to one another. This allows an a priori relation of Λ to the total mass of the universe to be eschewed. But it is most remarkable that the singularity of an empty world is transformed away by the choice of a non-static metric, not by the postulate of cosmic time. This is probably what hinders Weyl from analytically investigating the fourth case: cosmic time in a non-static metric. True, Weyl does not even speak of non-static *coordinates*, but only of a world that would no longer be static were it regarded "as a whole" (*als Ganzes*). (In his correspondence with Klein, Einstein had also made the distinction between two ways of regarding the De Sitter world, one in itself [*an sich*] which is singularity-free and one as static.) The a priori reasoning carried out by Weyl has enabled him to realize that the first and the third cases are in fact identical, the absence of cosmic time in the former being due only to the local character of the metric. Provided the metric is static, cosmic time allows matter to be inserted in an apparently empty world; the rejection of the static metric is also the rejection of cosmic time (second case), in which case the world remains empty. We see how the truly speculative supremacy of the idea of cosmic time determined here both the occurrence of a non-static representation and the impossible mathematization of this representation. Weyl had fixed his task as one of spelling out this supremacy which was already present in Einstein's arguments, by looking beyond their rather simple reduction of 'physical possibility' to the properties of the cylindrical model alone. The mathematics of the missing fourth case was developed some ten years later by Lemaître: tackling the problem of the remote singularity at $r = \sqrt{3/\Lambda}$ predicted by Eqn (4.17), Lemaître (1933, pp. 80–2) showed its fictive character as due to the static coordinates.

Weyl began by rejecting the consequences of the statistical interpretation which he himself had constructed. This opened up the vast fields of possibilities. In the first edition of his book, he expressed the opinion that a priori considerations could only specify what relations were possible in nature, but that the choice of actual physical laws had to be based upon empirical considerations (1918b, pp. 226–7. This statement was suppressed in later editions.) As a matter of fact, the resolution of the Einstein–De Sitter opposition also demonstrates that the actual relation of Λ to the total mass is now open to empirical determination; it is not the value of the constant that fixes the mass, but the other way around. This position embodies Weyl's new approach to the broader issue of the dualism matter/metric. While in Einstein's model, matter generates the metric field, it is the metric field which seems to pre-exist matter in the original form of the De Sitter model. Weyl's resolution sends both models back to back on this issue, but it yields a new, more general form of conflict: either the universe is empty but homogeneous because it is in motion, or it has matter at rest but this cannot be homogeneous. Quite significantly, Weyl did not identify this ultimate conflict arising from the resolution as static versus non-static. He proceeded to annihilate it by incorporating the conceptual framework of his own unified theory (1922, pp. 298–300. See also pp. 261–2). In his view, a natural calibration of the system of gauges as far as time is concerned was unquestionably offered by the static nature of the world. From this uncritical position, it was easy to show that the additional curvature required by the new geometrical unification of electromagnetism and gravitation is constant. The true conflict, therefore, Weyl identified as existence of a material particle versus mass-horizon. It was the purpose of the final analysis of cosmology in the fourth edition of *Raum, Zeit, Materie*, to show that the addition of the electrical term to the cosmological constant removes the conflict. Inserting the electrostatic potential into the radially symmetrical and statical solutions as hitherto developed, he now concluded that the equatorial singularity occurs just when $r = 0$. This was the ultimate bridge between local and cosmic considerations, for it is not the removal of the mass-horizon that also removes the conflict, but the very fact that "matter is. . .a true singularity of the field" (p. 300).

(c) Lanczos and the idea of a new type of combination of space and time

In October 1922, almost the same time as the first of Friedmann's paper was released and passed virtually unnoticed, Cornelius Lanczos published an important paper which can be taken as the first analytic treatment, if not of the non-static coordinates, at least of the *limits* of static coordinates as hitherto worked out. The influence of Lanczos's work cannot be belittled. While that of Friedmann failed to capture attention, Lanczos's articles, on the contrary, were to stir up many controversial discussions, notably one between Lanczos and Weyl; this seems to have prompted Weyl to reconsider the nature of geometrical foundations of cosmology. Furthermore, Lemaître acknowledged his debt to Lanczos for he, rather than the then unknown Friedmann, had a strong influence on the path towards the idea of primeval atom (see Lemaître 1927, p. 50). Also, Lanczos had investigated and published in a 1923 article, the relationship between Doppler redshifts and distance, and found a linear one.

The work of Lanczos can be characterized as a reformulation of the geometrical approach championed by Klein. On the fundamental issues, and right from the outset, Lanczos's position was never ambiguous. As he recalled it much later, in a series of lectures delivered at the University of

Michigan in 1962, his basic conviction was that the physics of the twentieth century offered "a new physical world picture in which matter appeared as a certain curvature property of the space–time world" (1962, p. 110). With an extraordinarily high degree of admiration for Einstein's achievements, Lanczos also said (pp. 105–6) that, in retrospect, Einstein certainly had very good excuses for not returning to cosmological considerations after 1917, except in brief remarks. Much more urgent were the problems of the nature of electricity and the quantum phenomena; cosmology should follow only as a natural extension of the unified view yet to come. And indeed, Lanczos's 1922 discussion of the puzzling properties of the De Sitter universe are driven by a very meticulous analysis of the determining character of geometry in relativistic concepts.

First of all, quite irrespective of all interpretations, Lanczos is struck by the very occurrence of a singularity in the De Sitter model. As he puts it, the model was designed to be to the differential equations of gravitation what Euclidean geometry is to $R_{\mu\nu} = 0$ (1922, p. 539). The whole problem with such an occurrence is thus connected with time; it is not something which can be derived from the spatial geometry alone. Lanczos refers to the transformation of coordinates that Klein proposed so as to deprive this model from matter at the singularity. He notes (p. 539, n. 3) that this transformation is subject to a restrictive condition, namely, $\omega^2 > \nu^2$. In order to cover the whole range of variability for the five coordinates, Lanczos suggests the following transformations:

$$x = \cos it \cos\phi \cos\psi \cos\chi$$
$$y = \cos it \cos\phi \cos\psi \sin\chi$$
$$z = \cos it \cos\phi \sin\psi \qquad (4.28)$$
$$iv = \sin it.$$

For the very first time, it appears that if the spatial coordinates are expressed so as to depend on time, no trace of singularity is to be found. Indeed, the resulting ds^2 can be written,

$$ds^2 = -dt^2 + \left(\frac{e^t + e^{-t}}{4}\right)^2 (d\phi^2 + \cos^2\phi d\psi^2 + \cos^2\phi \cos^2\psi d\chi), \quad (4.29)$$

in which $\cosh t = (e^t + e^{-t})/4$. Apart from clearly exhibiting the dependence of ϕ, ψ, χ on time, this transformation is nowhere singular. On our intuitive model (Fig. 4.2), it corresponds to $ds^2 = -R^2\cosh^2 t\, d\chi^2 + R^2 dt^2$, resulting from the change of variables $x = R\cos\chi \cosh t$, $u = R\sin\chi \cosh t$, and $v = R \sinh t$ in the equation (4.9) of the reduced model. This is the ds^2 of a spatially closed and homogeneous universe, with a radius of curvature that increases up till infinity, either in the positive or negative direction of time.

However, this development is just a side issue in Lanczos's paper. It is found in a rather brief footnote, which ends with a remark on the interesting possibility that very different physical interpretations are allowed by one and the same geometry. Each interpretation seems to derive rather directly from the choice of coordinates. Lanczos's main interest is geometrical indeed—he is not concerned with an attempt to find out the correct representation of a 'really' empty world. Rather, the core of his paper is a new parametrization of the De Sitter universe, by which it will be seen that matter cannot 'explain' or 'account for' the singularity. In this sense, Lanczos maintains that the singularity is an intrinsic feature of this universe. In his view, only *Einstein*'s model lends itself to a radius of curvature which may be 'really' variable with time; this is his main point, which he tried to prove soon afterwards in a 1924 paper.

How does this come about? The spatial coefficients of the cylindrical universe are independent of time, but the specification of the coordinate system, Lanczos now emphasizes, is essential, "for time does not represent by itself an invariant concept, it is meaningful only as proper time" (1924, p. 73). A system of coordinates in which all proper times are identical is called by Lanczos a *stationary* system, as distinct from the *static* systems of De Sitter and Schwarzschild's exterior metric. (On the different and confusing senses given by various authors throughout the 1920s, such as Friedmann and Robertson, to the terms static, stationary, and non-static, see J. North 1965, pp. 112ff.) Furthermore, since the concepts of differential geometry alone cannot characterize the global properties of some given configuration, the range of variability of the coordinates cannot be derived uniquely from the field equations either. On this basis, Lanczos claims that the time coordinate, as distinct from the space coordinates, is not necessarily concomitant with a range of variability from $-\infty + \infty$ (1924, p. 77). The time coordinate could well represent a *periodic* coordinate, a sort of angular coordinate that would be a function of some period of time. Such a periodicity offers the advantage of truly eliminating any kind of singularity at infinity, that is, *both* the boundary conditions at infinity of the field equations in their original form, and the singularity at finite distance but physically unattainable of the De Sitter type. In fact, the relevant and distinctive property of Einstein's universe is, Lanczos argues, that its period can be taken arbitrarily large. By contrast, the transformation of coordinates (4.28) which allows the De Sitter universe to be made variable in time involves a radius of curvature which is a function of *imaginary* time. In this sense, a *real* time in the De Sitter universe simply deprives it of the possibility of becoming periodic. This brings out a most remarkable view of the significance of a comparison between the two models. Originally, as we remember, De Sitter had begun by writing his metric with both space and time spherical; he changed to the hyperbolic form just to restore the physical meaning of time. According to Lanczos, the root of the problem with this universe is that this change is still insufficient to yield comparison with the possible senses of reality implemented by Einstein's universe. If Lanczos does not seem to perceive the implications of his non-static metric (4.29) in today's sense of the term, it is probably because he has in mind that only Einstein's universe is compatible with a 'real' periodicity.

With this type of argument borne in mind, let us return to the first article of 1922. In the equation of the pseudo-sphere $x^2 + y^2 + z^2 + u^2 - v^2 = 1$ (Λ is taken to be equal to 3), Lanczos introduces the following parametrization of the De Sitter *static* universe:

$$x = \cos\phi \cos\psi \cos\chi$$
$$y = \cos\phi \cos\psi \sin\chi$$
$$z = \cos\phi \sin\psi \qquad (4.30)$$
$$u = \sin\phi \cos it$$
$$iv = \sin\phi \sin it.$$

With these coordinates, the line element is

$$ds^2 = \cos^2\phi(d\psi^2 + \cos^2\psi d\chi^2) - \sin^2\phi dt^2 + d\phi^2. \qquad (4.31)$$

All points of the pseudo-sphere are equivalent, so that a singularity can arise only through the chosen coordinate system. The equatorial sphere is given here by $\phi = 0$, that is, the sphere $x^2 + y^2 + z^2 = 1$. According to Weyl, this singularity cannot be transformed away because the velocity of light is zero on that sphere. But does the interpretation of the singularity in terms of matter exhaust its nature? Lanczos seems to agree with Weyl that there is no other way of viewing matter as being already by itself some kind of singularity of the pre-existing field, but he does not conceive of this as a way of resolving the nature of the singularity occurring in the De Sitter model. Quite the contrary, one must still distinguish between apparent and intrinsic singularities, for the $g_{\mu\nu}$ describing matter as a true singularity remain everywhere finite and continuous, while a true singularity such as that of the De Sitter metric is a discontinuity (Lanczos 1922, p. 541). For the latter case, Lanczos speaks of the occurrence of an 'edge' in the otherwise regular metric structure. This is obtained by removing a thin portion of the equator of the sphere, situated between $\phi = \epsilon$ and $\phi = -\epsilon$. By rejoining the two half-spheres thus detached from one another, an 'edge' appears, and it is this sort of singularity which is said to be due to matter according to Weyl. In order to compute the density of this matter, Lanczos writes De Sitter's metric under the form that rules out all 'apparent' singularities, and in which $\phi = \pm\epsilon$ is substituted for ϕ. Long and tortuous calculations, which need not be reproduced here, lead him to the result, $R = 4/\epsilon$. Thus, if $\epsilon = 0$, the scalar of curvature R is infinite and the quantity of matter is infinite as well. When ϵ tends towards zero, the density tends towards infinity. This result contradicts Weyl, since the density does not reach a maximum when the thickness of the equatorial singularity is made to tend to zero. The density overtakes any limit, a result which Lanczos claims to be indisputable since it is expressed in terms of a scalar, independently of all reference systems (pp. 542–543).

Lanczos goes on to investigate the properties of the equator itself, i.e. the middle surface of the zone of matter. He discovers that the $g_{\mu\nu}$ do not undergo a discontinuous jump until the equator itself is reached; they remain continuous throughout the extended zone and break down at the equator only. Lanczos describes this fact as revealing an astonishing result, namely, the equator alone contributes to the matter; the extension of the zone does not contribute at all, for any increase of its density would be compensated by contraction of the zone. As a consequence, the singularity under consideration is really of a peculiar type. The above-mentioned construction of an edge does not portray it quite correctly, since in this construction the singularity stands out less as the piece removed is small. The true situation is just the inverse. Lanczos compares it with two congruent spheres lying on each other, with their poles as tangent point. By removing a thin cap around each one of the two poles, and then joining the two spheres again, one gets an ever greater edge as the removed portion is small. The singularity increases as ϵ becomes smaller, even though $\epsilon = 0$ is a perfectly regular point.

In his reply to Lanczos, Weyl (1923a) provides means to effectively derive the radius of the zone of matter. As he argues, it is not the case that the point $\epsilon = 0$ remains regular if, by a method of step-by-step approximation, one *begins* with an extended zone of matter and tries to infer its density. Then, it is found that ϵ cannot at all tend towards zero. Rather than further disputing the calculations, Lanczos's new reply returns to basic principles and discusses the very meaning of the supposed 'malleability' or extendibility of the zone of matter (1923, p .180). In fact, it is not consistent to argue that the zone may be extended at will (for instance, so as to cover the totality of the world): the zone must be taken to cover an infinitesimal portion around the equator, for otherwise the general assumption of incompressible fluid is found by Lanczos to be no longer applicable. Here too, and more importantly, this critique clears the ground for Lanczos's later development. For Lanczos conceived of the periodic radius of curvature as that which also allows understanding of motions directed towards evolutionary processes within the universe. In Einstein's universe, the variation of the radius of curvature could explain no less than the emergence and formation of nebulae from homogeneous and uniform matter, at least if the density is taken to be increasing in those phases when the radius is contracting (Lanczos 1924, p. 86). But such a process of condensation involves terms of pressure, so that in the last resort the only viable model, in which the metric and the matter distribution are everywhere regular, is simply this refit cylindrical universe. By contrast, the actual knowledge of the physics prevailing within the horizon of the De Sitter universe seems to have been finally regarded as impossible by Lanczos. A result of that kind was implicitly confirmed by Max von Laue and N. Sen (1924). These two authors began with the assumption that the De Sitter universe may be identified with the limiting case of the gravitational field created by uniform matter within a hollow sphere. Applying conditions of discontinuity for the solutions of the field equations with spherical symmetry, and taking the radius of curvature to be that of the De Sitter universe, it was possible to prove that the density of matter on the surface of the sphere remains undetermined and undeterminable. It is thus impossible to calculate the total mass of the mass-horizon in that universe.

More than simply testifying to the divergence from Weyl's viewpoint, Lanczos's radical attitude reveals the philosophical insecurity of general relativity with regard to its global implications. While Lanczos goes from the geometrical to the physical, Weyl deduces the geometrical properties from what he claims as the only possible physical basis of De Sitter's universe. At the other extreme, Klein believed that the purely geometrical transcription of allegedly physical properties posed a further challenge to "our ordinary way of thinking" (1918c, p. 423), in that Einstein's model is flatly contradicted. Of course, this remark could highlight the mystery, but did not plumb the depth of those intuitions about physical nature that underlie all natural science. In fact, the first genuine clarification of the physical character of cosmological considerations was to come from an unusual genius, who could combine expertise in mathematics and physics with practising astronomy at the highest level: it was Eddington who provided the needful synthesis in the first edition of his epoch-making textbook, *The Mathematical Theory of Relativity* (1923b).

6. Eddington's solution of 1923: its foundations and limits

(a) From physics to metaphysics and back

Eddington gave the debates we have been following their most lucid expression:

Is De Sitter's world really empty?. . .there is a singularity at $r = \sqrt{3/\Lambda}$ similar to the singularity at $r = 2m$ in the solution for a particle of matter. Must we not suppose that the former singularity also indicates matter—a 'mass-horizon' or ring of peripheral matter necessary in order to distend the empty region within? If so, it would seem that De Sitter's world cannot exist without large quantities of matter any more than Einstein's; he has merely swept the dust away into unobserved corners (1923b, p. 165).

As we have seen in Section 4a, the values $2m$ and $\sqrt{3/\Lambda}$ are, in first approximation, two roots of Eqn. (4.17): the first defines the Schwarzschild sphere which, as Eddington says, gives a particle "the appearance of impenetrability", while the second "may be described as the *horizon* of the world" (p. 101). Is this horizon just as impenetrable? To what extent can the analogy be carried on? Here we hit on Eddington's major problem; he will make a point of carefully distinguishing between the local and the global, which in his language is verging on a distinction between reality and appearance, and even between reality and illusion.

Eddington's work stands as both the end point of the Einstein-De Sitter controversy in its early form and the beginning of an entirely new approach to cosmological questions. There is a strong metaphysical commitment behind Eddington's strategy, and it is impossible to overlook this without impoverishing the significance of his scientific achievements. On the one hand, Eddington's contributions are well known for their impeccable pedagogical virtues for the scientist: as Dirac has emphasized, English science students "really had no chance to understand relativity properly until 1923, when Eddington published his book" (Dirac 1982, p. 82)—and the book can still be recommended to the contemporary relativist. But apart from its function as a popularization of relativity, the book was also, as Dirac goes on to say, "interspersed with a lot of philosophy". This philosophy grew from a more or less idealistic interpretation of the theory of relativity, strikingly at variance with the more pragmatic positions adopted by most of the physicists in Eddington's day.

Eddington's philosophy of science is not so far from that of Kant. The most sustained account of his position is contained in his much later book, *The Philosophy of Physical Science* (1939), where he argues that the physical world is not quite identical with the real world, for the former is only what the physicist can describe in terms of the operational means available. So the laws of the physical world have a subjective origin which is reflected in the way the physicist *selects* those aspects of reality amenable to his own way of working. As Eddington puts it: "By defining the physical world and the physical objects which constitute it as the theme of a specified body of knowledge, and not as things possessing a property of existence elusive of definition, we free the foundations of physics from suspicion of metaphysical contamination" (p. 3). In fact, Eddington's philosophy of science is a subtle equivocation between realism and idealism, for the properly physical knowledge was never thought of in this system as ruling out other forms of the 'real' (see J.W. Yolton 1960, pp. 88-108 and J. Merleau-Ponty 1965, pp. 49ff.).

Eddington wrote *The Mathematical Theory of Relativity* in 1922, almost as an afterthought to providing a mathematical appendix for the French, 1921 edition of *Space, Time, and Gravitation* (see Eddington 1921a). Commentators seem to agree on the way Eddington's ideas developed in the three years that passed between the publication of these two books: Eddington moved from an outright 'realism' about space-time (that is, space-time regarded as an absolute, an ontological property of the world uncovered by physical science) to the subtle notion of structure, in which a more abstract algebra supersedes this realism. I say supersedes, because the realism is not done away with completely: Eddington's fundamental problem remains the same from the outset of his career—how can an independent reality end up as an aspect of the form of our physical laws, however abstract this form may be? In the own terms used by Eddington, the fundamental problem is one of attacking "Nature from opposite ends" (1921b, p. 105), that is, the purely deductive and general process of developing a world-geometry should ultimately combine with the particulars identified by induction in experimental measurement. As to the laws of relativity physics in particular, Eddington finally shows (in *The Philosophy of Physical Science*) that they are universally valid since they can never be confuted: they simply reflect the manner in which the selective process of the real world can be effected; but of course, this selection is pressed upon us not only by ourselves but also by nature as an independent entity. Clearly, it is in the context of Eddington's fundamental problem about this interaction that we can begin to make sense of the reference (1923b, pp. 2-3) to mysterious "world-conditions", i.e. conditions the knowledge of which is attained only indirectly, through the medium of mathematical symbols. And in the rest of this section, we shall see how in 1923 Eddington's interest in the formal development of the known models of the universe echoes this highly original and independent philosophical journey; cosmology will appear as the ideal ground where all "metaphysical contamination" can be exorcized.

(b) The four themes of Eddington's research

In his early career Eddington's cosmological research comes across as the culmination of his struggle with the metaphysical foundations of natural science. He seems to have arrived at cosmology after having exhausted three other lines of investigation. In this subsection, I shall examine each of them in turn (taking J. Stachel 1986 as basis) and show how, in terms of cosmology, the four themes are deeply interwoven.

In the first place, there is the famous participation of Eddington in the 1919 eclipse expedition (see D. Moyer 1979). In fact, Eddington's interest in Einstein's theory of relativity and the eclipses had been kindled well before this expedition, and even prior to De Sitter's 1916 publications on the theory (which I have discussed in Chapter 2). There was a Brazilian expedition in 1912 under leadership of the American astronomer C.D. Perrine, and one of its aims was to test Einstein's 1911 prediction of a deflection in the apparent position of a star when its light passes in the vicinity of the gravitational field of the sun. The expedition was a fiasco, but Eddington seems to have first become acquainted with Einstein's work via Perrine (see for instance Eddington 1913a; at any rate his familiarity with both the special and the general theory is clear from his 1915 paper).

Eddington's first paper to be entirely devoted to the general theory dates from December 1916. In it, he favours Poincaré's conventionalism: that is, as he had described it a little earlier in his obituary for Poincaré, "space is neither Euclidean nor non-Euclidean...But a certain geometry may be more convenient than any other, because it enables the facts of nature to be expressed with greater simplicity" (1913b, p. 227). And he goes on to adapt this conventionalist thesis to the needs of general relativity. He understands the requirements on the immateriality of the world-coordinates as implying that "the space and time of physics are merely mental scaffolding" (1916, p. 328). In particular, his central claim is that, if the conventionalist view of the laws of geometry is to be maintained when gravitation is made part of these laws, then Einstein's law of gravitation cannot be used as an explanation. If we follow Einstein, Eddington says, then "although we do not seek a *cause* of gravitation in the properties of space, it may well happen that the *law* of gravitation is determined by these properties" (Eddington 1916, p. 329). In Eddington's eyes, this distinction between the explicable and the actual nature of the law was founded in the rejection of the earlier mechanistic model. In his first comprehensive exposition of the theory, the *Report on the Relativity Theory of Gravitation* written in 1918, he stated quite unambiguously that although, according to this theory, "a certain connection between the gravitational field and the measurement of space has been postulated,...this throws light rather on the nature of our measurements than on gravitation itself" (p. 91). The consequence is that the fundamental aims of physical science are like those of pure mathematical theory. All we have to do, Eddington concludes in his *Report*, is to "accept some mathematical expression as an axiomatic property which cannot be further analyzed" (p. 91). If the law of gravitation in its original, Einsteinian form is to constitute a true explanation, then it should follow this model of pure mathematics. This quickly led Eddington to develop two simultaneous lines of research, one on the relevance of dimensionless numbers in physical theory and the other on the possibility of unifying gravitation and electromagnetism with concepts borrowed from atomic physics.

The connection between the two was made immediately clear in the Prologue of *Space, Time, and Gravitation*. For if space-time is not absolute and independent of measurements according to the new theory, "is there some absolute quantity in nature that we try to determine when we measure length? When we try to determine the number of molecules in a given piece of matter, we have to use indirect methods...but no one doubts that there is a definite number of molecules". Therefore, "counting appears to be an absolute operation" (1920a, pp. 7-8). In an article written in the same year, Eddington put it pithily enough when he said that this operation reflected the search for explanations: "Are there...no genuine laws of the external world? Is the universe built from elements which are purely chaotic? It can scarcely be doubted that our answer must be negative. There *are* laws in the external world, and of these one of the most important (perhaps the only law) is a law of atomicity" (1920b, p. 156). The determining influence of the human mind on the description of nature loses its authority precisely in respect of quanta, for "the laws of quanta do indeed differentiate the actual world from other worlds possible to the mind" (1920a, p. 200). Thus, by associating the constant of gravitation with two other fundamental constants of nature, the velocity of light and the quantum, Eddington was able to discern a fundamental unit of length which he thought must play some fundamental part in any complete interpretation of gravitation.

Eddington's views on the process of explanation had many implications on *what* is and must be explained, and these were summarized in the preface to the second edition of the *Report* (1920c, p. xi) where he wrote: "Matter does not cause the curvature of space–time; it is the curvature". There is a fundamental unity between gravitation and geometry, in the sense that "we need not regard matter as a foreign entity causing a disturbance in the gravitational field; the disturbance *is* matter" (1920a, p. 199). So that what we can ultimately accept as an explanation in physical science takes the form of identities, that is, mathematical equations. In the context of Minkowski's programme, this means that symmetry in the largest sense does not simply dictate causality as if a mere inversion had occurred. For if curvature is now seen as the 'cause' of those phenomena that we interpret as features of gravitation, we may ask where the curvature itself comes from. This must in the final analysis be 'caused' by whatever is responsible for the occurrence of a spatio-temporal interval in the universe (understanding spatio-temporal in the metric sense). What must be explained is not cosmic order as springing from some underlying cause, but rather this order itself 'causing' all that may be observed (1923a, p. 314). And by taking the field equations without matter but with the cosmological constant, we shall at least have (as a starting-point) the representation of this order:

The interpretation of the equations $R_{\mu\nu} = \Lambda g_{\mu\nu}$ is that space–time has a homogeneity and isotropy as regards a certain characteristic. The order, which we are seeking to explain, arises because actual space–time is not the most general type of metrical continuum imaginable by a mathematician, but is limited in this way (1923a, p. 322).

So a constant like Λ seems to provide the needed key. Gravitation, if it is to accord with the proposed scheme, must be thought of in the terms first disclosed by Weyl. Indeed, it was primarily in the course of extending Weyl's original proposal for unification that Eddington came to feel that he had found an *explanation* of gravitation, rather than the mere *description* given in Einstein's 1915 theory.

For explanation, in Eddington's sense, goes hand in hand with unified physical theory. As he put it, "the geometry of actual space and time" still does not get off the ground of *pure* geometry and its mental constructions; the only way of bringing it into touch with physics is by developing "the geometry of the world-structure, which is the common basis of space and time and things" (1921b, p. 121). Now, Weyl's theory of the system which sets up a separate unit of length at every point of space and time (the gauge-system) lies precisely at the junction which makes pure deduction open itself to external reality. For without a *natural gauge* (or metric), we would never actually compare lengths as we do, for instance, when we interpret the behaviour of atoms on the sun by comparison with those on the earth. Eddington wanted to develop Weyl's theory by starting off from "the minimum degree of comparability which permits of any differentiation of structure", and this is the "comparability of proximate relations" (p. 121).

Weyl's theory took advantage of a mathematical property that was first uncovered by Levi-Civita in 1915, the logical independence of metrical and affine structures on a space-time manifold; the only relationship between the two is the unique symmetric affine connection associated with an otherwise specified metric. In Weyl's theory, the electromagnetic potentials were identified with the additional, geometrical element (a vector field) needed to specify the affine structure of space-time. Eddington, in developing Weyl's proposed unification, traced a quite remarkable convergence between the purely mathematical cogency of the theory and the conventionalist philosophy of geometry. At the outset he noticed that the curvature (or Riemann) tensor could be formed without any metric, starting from the affine connection alone. Now, while the affine connection is symmetric, the curvature tensor generally is not. Which is what led Eddington to think of the connection as the minimum requirement for any possible physics. The connection allows comparison between quantities at neighbouring points in space-time: "For if there were no comparability of relations, even the most closely adjacent, the continuum would be divested of even the rudiments of structure and nothing in nature could resemble anything else" (1921b, p. 106). Following Weyl, Eddington suggested that comparison of lengths was provided by a gauge, but insisted that a natural gauge could be somehow already contained in the (affine) world geometry. In a letter to Weyl of 10th July 1921, he explained his reasons for preferring to start from the affine connection: this expresses the inevitable character of the equation $R_{\mu\nu} = \Lambda g_{\mu\nu}$, in the sense that "symmetry must be relative to something else; and the only thing it can be symmetrical to is the shape of the world at that point, that is to say, its radius in any direction must be a particular function of the radius of space in that direction". Eddington then went on to state the crucial conclusion: "The equality of the radii of the curvature of the world in different directions and at different places is inherent in our definition of equality of length in different directions and in different places". In other words, a quantity like the cosmological constant is simply an expression of the proportionality between the symmetric parts of the curvature tensor and the metric tensor.

Clearly enough, it is precisely this conception of "the world as self-gauging" (1921b, p. 111) which gives epistemological analysis its implications at the level of cosmology. But before turning our attention to this aspect of Eddington's thought, it is worth emphasizing that the mathematical height of physics attained at that point was a source of contention between Einstein and Eddington. As Einstein was to explain, for Eddington "the metric was to appear as a deduction from the theory" (Einstein 1923, p. 448). But this deduction, however elegant and mathematically seductive, still fails to explain the structure of electrons. In short, Eddington's work was "beautiful but physically meaningless" (Einstein to Weyl, 5 September 1921).

But it is Eddington's growing interest in the promises of cosmology throughout these years that most strengthened his convictions. While Eddington insisted to Einstein that his work on unification with its large-scale implications "is an *extension*, not an *emendation* of yours" (letter of 12 June 1921), he could not accept Einstein's commitment to Machian arguments. So much so that Eddington's very first reception of the cosmological constant (1918, pp. 90–91), which soon proved crucial to the proposed unification, was somewhat cool because it was originally associated with Mach's principle. In fact, with the cosmological models themselves, Eddington's first reaction was generally skeptical.

Eddington's letter to De Sitter on 16th August 1917 is a reply to De Sitter's second paper for the *Monthly Notices* which he had just received. He immediately levels an objection at the world-matter of model A: "That is simply the ether coming back again. . .an idle substance whose only function seems to be just to be there". It is impossible to think consistently of this new medium. Here is Eddington summing up Einstein: "Only mind that you don't go sticking light-waves and things into it this time, or you will be trying to find our motion relative to it, and landing us in the same difficulty as before". He finds model B much less open to objection because it unveils "some distinctive property of space when we use 'natural measure' which does not appear when we use any other kind of measure". Could this lead to a new interpretation of the nature of rotation and space–time structure? Eddington asks De Sitter if, according to his model, it is true that "if we use axes in absolute rotation, a fluid globe at 'rest' takes a spheroidal form; is this spheroid a sort of reflection of the shape of space referred to the same axes?". This is how Eddington first touched on the question of a relationship between the local and the global in terms of a given cosmological model. But when he sent De Sitter a copy of the first edition of his *Report*, he felt little confidence in his understanding of Einstein's theory of gravitation and wrote that "it is a subject in which it is very easy to go astray". In the *Report* itself, he still said about the two models that they are "independent speculations, arising out of, but not required by, the theory [of general relativity] hitherto described" (1918, p. 84). It was only in the preface to the second edition of his *Report* that he ventured to argue that Weyl's unified theory makes cosmology much more natural (p. xi). As he wrote to Einstein on 1 December 1919, "Weyl's work. . .removes some of my prejudices aginst your 'cosmological views' of space-curvature. . .Weyl's mathematics. . .seems to lead almost inevitably to your cosmological terms".

The argument of the *Report* is that cosmology is the theory of the inertial frame. It is an attempt to come to terms with the daunting fact that "the rotation of the earth detects something of the nature of a fundamental frame of reference" (p. 83). A privileged frame manifests itself by the occurrence of pseudo-Euclidean $g_{\mu\nu}$ at infinity, which naturally leads Eddington to ask where the ascription of the inertial frame comes from. It is the great merit of Einstein's insight, he argues, to reveal that such a question has physical significance only if we say that the $g_{\mu\nu}$ tend to zero at infinity, for this allows us to assert that "the property of the inertial frame arises from conditions within a finite distance" (p. 84). And an elegant way of doing that is to abolish boundary conditions at infinity altogether. But in Einstein's hypothesis, the old differentiation between space and time is restored and this, says Eddington, "throws away the substance for the shadow" (p. 87). Moreover, "the solution involves a very artificial adjustment", which is the direct relationship of the constant Λ to the total amount of matter in the universe. Model B is to be preferred despite its singularity which Eddington, as early as this *Report* of 1918, understands in terms of complete relativity. That is, the singularity means that "at any fixed point ds is zero however large dt may be" (p. 88), but this is relative to the viewpoint of the observer at the origin of coordinates for "all parts of this spherical continuum are interchangeable". In the De Sitter model, the inertial frame is a property of the modified law of gravitation, not of the matter distribution, since the model involves "no assumption of the existence of vast quantities of matter not yet recognized" (p. 88). As for the redshift, Eddington is of course aware that only model B is compatible with it. He describes it as a metric effect, and notes how it "would in practice be attributed to a great velocity of recession" (p. 89). The result is that the systematic radial velocity is spurious.

(c) The challenge of a compromise: 1. Time and becoming

In the following years, Eddington went further and further in teasing out what looks like an innocuous confusion between the practical and the theoretical aspects of the redshift. The argument of *The Mathematical Theory of Relativity* is based on the fact that this identification does not follow as a matter of course; and on the basis of such careful reconsideration, the other themes of cosmological interest—space-time structure, the status of world matter, the interpretation of Λ—all appear to be articulated into one integrated scheme.

The argument arises from a reflection on the relationship between the two cosmological models then known. Eddington thought the De Sitter model was the natural form of any empty space-time, while Einstein's was "a world containing as much matter as it can hold", and therefore "it would seem natural to regard De Sitter's and Einstein's forms as two limiting cases, the circumstances of the actual world being intermediate between them" (Eddington 1923b, p. 160). The true model would then simply depend on the amount of matter actually present in the universe. This in fact, was a view that Eddington himself had entertained for some time, when the first systematic presentation of his generalization of Weyl's theory came out (1921b, p. 112); that is, he had identified the equation $R_{\mu\nu} = \Lambda g_{\mu\nu}$ as the gauging-equation for empty space and argued that the difference between Einstein's and De Sitter's universes could not be theoretical; as Weyl himself had thought, it depended on the total amount of matter. But Eddington highlights the central part of his new argument by saying that "this compromise has been strongly challenged, as we shall see".

Of course, in retrospect, we can see that it was not just the explicit reasons he gave which blocked Eddington's path to such compromise. A powerful implicit consideration is that as thinker Eddinton was critical of any view on the nature of time which (as it was later recognized) could make the compromise feasible. Before turning to the explicit part of this challenge ofcompromise, I shall examine the preconceptions behind Eddington's philosophy of time in the decade of interest to us.

In his 1927 Gifford Lectures, which were the basis of his book *The Nature of the Physical World*, Eddington says in a clear-headed way that the very idea of time as an intrinsic aspect of the world is something which cannot be touched by any physical theory. What we understand by 'becoming' is quite subjective: "there is no 'becoming' in the external world which lies passively spread out in the time-dimension as Minkowski pictured it" (1928, p. 92). In this sense, the theory of relativity captures this subjective dimension of time. A formal proof of this is provided by the Minkowski ds^2, in which the time interval is distinguished from the three spatial intervals by the fact that it is multiplied by the imaginary number i, and we "must not think of [this number] as occurring with some mystical significance in the external world" (1923b, p. 12). Therefore, the only 'physical' way of understanding time on a cosmic scale is just that: "There is no radius of curvature in a real timelike direction" (p. 154). From the physical viewpoint, this impossibility is concomitant with the eternal existence of the world; one implication at the atomic level is that an electron

> would not know how long it ought to exist, unless there existed a length in time for it to measure itself against. But there is no radius of curvature in a timelike direction; so that this electron does *not* know how long it ought to exist. Therefore it just goes on existing indefinitely (p. 155).

The world is self-gauging spatially only, not temporally. This is not to mean, by the same token, that Eddington rejects any religious idea of creation. His strong religious background was that of a Quaker, as he explained at length in his 1929 Swarthmore Lectures (1929), where he suggests that acts of faith and rational knowledge are not necessarily connected. Not only is religious experience not irrational but it is also compatible with the scientific search for truth.

The language of physical science is that of symbols, so much so that "the external world of physics has thus become a world of shadows" (1928, p. xvi); the relation of these shadows to the external world is registered by our recognition of time, which indeed is dual: one is physical time, the other is the time of consciousness (pp. 99ff.). The correct distinction imposed by this duality must be kept uppermost in mind when we try to interpret physical laws. Thus, the second law of thermodynamics seems to predict a well-defined direction of time at some global scale of nature. But this, so Eddington argues, is produced by the physico-mathematical quantity involved in the law (the entropy-gradient) and would be wrongly interpreted as indicating some intrinsic becoming in the universe:

> Whilst the physicist would generally say that the matter of this familiar table is *really* a curvature of space, and its colour is *really* electromagnetic wavelength, I do not think he would say that the familiar moving on of time is *really* an entropy-gradient . . . there is a distinct difference in our attitude towards the last parallelism. Having convinced ourselves that the two things are connected, we must conclude that there is something as yet ungrasped behind the notion of entropy. . . In short we strive to see that entropy-gradient may *really* be the moving on of time (instead of *vice versa*) (p. 95).

As a result, Eddington's main problem was that 'the scientific world is. . .a shadow-world. . .Just how much do we expect it to shadow?" (p. 109), and this turned into a fundamental dilemma in the face of a physicist's view of creation. For the laws of thermodynamics impose more and more organization as we travel back into the past. However, Eddington wrote, "as a scientist I simply do not believe that the present order of things started off with a bang; unscientifically I feel equally unwilling to accept this implied discontinuity in the divine nature. But I can make no suggestion to evade the deadlock" (p. 85).

The search for unification and the attempt to found cosmological models on this basis gave Eddington many opportunities to put the merits of this philosophy of time to the test. Thus, in Weyl's theory, the result of measurements depends on the electric and magnetic forces which have acted not only on the scales but also on the clocks since they were last compared with some standards. The non-integrability of temporal intervals would, as Weyl immediately pointed out, have important observable consequences. For according to Einstein's theory, the time kept by an atom on the sun can be compared immediately (at least in theory) with that kept by a similar atom on the earth. In its original form this is no longer true of Weyl's theory: we must trace back the history of the two atoms step by step till they come together in some primitive medium, and we must allow for the different electromagnetic forces which have acted on them. In other words, the atom now at rest on the Sun has had an experience which differs in one crucial respect: its great velocity of fall has been destroyed by encounters with other atoms, and the electromagnetic forces of these encounters should make a systematic contribution to the spectral lines emitted by the Sun, so that the average time-keeping of solar atoms may differ systematically from that of terrestrial atoms. (Obviously, Einstein's explanation of the motion of Mercury's perihelion would not be affected by these modifications Weyl proposed, since there seems to be no intervention of the electromagnetic field in that case.) Failure to detect any such effect prompted Weyl to introduce a new theoretical apparatus (1922, p. 308). (On the astronomical evidence, see J. Earman and C. Glymour 1980.) The redshifts of the light emitted by the Sun do depend on its gravitational field, not on the past history of the Sun and the Earth. So, Weyl now argued, a distinction must be made between two modes of determining a quantity in nature—persistence and adjustment. The dimension of material objects is determined by the fact that they adjust themselves to the field in which they are embedded, not by the persistence of anything in their past history. Eddington quickly seized the opportunity to strenghten his own development of Weyl's theory. He saw the possibility of equating the embedding field with the *spatial* radius of curvature of the whole world, so that the adjustment of, say, the size of an electron would occur in proportion to this radius (1923b, p. 208). In this way, the very idea of an oriented history of the universe in terms of relativistic notions was discarded as soon as it emerged.

This had quite an effect on Eddington's perception of the nature of time in the cosmological models. In the early *Report*, he began by raising difficulties with the cosmic time function in Einstein's model: "time is not symmetrical with respect to the other coordinates; in general matter moves with small velocity, so that the different components of the energy tensor $T_{\mu\nu}$ are not of the same order of magnitude" (1918, p. 84). In *Space, Time, and Gravita-*

tion, a distinction began to emerge between the real which falls within the province of relativity, and the real which transcends it, so that Einstein's model might represent something of the sort that cannot be reached by any known experimental route. This would involve some super-observer:

> It need not perturb us if the conception of absolute time turns up in a new form in a theory of phenomena on a cosmical scale, as to which no experimental knowledge is yet available. Just as each limited observer has his own particular separation of space and time, so a being coextensive with the world might well have a special separation of space and time natural to him. It is the time for this being that is here dignified by the title 'absolute' (1920a, p. 163).

But in *The Mathematical Theory of Relativity*, Eddington was far more radical just as the distinction between symbols and reality was far sharper. He rejected the construction of time-partitions throughout the four-dimensional world, because he thought that by doing so, physics was simply borrowing the idea of world-wide instants from the mere time succession of our consciousness. That is why

> the original demand for a *world-wide* time arose through a mistake. We should probably have had to invent universal time-partitions in any case in order to obtain a complete mesh-system; but it might have saved confusion if we had arrived at it as a deliberate invention instead of an inherited misconception. . . It is important for us to discover the exact properties of physical time; but those properties were put into it by the astronomers who invented it (1923b, pp. 24–5).

This classical conception of universal time and the time function of Einstein's cylindrical universe (p. 167) have something in common—they each exemplify the purely inventive part of physical science; they are necessary for science as a construct but they have no counterpart in reality.

(d) The challenge of a compromise: 2. Astronomical evidence versus theory

Eddington's difficulties in finding a compromise between the two models was also strongly influenced by an apparent conflict between astronomical evidence and theory. On the one hand, Einstein's model seemed to be in keeping with Eddington's views on the nature of any possible mathematical basis for physics; this tallied well with the inventive aspect just referred to. But on the other hand, only the De Sitter model complied with the new facts about extra-galactic nebulae.

In his final hesitation between the two models, Eddington says that only Einstein's universe is compatible with the occurrence of a remarkable coincidence between very large pure numbers: the ratio of the radius of an electron to its gravitational mass and the total number of elementary particles in the universe (1923b, p. 167). It is a theoretical advantage, for such numbers were assumed by Eddington to be the key to the determination of the (absolute) constants of Nature. That is, a number such as the total number of particles should have a certain relationship to the quantity of matter in the universe—this quantity being fixed in the model in accordance with a definite law. Therefore, while he was initially very critical of the large amount of matter in the cylindrical universe, identifying it with the old ether of nineteenth-century physics, Eddington now became much more wary and wrote of this as just "ordinary stellar matter" and continued that "the formulae are only intended to give an approximation to the general shape [of space]" (p. 168). He did conclude on a note of hope, saying near the end of his book that the law yielding the quantity of matter "at present seems mysterious, but it is perhaps not out of keeping with natural anticipations of future developments of the theory" (p. 240).

If this hope of finding a physical account for the occurrence of large pure numbers had not motivated Eddington, the astronomical virtues of the De Sitter model would no doubt have prevailed. In fact, Eddington was the first to appreciate both the theoretical and the practical incidence of the newly detected redshifts of remote spiral nebulae. It all began with a letter he wrote to the astronomer Harlow Shapley, on 30 December 1918:

De Sitter's hypothesis does not attract me very much, but he predicted this (spurious) systematic recession before it was discovered definitely; and if, as I gather, the more distant spirals show a greater recession, that is a further point in its favour (quoted in R.W. Smith 1982, p. 174).

In the following years, the measurement of ever greater recession velocities in proportion to the estimated distances was a major astronomical breakthrough. In his 1923 book, Eddington published the most comprehensive table of radial velocities of spiral nebulae so far achieved. The table had been prepared by V.M. Slipher at the Lowell Observatory and was released there before it appeared in specialized astronomical journals. It comprised no less than 41 measurements, 36 of which did indicate positive (receding) velocities while only 5 showed a velocity of approach (violetshift) (1923b, p. 162). Despite the huge preponderance of receding velocities, Eddington was extremely cautious in his conclusions; for instance, the great Andromeda nebula, one of the closest, was found to be approaching at rather high velocity. Basically, what Eddington retained from this was the need for a reconsideration of the relationship between the metric and Doppler (kinematic) redshifts. Not that he was led in any way to formulate clearly a velocity/distance relation (such a relation was, in fact, only implicit in his calculations): his problem was rather the theoretical interpretation of motion. Other astronomers and theoreticians were already working actively at that time on a velocity/distance relation, and Weyl integrated it in his own work which followed soon after the first edition of Eddington's book (and which I shall discuss in the next chapter).

As Eddington was later to confess (1933, p. 15), the phenomenon of receding motion for the nebulae did not appear to him, already at that early stage, as something that could be looked at independently of theoretical considerations. The very structure and the dimension of atoms and electrons had to be seen, he thought, in the physical determinants of De Sitter's model. No doubt the thought of a connection between the submolecular and the cosmic scales (in terms of the fundamental concepts of physics) would lead Eddington to find out the physical significance of this paradoxical model.

The motivation for such a connection is best expounded in a non-mathematical way in *Space, Time, and Gravitation* (1920a, pp. 30-31, 80-81, and 182-183). Eddington is here moving away from the reservations expressed by Einstein in his 1920/1921 conferences as to generalization of field concepts to all dimensions (see Chapter 3). How can we form a world picture which would be independent of any particular observer, i.e. an 'absolute' picture? Three factors have an influence on the way the picture is constructed: position, motion, and gauge of magnitude of objects. The problem can thus be formulated as follows: "Can we form a picture of the world which shall be a synthesis of what is seen by observers in all sorts of positions, having all sorts of velocities, and all sorts of size?" The synthesis in terms of position is almost immediate, since it is a matter of intuition that one and the same thing can be perceived from different positions; the integration of all possible motions is achieved by the theory of general relativity. Similarly, "space is not a lot of points close together; it is a lot of distances interlocked" (1923b, p. 10) so that an infinitesimal interval is absolute in so far as it *contains* position. The synthesis in terms of dimension is the final step, which is supposed to actually seal our 'absolute' understanding of things. It must show how the interval itself is *contained* in something even more basic (1920a, p. 33 and pp. 167ff.).

This final step implies an original combination of Weyl's gauge-invariance theory and De Sitter's model of the universe, for despite their different methods they both arrive at the same determination of the absolute. Eddington sees Weyl's theory as providing a natural way of conceiving space curvature: "no part of space-time is flat, even in the absence of ordinary matter, for that would mean infinite radius of curvature, and there would be no natural gauge" (p. 177). Now, in the De Sitter model, the law of gravitation remains one of material structure if it is taken as showing "what dimensions a specified collection of molecules must take up in order to adjust itself to equilibrium with surrounding conditions of the world" (1923b, p. 153). This adjustment results from an epistemological analysis of the finite space:

all spatial measurements being made from within the universe, the universe cannot have an arbitrary radius of curvature, so that the very possibility of measurement already implies the existence of a homogeneous and isotropic length; a purely dimensional analysis (pp. 153-5) leads Eddington to the remarkable conclusion that this length is quite simply identical to the radius of curvature of the De Sitter metric, that is, the quadratic of curvature in every direction and at every point in empty space is $ds^2 = 3/\Lambda$.

Clearly, then, it is via this interpretation of the cosmological constant that the De Sitter model can be regarded as a truly physical one. Eddington draws two conclusions from this fact.

Firstly, the over-arching connection between the local and the global which would transcend any difference between the two models is achieved by means of the concept of *symmetry* precisely to the extent that it supersedes that of *causality*. Indeed, one condition for the very existence of a particle of matter is that it has symmetrical properties (pp. 125-6). But the identification of length of the natural gauge with radius of curvature of the De Sitter universe provides an implicit justification for this symmetry, since now "we have introduced a new and far-reaching principle into the relativity theory, viz., that symmetry itself can only be relative; and the particle, which so far as mechanics is concerned is to be identified with its gravitational field, is the standard of symmetry" (p. 155; see also 1923a, p. 322). One of the powerful arguments in favour of *Einstein's* model is thereby re-interpreted: if there is no directional variation of inertia in the universe, it is not because distant masses would be distributed uniformly and cause inertia to be isotropic, but because of these properties of symmetry that, in fact, are most naturally accounted for by the De Sitter model.

The other and very different implication of the physical character of the cosmological constant relates to the redshift. Given the physicality of the constant, discussed above on epistemological grounds, as an integral part of the metric, how is it implicated in the account of the phenomenon? Or, more precisely, if the metrical properties of the De Sitter model are compatible with the redshift, would it be possible to account for its occurrence by means of the physical 'cause' constitutive of all 'natural' intervals, the constant itself? What is at stake here is the verification, the observable consequences of the De Sitter metric which is now taken to reflect the implicit presuppositions about the basic homogeneity of the universe.

Before grappling with this central issue, Eddington derives the equation of motion for material particles and light rays in the De Sitter universe in a very simple and elegant way. He finds that all properties of motion depend on the cosmological constant alone (1923b, p. 161; J. North 1965, p. 96). The equation of motion in the empty space-time is

$$\frac{d^2r}{ds^2} = \frac{1}{3}\Lambda r. \tag{4.32}$$

This is obtained quite straightforwardly from Schwarzschild's exterior metric written with the cosmological constant. For slow orbits, the equation of motion in the gravitational field of a particle yields, instead of (4.17), the Newtonian orbit subject to a central potential:

$$\phi = -\frac{m}{r} - \frac{1}{6}\Lambda r^2. \tag{4.33}$$

The term in Λ corresponds to a central force of repulsion, whose magnitude is $(1/3)\Lambda r$.

Equation (4.32) has these very clear implications, which make the De Sitter universe quite distinct from the Einstein one:

... a particle at rest will not remain at rest unless it is at the origin; but will be repelled from the origin with an acceleration increasing with the distance. A number of particles initially at rest will tend to scatter, unless their mutual gravitation is sufficient to overcome this tendency (p. 161).

The derivation is as simple as its implications are inescapable, yet the clarity of Eddington's statement is wholly new; in De Sitter's calculations of 1917, receding motions were not favoured above approaching motions. True, Eddington notes the strange fact that strictly speaking no approaching motion is compatible with (4.32), yet what can be observed of the radial velocities of spiral nebulae is not so conclusive as to rule out the possibility of such motion altogether. Characteristically, however, Eddington is more interested in the conceptual progress:

It is sometimes urged against De Sitter's world that it becomes non-statical as soon as any matter is inserted in it. But this property is perhaps rather in favour of De Sitter's theory than against it (p. 161).

By using the expression "non-statical", Eddington means something which does not depart from a classical conception—motion in an otherwise static space-time. He is still very far from a non-statical metric, yet Lemaître in his 1925 paper (p. 41) will cite this sentence of Eddington's as an argument in favour of his own theory, the non-statical formulation of the De Sitter metric. Here, then, we find a critical moment in the whole story, and its issue hinges on Eddington's way of combining motion and redshift under the aegis of the cosmological constant as a physical agent.

(e) 'Reality' and 'appearance' of the redshift

The De Sitter model would now appear to offer two competing explanations for the redshift: the slowing down of atomic vibrations, and the kinematic tendency to scatter. Eddington warns against a confusion between the two:

"the slowing down of atomic vibrations...would be erroneously interpreted as a motion of recession" (1923b, p. 161). He then proceeds to a rather simple calculation establishing the numerical value of the redshift due to the Doppler effect (p. 164). In the De Sitter metric, $ds^2 = -R^2d\chi^2 - R^2\sin^2\chi\,(d\theta^2 + \sin^2\theta d\phi^2) + R^2\cos^2\chi dt^2$, $R\chi$ will be the time taken by the light emitted from an object to reach the origin. From Eqn (4.32) which gives the acceleration, we can find the 'velocity' ($t\,\dfrac{d^2r}{ds^2}$) which is equal to $(1/3)\Lambda rt$. If the acceleration $(1/3)\Lambda r$ is continued for the time $R\chi$, the change of velocity over that period will be $(1/3)\Lambda rR\chi$, that is, in natural measure, $(1/3)\Lambda r^2$. Thus, this velocity results in a Doppler effect whose numerical value is precisely what is anticipated in terms of the metric itself (Eqn 4.18).

To Eddington, this equality has some crucial implications. In the first place, it seems to allow him to resolve a paradox: in virtue of the metrical properties of the shift, two observers A and B, separated by a large distance, will not account in the same way for a difference in the frequency of light. In terms of the metric, the difference should be something absolute: if A observes a displacement towards the red of the vibrations of an atom at B, B will observe a displacement towards the violet of the vibrations of an atom at A. This is absurd, Eddington says, because B could equally well be chosen as origin and then the light from A would exhibit a redshift.

As far as we can trace it back, the paradox seems to have been formulated by Weyl in 1921. In a paper which contains his very first findings about the impact of the redshift measurements, Weyl argued that if the horizon of the De Sitter universe was actually made up of matter, then the redshift was to be ascribed neither to the metric nor to some kinematic effect, but rather to the gravitational effect of the matter amassed at the horizon. And instead of favouring the De Sitter model on the basis of this fact of observation, Weyl sees in the redshift a way of consolidating his earlier view of matter in *Einstein*'s universe as a sort of extended horizon of the De Sitter model. He believes that the linkage of the redshifts with the existence of separate stellar systems comparable to the Milky Way tends to undermine the validity of Einstein's assumption that matter was uniformly distributed. If the world were spatially closed in keeping with the cylindrical model, then the redshifts would reveal that matter, instead of being distributed uniformly, "moves freely in the form of single islands like the Milky Way" (Weyl 1921b, p. 478). In other words, the redshift could be interpreted in the cylindrical universe, just as it is in the De Sitter model, in terms of gravitational effect produced by accumulated matter.

And indeed, the interpretation of redshifts in terms of metric effect in the De Sitter model is complicated by a difficulty which Weyl takes to be insuperable. Combining Eqns (4.1), (4.2), and (4.4), Weyl writes down the mathematical expression of the redshift of a star at S measured from a centre O;

the redshift increases as r increases. But a consistent interpretation, as Weyl emphasizes, is subject to the essential condition that *"the light emitted by the star has in the whole space the same frequency as that measured in the 'static' time t"*. By static time, Weyl means here that in order for the measurements to be consistent, there must be a contemporary space between O and S. Not surprisingly, then, the paradox is that an observer at O will detect a violetshift coming from a star at S. Weyl finds that this asymmetry rests upon the fact that non-singular solutions of Maxwell's equations are possible only within one half of the hyperboloid, in which $x_4^2 - x_5^2 \geq 0$, and he adds that "this is of course correct only if the world consists of this portion alone, i.e. if it is completely covered when static coordinates are used". As long as each observer remains within his 'static' reference frame, the paradox stated by Weyl does not occur; thus, if the centre O is transported to the star at S, there will be no ambiguity in O's observations of the neighbouring stars (including what was formerly point O)—they will all have their light shifted towards the red. But if a given observer seeks to *extend* his inertial reference system (at rest) and to deduce (or "anticipate", as Eddington would say) the measurements made by another observer in this same reference system, then the paradox does come into play.

This tentative integration of the redshift of extra-galactic nebulae into the cosmological theory which was being developed at the same time in the fourth edition of *Raum, Zeit, Materie* is certainly a decisive development in Weyl's thought. The construction of the cosmic time function, which was largely established on an a priori basis, is now confronted with the question of measurements effectively made by observers. The idea of contemporary space linking any two observers in order for a redshift measurement to be consistent, no matter how far apart they may be, implies something more than identical separation of space and time at any given point; there is a need to think of a time function linking any two points which may be separated by a large distance. Although the implication is perceived clearly enough by Weyl in his paper, no solution in terms of the metric is proposed because the only possible contemporary space is identified with a static frame of reference. Instead of reflecting on the metric, Weyl reiterates his view of the horizon as a mass-horizon and accounts for the redshift in the *two* models as a gravitational effect.

Eddington's way out is that there is a fallacy in the argument that is responsible for the asymmetry:

The fallacy lies in ignoring what has happened during the long time of propagation from A to B or B to A; during this time, the two observers have ceased to be in relative rest, so that compensating Doppler effects are superposed (1923b, p. 164).

At the very least this solution seems to raise a lot more questions than it solves. Eddington supposes that the (relative) Doppler effect is superposed to the (absolute) metric effect in such a way that the former compensates for the latter. It is a quite puzzling solution: since a particular phenomenon may have two explanations, and as the effects anticipated in each case may be identical, Eddington argues that the explanations themselves must be identical. At the very best, what we have here is a sort of 'practical' solution and a 'practical' identity; but the asymmetry is still there as a matter of principle.

Again, of course, the difficulty with the solution hinges on the static coordinates. But, quite apart from that, the equality between the Doppler and the metric effects is, in consequence, such that

> we may thus regard the redshift for distant objects at rest as *an anticipation* of this motion of recession which will have been attained before we receive the light.

In effect this corrects Eddington's earlier position, according to which the radial velocities were simply spurious:

> If De Sitter's interpretation of the redshift in the spiral nebulae is correct, we need not regard the deduced large motions of recession as entirely fallacious; it is true that the nebulae had not these motions when they emitted the light which is now examined, but they have acquired them by now.

An interpretation like this clearly pushes the constraints imposed by the static coordinates to their limits. A process of anticipation is effective each time an observation is made, so that all observers are somehow led to 'fix' the accelerating course of the universe. The actual observation of the universe in a dynamic state is made impossible by the fact that this state is, always and on every occasion, anticipated. So from the position of observability, the universe remains at rest.

An implicit distinction begins to come to light here, between the universe as it may be 'observed' and the universe as it is 'in itself' (in its causal realm as it were). Is not such a distinction strikingly reminiscent of the epistemological point we have already looked at about the necessarily dual nature of physical knowledge, observer-dependent and observer-independent? Casting an eye back on his work of 1923, Eddington said some ten years later, when the controversy about non-static coordinates had settled, that the question of whether the receding motion was a "true" or "fallacious" phenomenon had turned out to be a matter of definition (1933, p. 49). In Eddington's philosophy, the term definition is certainly not an innocuous one. The point is that he may well have regarded his present reconstruction of the motion properties in De Sitter's world as already settling the question of what this universe is like, the ambiguity on the interpretation of motion being just a matter to be resolved by a correct understanding of the definition. Thus in the last of the supplementary notes that Eddington added at the time of the second edition of *The Mathematical Theory* (1924), he had already suggested that "those laws of nature which we discover must be implicit in the definition of the quantities which obey them" (p. 262). In particular, the task of physical science would come to an end when (all the equations that held sway in the physical world having been written down) the number of independent equations did not exceed the number of definitions.

Such an epistemology must have dictated Eddington's views, since it rather obviously falls short of the more immediate needs of physics. Some of Eddington's contemporaries were quite upset, and not without reason endeavoured to look for entirely different explanations (see Chapter 5). How is it possible for the acceleration to be a 'reality' which nevertheless cannot be observed? As John North has rather caustically remarked of this phase of Eddington's cosmology, "a consistently accelerating particle cannot remain perpetually at rest" (1965, p. 97). The difficulty, once more, is the nature of time. If the scattering of particles is never actually observed, but always anticipated, those instants of time that 'fix' the dynamic state of the universe do not express anything intrinsic to the universe. Those instants ignore what happens physically during the time separating the emission and the reception of light; Eddington's solution is even intended to show that what has happened is immaterial to the metric. It is, however, an answer to just such a question which was to motivate both Weyl's and Lemaître's later works.

(f) The mass-horizon as a necessary illusion

In *Space, Time, and Gravitation*, Eddington eloquently describes the relativity of temporal phenomena in the De Sitter universe. At the horizon, time does not flow any more,

but if attracted by such a delightful prospect, we proceeded to visit this scene of repose, we should be disappointed. We should find nature there as active as ever. We thought time was standing still, but it was really proceeding there at the usual rate, *as if* in a fifth dimension of which we had no cognizance. Casting an eye back on our old home we should see that time apparently had stopped still there (p. 160, emphasis added).

It is as though one true cosmic time exists, but that it remains impervious to our measurements, in contrast with the multiplicity of actually measurable times in the static metric. Although Eddington does not say it, the fictional fifth dimension could be brought in by the super-observer coextensive with the natural separation of space and time in Einstein's universe. Together with the notion of anticipation developed in *The Mathematical Theory*, this view of the singularity was to be expanded in terms of the newly established properties of motion.

If, at the very horizon, the velocity of light is zero, the reason is that those objects located there *have always had* that velocity:

We are supposing to be observing a system which has *now* the velocity of light, having acquired it during the infinite time which has elapsed since the observed light was emitted (1923b, p. 164).

In terms of proper time, material particles and light rays need an infinite time to reach the periphery from their starting-point. If they were actually to reach it, they would regress to an infinite past, which is their origin, and therefore obviate that indefinite existence which Eddington had identified as a prerequisite to any relativity physics. In this sense, the horizon of observability in the De Sitter universe is certainly not something which shares the status of various other physical properties.

And indeed, the source of the difficulty with the singularity in this model is epistemological. Eddington begins by arguing that "a singularity of ds^2 does not necessarily indicate material particles, for we can introduce or remove such singularities by making transformations of coordinates" (p. 165). Eddington's earlier position, as we have seen, was much less ambiguous since he seemed to ascribe the singularity to the coordinates alone. Yet here he bemoans the lack of available means for distinguishing a real singularity from a coordinate singularity: "It is impossible to know whether to blame the world-structure or the inappropriateness of the coordinate-system". It would be incorrect to interpret this passage as saying that the blame *is* to be borne by the coordinates and that our inability to make the distinction is only due to a lack of suitable method (see Tipler, Clarke, and Ellis 1980, p. 101). In talking about inappropriate coordinates, Eddington has something of profound significance in mind: the reunion of De Sitter's cosmology with his prior considerations of epistemological order. For he goes on to say that, 'if De Sitter's form for an empty world is right, it is impossible to find any coordinate system which represents the whole of real space–time regularly" (1923b, p. 166). Thus, the difficulty of distinguishing our own inability to understand from a genuine singularity is not a (temporary or accidental) matter of method; it is anchored in the prior conditions of any conceivable knowledge of the world as a whole.

Eddington supports his argument by using a geometrical reconstruction of the De Sitter universe in which the properties of motion stand out in relief. We can use our reduced model to see what Eddington is actually doing; the constraints imposed by the static reference system, which Eddington is aware of, should be clear enough (pp. 164-6). (The following is based on Schrödinger 1956, pp. 16-22.)

Eddington defines the variable t in the same way as Weyl, that is, via the transformation (4.2). Time being a function of the ratio v/u, the coordinate x

EDDINGTON'S SOLUTION OF 1923   303

or any function of x will be a space coordinate in the pseudo-Euclidean space–time. Eddington chooses $\sin\chi = x/R$, so that on the reduced model the three coordinates are transformed like this:

$$x = R\sin\chi, \quad u = R\cos\chi\cosh t, \quad v = R\cos\chi\sinh t. \tag{4.34}$$

In the new system of coordinates (χ, t) on the hyperboloid, an infinitesimal interval is

$$ds^2 = -R^2 d\chi^2 + R^2\cos^2\chi\, dt^2. \tag{4.35}$$

This is nothing other than the static form of the De Sitter universe, which is more familiar in the (x, t') coordinates,

$$x = R\sin\chi, \quad t' = Rt, \tag{4.36}$$

since these coordinates give the infinitesimal interval, $ds^2 = -(1-x^2/R^2)^{-1}dx^2 + (1-x^2/R^2)dt'^2$. With the transformations (4.34), the space-partitions are cut out by planes perpendicular to the x-axis, while the time partitions are cut out by planes which contain the x-axis. The intersection of the hyperboloid with any such plane which contains the x-axis is what Eddington calls a "lune". He then goes on to say (see Fig. 4.10):

The light-tracks, $ds = 0$, are the generators of the hyperboloid. The tracks of undisturbed particles are (non-Euclidean) geodesics on the hyperboloid; and, except for $x = 0$, the space partitions will not be geodesics on the hyperboloid, so that particles do not remain at rest (Eddington 1923b, p. 165).

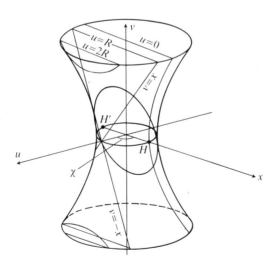

FIG. 4.10. Eddington's reconstruction of the De Sitter metric (adapted from Rindler 1977, p. 186).

All geodesics must pass through O in the bottle-neck of the hyperboloid; for the space-like geodesics, this is true only when the plane perpendicular to x is $x = 0$. The properties of motion are thus reconstructed here in such a way that they appear to result from non-inertial paths.

The other interesting features are related to time and observability. On Fig. 4.10, the points H and H' stand for $x = \pm R$ and $\chi = 0$. All contemporary spaces intersect each other at these two points (which are two-dimensional spherical surfaces in the complete five-dimensional world). That is why the time t is indeterminate there, as can be seen from (4.34) when $\chi = \pm 90°$. In other words, the two points H and H' represent the mass-horizon as it was already described by Weyl. Even for all real values of t and the complete range of variability of χ from $- 180°$ to $+ 180°$, the new coordinate-frame only covers the part of the world for which $u \geq v$. Its boundaries on the surface are the four generators (null geodesics) issuing from the points $v = 0$ and $x = \pm R$, that is, the 'double-wedge' of the hyperboloid; this accounts for the fact that, for an observer spending an infinite time on his geodetic line, only a lune of 90° will be accessible to observation. The two domains $|\chi| > 90°$ and $|\chi| < 90°$ seem to be completely separated from one another, and this is apparently due to the mass-horizon. As Eddington writes, "two observers cannot communicate the non-overlapping parts of their experience, since there are no light-tracks (generators) taking the necessary course" (1923b, p. 165). The facts had already been established by Klein and Lanczos. But what about a change of origin such that the experience of another observer covers a different, but partly overlapping, lune? In any one half of the world, all time-like geodesics are hyperbola branches which are cut out by the bundle of planes passing through O. In fact, most of these geodesics spend only a finite interval of proper time inside the observable world of a given observer. The sole exception is precisely the hyperbola branch $x = 0$ which remains permanently inside: this represents the observer permanently at rest at the spatial origin ($\chi = 0$) of the static frame. Two groups can be distinguished among the other free particles which are asymptotic to one or other branch of the hyperbola $x = 0$: those that have been inside from $t = -\infty$ (they will leave this region later) and those that enter it to remain insider forever. Of course those phenomena of going in and out are quite exceptional if they are irreversible, for most of the particles spend just a finite interval of their proper time within the region. According to (4.36), the proper time of the observer at O is given by Rt, i.e. it determines the variable t. But this determination of the world time appears to refer to what Eddington calls a matter of definition, for there will be as many different cosmic time functions as there are different origins. So, if the mass-horizon does not in the least hinder most time-like and null geodesics from freely entering and leaving a given lune at different points, this is because the 'reality' of the mass-horizon is itself dependent on the adopted 'definition' of world time. Eddington seems to content himself at this stage with an identification of the static frame with that of the observer at rest at the spatial origin; but we shall see in Chapter 5 that, in the second edition of his book, he began to go more thoroughly into the limits of this identification.

Given this geometrical structure, in what sense is the De Sitter model representative of the physical universe seen as a totality? A step-by-step process is implied here, for

the whole of De Sitter's world can be reached by a process of continuation; that is to say the finite experience of an observer A extends over a certain lune; he must then hand over the description to B whose experience is partly overlapping and partly new; and so on by overlapping lunes. . .we arrive at De Sitter's complete world without encountering any barrier or mass-horizon (Eddington 1923b, p. 166).

This leads to only one possible conclusion about the nature of the horizon itself:

I believe then that the mass-horizon is merely an illusion of the observer at the origin, and that it continually recedes as we move towards it (p.166).

Again this is baffling. The horizon is described as an illusion because it is swept along by the properties of motion. But it is hard to say if those very same properties work to deprive the horizon of its very existence. The horizon is not impenetrable, as the geometrical construction shows, but this too seems to be a consequence of motion and we still do not know whether the illusion is created by matter or not. Eddington only succeeds in offering the conjecture that the gravitational flux right up to the boundary vanishes, and he finds this "inconsistent with the existence of a genuine mass-horizon"; this last calculation, however, was suppressed in the second edition of the book.

In the course of infinite existence, no observer can have a view of the whole of De Sitter's universe without resorting to some other observer's experience. Therefore, if the model is a real one, then the mass-horizon is also a *necessary* illusion. In the context of the argument over the nature of the singularity, the consequence is that, whatever this nature may be (our coordinates or space–time structure), the horizon remains an illusion. Despite our freedom to create or remove a singularity by using transformations of coordinates, the singularity that has been found to hold in De Sitter's model reflects some *unknowable but intrinsic* property of the universe. This parallels the world time postulated in Einstein's model, but it also sanctions the ultimate difference between De Sitter and Einstein.

7. Solutions of old problems, problems of new solutions

The idea of a compromise between the two models had appeared from the outset quite natural to Eddington (1923b, p. 160), and it was with a good deal of reluctance that he resigned himself to the arguments that mitigated it. But in doing so, Eddington was led to discover that the conflict between the two models illustrated the force of a much more profound conflict between two alternative conceptions of the universe.

This discovery occurred when Eddington questioned what it means to speak of harmony between physics and geometry in a dual way, with reference to both the general conditions of knowledge and the world-conditions operative in the concrete universe. The De Sitter model reflects the natural space geometry of the world, and the insertion of matter into it imposes all sorts of constraints; the most notable being the fact that while the world becomes non-static as soon as matter is allowed, the geometry, just because it is natural, remains static. At the other extremity, the Einstein model is one in which matter is a pre-condition for the model's existence; that is not to say that the material content of the universe determines its laws (Eddington rejects the Machian background of Einstein's early thought), but what it does determine is the constants of nature. In short, Eddington writes: "Einstein's world offers no explanation of the redshift of the spectra of distant objects; and to the astronomer this must appear as a drawback. For this and other reasons I should be inclined to discard Einstein's view in favour of De Sitter's, if it were not for the fact that the former appears to offer a distant hope of accounting for the occurrence of a very large pure number as one of the constants of nature" (p. 168).

One common feature of both models is that each in its own way implies a distinction between space and time. It is the difference in the way the distinction is carried over that contains the seeds of a conflict between two conceptions of the universe. In Einstein's model, the totality of space–time can be described (at least in principle) in terms of observations, because the static coordinate system covers this totality. In other words, in this model, "there is no. . .barrier of eternal rest, and a ray of light is able to go round the world" (Eddington 1920a, p. 161). This imposes a cosmic time function quite distinct from the three space coordinates. But in De Sitter's model the distinction rests upon epistemological considerations. Its space geometry is the natural geometry of the world, since it provides what is required in the case of "an electron [that] could never decide how large it ought to be unless there existed some length independent of itself for it to compare with" (1923b, p. 154); but there is no such standard of comparison in the case of time, and the model simply reflects what Eddington takes to be the naturally non-cosmic dimension of time. Hence the language of illusion when the temporal, and so also the motion properties of De Sitter's model, are discussed in terms of extended inertial frame. Instead of the ideal observer equipped with a linear time function throughout space–time, we have the step-by-step process which alone enables any observer (by means of real communication with other observers) to restore the unity of the universe. The process illustrates the limits which are inherent in the anticipation of receding motions, for there are physical events which are perfectly free to cross the boundary of the observer at the origin, that is, which cannot be anticipated.

Weyl's original view of the cosmological debates in 1918 was that the De Sitter metric is quite ill-adapted to the kind of assumptions operative at the large-scale level. The singularity reflects a physical entity only at the cost of violating the uniformity with which distribution of matter should be depicted. Eddington's solution of 1923 does not provide a direct demonstration that the singularity is non-material; but by including the singularity in the general properties of receding motion, it transforms the paradoxical metric and makes it central to a concept in which the universe is taken seriously as a theme for theoretical physics—seriously, that is, without the postulation of some separate 'super' observer.

Of course, the dynamic behaviour of test bodies in a static metric is still a far cry from explaining the occurrence of such highly organized systems as the galaxies. As Lemaître was to understand from Lanczos, only the language of non-static metrics seems to lend itself to the resolution of such problems. Certainly it was clear from Eddington's work that the De Sitter metric is one of material structure in the sense that it expresses some constant ratio of spatial lengths; but nothing in his argument indicates how it can be related to the properties of motion. Eddington is simply not preoccupied by a change of coordinates. In his view, it remains for a unified theory of gravitation and submolecular phenomena to understand at one stroke "the possibility of the existence of an electron in space", as he called it (1923b, p. 153), and its observable effects. At the very least, the merit of the prior epistemological analysis of both models is that it reveals what a physics of any possible scale must comprehend.

On the other hand, the impact of the growing evidence that large and systematic redshifts did exist had prompted Weyl to change the terms of the sought-for compromise between the two models, since the assumption of uniformity of matter distribution could be dispensed with in either model. But Weyl, with his philosophy of cosmic time which is so different from Eddington's, could hardly remain silent in the face of what looked like (from his standpoint) obvious flaws in the proposed solution. Eddington had already invoked the language of illusion in *Space, Time, and Gravitation* when he identified a pseudo-cause for the relativity of the time singularity in the De Sitter model: what gives the impression of usual rate to the phenomena occurring at the singularity when the observer travels there, is that the phenomena appear to take place "as if in a fifth dimension", of which no knowledge is possible. At this point, rejecting Eddington's philosophy of cosmic time proved enough to form the basis of the most extraordinary episode in Weyl's career as cosmologist. In a new and decisive contribution, in which the status of super-observer is dramatically revisited, it was nothing less than a consistent physical representation of this mysterious dimension which emerged.

References

Bachelard, G. (1929). *La valeur inductive de la relativité*. Alcan, Paris.
Bondi, H. (1960). *Cosmology*. Cambridge University Press.
Cartan, E. (1922). Sur les équations de la gravitation d'Einstein. *J. Math. Pures et Appl.* 9ème série (t.1), **87**, 141-203.
Coxeter, H.S.M. (1943). A geometrical background for De Sitter's world. *Amer. Math. Monthly,* **1**, p. 227.
Dirac, P.A.M. (1982). The early years of relativity. In *Albert Einstein: Historical and Cultural Perspectives*, (eds., G. Holton and Y. Elkana), pp. 79-90. Princeton University Press.
De Sitter, W. (1917). Further remarks on the solutions of the field equations of Einstein's theory of gravitation. *Proc. Kon. Akad. Wet. Amst.* **20**, 1309-12.
Earman, J. and Glymour, C. (1980). The gravitational redshift as a test of general relativity. *Studies in Hist. and Phil. of Sci.* **11**, 175-214.
Eddington, A.S. (1913a). The Greenwich eclipse expedition to Brazil. *The Observatory.* **36**, 62-65.
Eddington, A.S. (1913b). Jean Henri Poincaré, Obituary. *Monthly Not. Roy. Astr. Soc.* **73**, 223-8.
Eddington, A.S. (1914). *Stellar movements and the structure of the universe*. Macmillan, London.
Eddington, A.S. (1915). Gravitation. *The Observatory.* **38**, 93-98.
Eddington, A.S. (1916). Gravitation and the principle of relativity. *Nature.* **98**, 328-30.
Eddington, A.S. (1918). *Report on the relativity theory of gravitation*. Fleetway Press, London.
Eddington, A.S. (1920a). *Space, time, and gravitation*. Cambridge University Press.
Eddington, A.S. (1920b). The meaning of matter and the laws of nature according to the theory of relativity. *Mind* **29**, 145-58.
Eddington, A.S. (1920c). *Report on the relativity theory of gravitation*. 2nd ed. of 1918.
Eddington, A.S. (1921a). *Espace, temps et gravitation*. (French trans. of (1920a), J. Rossignol). Gauthier-Villars, Paris.
Eddington, A.S. (1921b). A generalization of Weyl's theory of the electromagnetic and gravitational fields. *Proc. Roy. Soc.* **A99**, 104-22.
Eddington, A.S. (1923a). Can gravitation be explained? *Scientia.* **33**, 313-24.
Eddington, A.S. (1923b). *The mathematical theory of relativity*. Cambridge University Press.
Eddington, A.S. (1924). *The mathematical theory of relativity*. Second edition of (1923b).

Eddington, A.S. (1928). *The nature of the physical world*. Cambridge University Press.
Eddington, A.S. (1929). *Science and the unseen world*. Allen and Unwin, London.
Eddington, A.S. (1933). *The expanding universe*. Cambridge University Press.
Eddington, A.S. (1939). *The philosophy of physical science*. Cambridge University Press.
Einstein, A. (1918). Prinzipielles zur allgemeinen Relativitätstheorie. *Ann. Physik* **55**, 241-4.
Einstein, A. (1920). Ether and relativity, (trans. G.B. Jeffery and W. Perrett) In *A. Einstein: sidelights on relativity*. 1922, pp. 3-24. Methuen, London.
Einstein, A. (1921). Geometry and experience, (trans. G.B. Jeffery and W. Perrett) 1922, pp. 27-56. In *A. Einstein: sidelights on relativity* Methuen, London.
Einstein, A. (1923). Theory of the affine field. *Nature*. **112**, 448-9.
Gonseth, F. (1926). *Les fondements des mathématiques*. Blanchard, Paris.
Husserl, E. (1913). *Ideen zu einer Reinen Phänomenologie und Phänomenologischen Philosophie*. (ed., K. Schuhman) 1976. M. Nijhoff, The Hague.
Klein, F. (1872). *Vergleichende Betrachtungen über neuere geometrische Forschungen*, Erlangen. Reprinted in *Math. Ann*. Bd. XLIII, 1893, 63-100.
Klein, F. (1873). *Uber die sogennante Nicht-Euklidische Geometrie, Math. Ann*. Bd VI, 112-45.
Klein, F. (1918a). Bemerkungen über die Beziehungen der De Sitterschen Koordinatensystems B zu der allgemeinen Welt der positiver Krümmung. *Proc. Kon. Akad. Wet. Amst.* **21**, 614-15.
Klein, F. (1918b). Über die Differentialgesetze für die Erhaltung von Impuls und Energie in der Einsteinschen Gravitationstheorie. *Nachr. Ges. Wiss. Göttingen*, 171-89.
Klein, F. (1918c). Über die Integralform der Erhaltungssätze und die Theorie der räumlich-geschlossenen Welt. *Nachr. Ges. Wiss. Göttingen*, 394-423.
Kockelmans, J.J. and Kisiel, T., eds. (1970). *Phenomenology and the natural sciences*. Northwestern University Press, Evanston.
Lanczos, K (1922). Bemerkung zur De Sitterschen Welt. *Phys. Zeitschr*. **23**, 539-43.
Lanczos, K (1923). Über die Rotverschiebung in der De Sitterschen Welt. *Zeitschr. für Physik*. **17**, 168-89.
Lanczos, K (1924). Über eine stationäre Kosmologie im Sinne der Einsteinschen Gravitationstheorie. *Zeitschr. für Physik* **21**, 73-110.
Lanczos, K (1962). *Einstein and the cosmic order*. Wiley, New York.
von Laue, M. and Sen, N. (1924). Die De Sittersche Welt. *Ann. der Physik,* **74**, pp. 252-4.
Lemaître, G. (1925). Note on De Sitter's universe. *J. Math. Phys*. **4**, 37-41.
Lemaître, G. (1927). Un univers homogène de masse constante et de rayon croissant, rendant compte de la vitesse radiale des nébuleuses extra-galactiques. *Ann. Soc. Sci. Brux*. **47(A)**, 49-59. English translation in *Monthly Not. Roy. Astr. Soc.* **91**(1931), 483-90.
Lemaître, G. (1933). L'expansion de l'univers. *Ann. Soc. Sci. Brux*. **53(A)**, 51-85.
Mehra, J. (1973). Einstein, Hilbert, and the theory of gravitation. In *The physicist's conception of nature*. (ed., J. Mehra), Reidel, Dordrecht.
Merleau-Ponty, J. (1965). *Philosophie et théorie physique chez Eddington*. Belles-Lettres, Paris.

Moyer, D. (1979). Revoluation in science: The 1919 eclipse test of general relativity. In *On the Path of Albert Einstein*. (eds., A. Perlmutter and L.F. Scott) pp. 55-101. Plenum Press, New York.

North, J. (1965). *The measure of the universe*. Clarendon Press, Oxford.

Pais, A. (1982). *Subtle is the Lord*. Oxford University Press.

Schrödinger, E. (1956). *Expanding universes*. Cambridge University Press.

Smith, R.W. (1982). *The expanding universe*. Cambridge University Press.

Stachel, J. (1986). Eddington and Einstein, in *The prism of science*. Israel Colloquium Vol.2 (ed., E. Ullmann-Margalit). Reidel, Dordrecht, pp. 225-50.

Tipler, F., Clarke, C., and Ellis, G.F.R. (1980). Singularities and horizons, a review article. In *General relativity and gravitation*, (ed., A. Held), Vol.2, pp. 97-206. Plenum Press, New York.

Torretti, R. (1978). *Philosophy of geometry from Riemann to Poincaré*. Reidel, Dordrecht.

du Val, P. (1924). Geometrical note on De Sitter's world. *Phil. Mag.*, **47**, 930-8.

Vizgin, A.P. (1984). Einstein, Hilbert, Weyl: Genesis des Programms der einheitlichen geometrischen Feldtheorien. *NTM-Schriftenr. Gesch. Naturwiss.* Leipzig, **21**, 23-33.

Weyl, H. (1918a). Gravitation und Elektrizität. *Sitz. Ber. Preuss. Akad. Wiss.*, 465-80.

Weyl, H. (1918b). *Raum, Zeit, Materie*. First edition. J. Springer, Berlin.

Weyl, H. (1918c). Reine Infinitesimalgeometrie. *Math. Zeitschr.*, **2**, 384-411.

Weyl, H. (1919a). *Raum, Zeit, Materie*. Second edition. J. Springer, Berlin.

Weyl, H. (1919b). *Raum, Zeit, Materie*. Third edition. J. Springer, Berlin.

Weyl, H. (1919c). Uber die statischen Kugelsymmetrischen Lösungen von Einsteins 'kosmologischen' Gravitationsgleichungen. *Phys. Zeitschr.*, **20**, 31-4.

Weyl, H. (1921a). Electricity and gravitation. *Nature*, **106**, 800-2.

Weyl, H. (1921b). Uber die physikalischen Grundlagen der erweiterten Relativitätstheorie. *Phys. Zeitschr.* **22**, 473-80.

Weyl, H. (1921c). *Raum, Zeit, Materie*. Fourth edition. J. Springer, Berlin.

Weyl, H. (1922). *Space, Time, Matter* (trans. H. L. Brose of 1921c). Methuen, London. Reprinted 1952, Dover, New York.

Weyl, H. (1923a). Entgegnung auf die Bemerkungen von Herrn Lanczos über die De Sitterschen Welt. *Phys. Zeitschr.* **24**, 130-1.

Weyl, H. (1923b). *Raum, Zeit, Materie*. Fifth edition. J. Springer, Berlin.

Weyl, H. (1927). *Philosophie der Mathematik und Naturwissenschaft*. In *Handbuch der Philosophie* (ed., R. Oldenbourg). *Philosophy of mathematics and natural science* 1949. (trans. O. Helmer) Princeton University Press.

Weyl, H. (1954a). Address on the unity of knowledge, delivered at the Bicentennial Conference of Columbia University. In *H. Weyl: Gesammelte Abhandlungen* (ed. K. Chandrasekharan) 1968, Vol. 4, pp. 630-9. J. Springer, Berlin.

Weyl, H. (1954b). Erkenntnis und Besinnung: Ein Lebensrückblick. In *H. Weyl Gesammelte Abhandlungen*. (ed. K. Chandrasekharan), Vol. 4, pp. 641-54. J. Springer, Berlin.

Whittaker, E. (1954). *A history of the theories of aether and electricity*. Vol.II. Philosophical Library, New York.

Yolton, J. W. (1960). *The philosophy of science of A.S. Eddington*. Nijhoff, The Hague.

5
The construction of a principle

1. Eddington's solution re-considered

In his Gifford Lectures of 1927, Eddington reiterated the interpretation of the redshift he had acquired four years earlier. More than ever he seemed to be convinced that the observed displacement of the light emitted by distant galaxies was a metric effect—the slowing down of the vibrations of light in its journey round the world—even though astronomers interpreted it as a Doppler effect, that is, by means of some purely kinematic mechanism (1928, pp. 166–7). On the basis of his theory, Eddington was able to conclude from the then-recent observations that the radius of curvature of the universe is something like 100 million light years. A year before, Hubble (1926) had calculated the first drastic revision of the size of the universe as a result of his discovery of Cepheids in extra-galactic spiral nebulae, and he had suggested a much bigger radius of the order of 10^{11} light years. What Eddington found unacceptable in this calculation was the fact that Hubble uncritically relied on the Einstein model of the universe, a model which he said had "generally been regarded as superseded" even though the De Sitter model remained "very speculative".

This apparent incommunicability illustrates conspicuously the increasing need for cooperation, felt by many throughout the second decade of the twentieth century, between theoreticians and observers. Even as late as September 1931, some two years after his announcement of the remarkable correlation between the distance of the nebulae and their redshifts, Hubble could write to De Sitter that the astronomers use "the term 'apparent' velocities in order to emphasize the empirical features of the correlation. The interpretation, we feel, should be left to you and the very few others who are competent to discuss the matter with authority" (quoted in R.W. Smith 1982, p. 192). In fact, Eddington was certainly not alone in this new wave of predilection for the De Sitter model, for he was soon followed by such eminent observers as Wirtz and Lundmark. Thus, as early as 1918, Wirtz had begun to expand Slipher's investigations of 1917 about the Sun's motion through the Milky Way by calculating this motion with respect to the spiral nebulae. Finding it impossible to accommodate purely random motions for the spirals, he introduced the so-called K-velocity term to account for a residual effect. In 1922 he went on to argue that this term might indicate something like a systematic velocity of recession. It was not until 1924 however, that he became aware of De Sitter's theoretical investigations, which prompted him to explore this velocity of recession as a function of distance. In his own interpretation, the Doppler shift was entirely subsidiary to the metric shift; but the distance measurements he was using remained extremely precarious until Hubble's momentous work of the following year.

In a remarkable book review of Eddington's *Mathematical Theory of Relativity*, Max von Laue was apparently the first to hit on a flagrant inconsistency in the proposed attempt to explain the occurrence of the redshift in the De Sitter model. He was surprised to see the clash between Eddington's statement that all points of the model are equivalent and the farfetched attempt to rescue the predominance of red- over violetshift from every point of observation. Compounding the metric and the kinematic effect in a static metric with cosmic time cannot really work, for von Laue suspected that the latter effect implies some kind of hypothesized "creation", namely, "at the beginning", as he said, all bodies must have been at rest with respect to the mass-horizon. Von Laue's point was that the subsequent attraction of the bodies by the mass-horizon could not be compatible with the prediction that *all* bodies would move *towards* it (1923, p. 384).

This review prompted Lanczos to re-open his 1922 investigations of the De Sitter model. He now argued that the bone of contention between von Laue and Eddington was very natural because the two authors were not, in fact, talking about the same things (Lanczos 1923, p. 168): von Laue was dealing with "static world lines" which, as Eddington had shown conclusively, are not geodetical in the De Sitter world. So, relying on his earlier claim that the mass-horizon could be no more than a product of special coordinates, Lanczos went on to ask if an objective criterion could be found here too in order to decide whether a given redshift is metrical or kinematical; all that results from the particular choice of coordinate system is apparent only and cannot count as a reality, the "real" being identified by Lanczos with the "invariant" (p. 169). Setting about the task of formulating an expression for the redshift which is independent of any coordinates, Lanczos discovered that the Doppler redshift was also totally independent of the De Sitter metric: all that remained was to formulate a simple proportionality ratio between the proper times of two atoms at two different places and the corresponding observed frequencies (pp. 170ff.). So there was a need to distinguish between two senses of the Doppler redshift: an apparent one, which resulted from a choice of coordinates 'forcing' the space–time structure (as when pseudo-Euclidean space is used to project the pseudo-spherical world), and a true one, where the coordinates were immaterial and the world-lines $t =$ constant 'naturally' diverging. Relying on the coordinates for the De Sitter world he had proposed in 1922, namely the set (4.28), Lanczos now found that these non-static coordinates provided the most natural counterpart of the observed redshifts (pp. 182–5). Indeed, with such coordinates, all points are not only 'truly' equivalent, but they also make it possible to consider the stars as being at rest, their spatial coordinates remaining unchanged through time. The resulting equation for the redshift showed that it would be determined by the two invariants of interest, namely, the distance separating two observers *and* the change of this distance through time. It is fascinating that all this agrees virtually word for word with the future expanding universe theory.

Certainly these new developments testify to Lanczos having gone some way towards a more physical interpretation of the time function associated with non-static coordinates. But his basic point was that *any* time function is a by-product of the chosen coordinates, and as these must remain immaterial he went on to describe mathematically how it could be possible to translate the redshift predicted by means of non-static coordinates in terms of the more ordinary static coordinates. What Lanczos did acknowledge however, was that the invariant formulation of the redshift was an idea he borrowed from Weyl's fifth edition of *Raum, Zeit, Materie*. And indeed, the decisive clarification of how to connect observed redshifts to distance in the De Sitter model of the universe came from this new and most fundamental contribution by Weyl. This edition shortly preceded the publication of yet another paper by Weyl, still in the year 1923, which seems to have come as a reaction to Eddington's mathematical theory of relativity. In effect, what has since become known as Weyl's principle is contained in these works. The paper, in particular, appears in retrospect to be no less than a way of physicalizing Eddington's hypothetical fifth dimension in which all paradoxes of De Sitter's singularity would supposedly normalize. Such a physicalization would solve many of the mysteries latent in Eddington's book and which von Laue and Lanczos had revealed each in their own way. Naturally such a leap from 'fiction' to 'reality' implies dramatic re-consideration of the status of a 'super-observer' in physical science; it also mounts a frontal attack against Eddington's epistemological concept of time as purely subjective which dominated the purported solution to the singularity. So much so that in revising the general introduction for the fifth edition of his book, Weyl showed growing awareness of the need to re-incorporate into physics the idea of becoming, an idea which had somehow been neglected ever since the rise to the theory of relativity. With regard to the basic couple action/passion which he had defined earlier as the source of the relativistic notion of space and time, he now explained that, as for the German words *tun* and *leiden*, "our grammar has only the verbal modes of the active and passive; there are no modes for the expression of a becoming, and much less for a state of affairs" (1923a, p. 5).

To be sure, both the new edition of the book and the subsequent paper contained quite a novel analysis of the conceptual and technical issues related to the De Sitter model. It comes then as a surprise to see Weyl re-stating more sharply than ever that this model is non-static when regarded as a totality (1923a, p. 294), and from there proceeding to spell out the technical difficulties again in the language of static coordinates. True, a few years later, after the non-static metrics had been re-discovered, Weyl explicitly acknowledged that such new coordinates faithfully reflected the geometrical construction of 1923 (1930, p. 937). The point is that the later non-static metrics offer a satisfactory technical solution—that we shall develop in the course of this chapter—to a sense of totality re-conceptualized in entirely new terms. Once more, the progression of Weyl's intellectual journey during these years is fundamental, for it is precisely this which determined the need of re-conceptualization in the first place.

2. From metric to topology and causality

In the first edition of *Raum, Zeit, Materie*, Weyl suggested an idea that proved wrong: he thought that the metric of the four-dimensional space-time of general relativity could be completely mapped out—and hence our actual space-time in which we live be discovered—by using light rays alone. But Weyl had to amend this idea in the later editions of his book, being pressed to do so as Lorentz had pointed out to him the flaw of the argument: two space-times with different metrics may have the same light-cone structure if they are *conformally* equivalent, that is, if the mathematical transformation from one to the other is only continuous and angle-preserving. Weyl then went on to argue that all we need to add to our observational data about the paths of light rays are the paths of free material particles. In other words, the preservation of time-like geodesics, which define the paths of such particles, would be sufficient to ensure that all space-times mapped from one to the other are metrically equivalent (1922, pp. 228-9 and 313-14).

So, having already conceived of the affine structure as being inherent in the metrical structure, there remained just one structure that still appeared to be fairly independent of the rest: this was the topological structure of space-time, which was also of the highest generality. The new cosmological arguments articulated by Weyl were undoubtedly regarded as an attempt to complete this story about how to fix uniquely *all* the structural properties of space-time. But the problem of how the topology of some space-time can be derived from its metric intervals is probably the most daunting for any geometry and geometrical physics. The issue, which has always been with us, also forms quite a vivid background in this early cosmological context. Thus, apart from developing Weyl's affine theory in his own way, Eddington had also shown interest, right from the outset, in the problem of relative independence of the topological and metrical structures of space-time. He argued in his *Space, Time, and Gravitation* that "the statement that the world is four-dimensional contains an implicit reference to some ordering relation": this is nothing other that the *interval* in the sense of the metric, but the question arises as to whether "that alone suffices without some relation corresponding to *proximity*" (1920, pp. 186-7). The concept of proximity referred to the local topological structure of space-time. Eddington left it as a possibility that the space-time topology may be interval-dependent, although just how the topology would be derivable from the knowledge of the interval remained much of a mystery, for "it must be remembered that if the interval between two points is small, the events are not necessarily near together in the ordinary sense". And in *The Mathematical Theory of Relativity*, he went on to argue that distance, rather than location, is in fact the concept which functions as a primitive in our physical theory of gravitation. Space is basically a network of intervals, and thus the topological difference (connectivity) between spherical and elliptical space cannot have any physical implication (1923, pp. 158-9).

Weyl then explored the relation of metric to topology in connection with its relevance to cosmology. The distinction between the topology and the metric is certainly not an easy one, yet the problem is given decisive momentum when the system under investigation is that of the total extent of the universe. Thus, as Weyl was keen to point out (in his 1927, pp. 108-9), there is no topological difference between the old Aristotelian, finite world and either the modern Euclidean, infinite one of Newton's mechanics or the pseudo-Euclidean one of special relativity; the difference is only metrical. Which provokes us, in turn, to ask what difference it makes to speak of a finite world in Riemann's sense. The closed sphere and the flat plane both have metrical homogeneity, but differ in the number of connected, infinitely distant fringes: the latter has only one, while the former has two, i.e. the infinitely distant past and the infinitely distant future. As for the physical character of the De Sitter world, which also extends from infinity to infinity in this sense, it becomes manifest by looking at some finer feature of its natural topology. For while nothing physical in the Einstein model favours the elliptical over the spherical topology—it seems to be no more than a matter of convenience to adopt the former because it avoids the universe being populated by an infinite number of 'ghost-images'—the De Sitter model "gives the world the double fringe of past and future but prevents the null-cone from self-overlapping" (1923a, p. 294). Thus, rather than an impediment to its physicality, Weyl now sees the topological perspective on the singularity of the De Sitter model as responsible for the whole model's superiority over Einstein's. The physical relevance of topology is therefore associated with a hypothesis as to the only possible meaningfulness of causality at the scale of the whole universe, namely, a closed time-like path must be causally anomalous in some way. This is quite a remarkable step, for this very expression of the relation of topology to causality still serves today as the most useful basis of discrimination between various candidates for the large-scale geometry of our universe (see in particular G.F.R. Ellis, 1975, p. 256). In his 1930 paper (p. 937), Weyl went on to distinguish between two senses of De Sitter's space which should be regarded as equally pertinent: this space is topologically closed with regard to its two fringes and topologically open with regard to the causal requirement.

So, not surprisingly, Weyl opens his 1923 article on De Sitter's cosmology by considering in the most general terms the meaning of causality at the astronomical level. Take any two stars A and B:

That both stars A and B belong to a common system, i.e. that they are causally linked together from the origin, means that their world lines have the same range of action (*Wirkungsbereich*) Σ (Weyl 1923b, p. 231).

There is here no explicit mention of this statement about causality as a principle or postulate, but obviously it works as such. For the point of the article is to articulate *causality* taken in this sense and the *observability* of the redshift in the De Sitter model. In other words: How does this statement about causality allow universal reciprocity of the observable redshifts in accordance with the necessary identity of all vantage points?

Clearly the definition of a common system serves as something like the definition of local common time. This is very different from Weyl's earlier argument for natural distinguishability of time from space. due to the Minkowskian signature at every neighbourhood of the pseudo-Riemannian space–time. Here, for the first time, we are dealing with talk about an *origin* (*Ursprung*). The term is quite difficult to grasp because the context does not tell us what kind of origin it may be. As he is talking about stars rather than galaxies, Weyl probably refers to origin in the sense of a dated or datable astrophysical event. But he carefully eschews the term time or scale of time in which this dating could be made, so that we are in limbo as to what "range of action" may possibly mean. In point of fact, the significance of this first statement will appear only in the course of the article, as the argument progresses towards another, more cosmic and all-encompassing sense of origin, namely, the chronogeometric one. Why the above definition of the range of action is just a stone's throw away from such an origin may be inferred from the point about topological openness with regard to causality. For this openness seems to imply that the topological structure in cosmology should be similar to that of special relativity, in which homogeneity is also given a priori. There is a way of preserving this homogeneity in the case of general-relativistic space–time (such as De Sitter's) where interaction with matter creates disturbance of this flat metric; or, to put it differently, there is a way of specifying a naturally suitable topology for the De Sitter cosmology, by looking at the implications of some hypothetical *undisturbed* state of matter. It is this peculiar state which will send Weyl back to the question of 'the' origin.

FROM METRIC TO TOPOLOGY AND CAUSALITY

For the time being, it is sufficient to note that causality as defined here stems from the demand of common time without which the very observation of redshifts would never be meaningful. Observability and factual evidence, rather than any prior epistemological concept which would constrain in some way the concrete universe, are to be taken as foundation. This is particularly true in the case of time, as Weyl is eager to emphasize that a fundamental *fact* about time, namely its unique direction from past to future, cannot lose its role of ultimate *foundation* even though it comes into conflict with the invariant character of field-laws (1923a, p. 287). Therefore, there can be little doubt that the common range of action of any two stars is defined in reference to a common origin assigning common time precisely in order to fix the meaning of observability.

When taken in this primordial sense, observation and fact include something which no mathematical analysis in terms of fields could possibly deduce. Rather, they lead us straight to some grand hypothesis. For all timelike geodesics belonging to a given range of action form a local bundle of world-lines diverging towards the future; the hypothesis connects the various bundles of this kind which we observe:

It is our hypothesis that all heavenly bodies which we know belong to a single system of that kind; this would explain the small velocities of the stars as a consequence of their common origin (p. 295).

Origin here explicitly stands for the actual beginning of the star system. And in his article Weyl explores the hypothesis in an analytical way.

He begins by positing metrical homogeneity of the four-dimensional hyperboloid $\Omega(x) = a$, in which $\Omega(x)$ is defined as above in Eqn. (4.1) in a five-dimensional pseudo-Euclidean space. The point is that some new form of cosmic time function should be found, such that homogeneity never breaks down as was the case with the transformation of coordinates (4.2). As Weyl puts it, "this hyperboloid with its two fringes of the infinitely distant past and the infinitely distant future gives a metrically exact representation of the world according to De Sitter's cosmology (1923b, p. 230). Weyl further explains that "the geodesics are cut out [of the hyperboloid] by the two-dimensional planes passing through the origin of the five-dimensional space". Clearly, the origin of the five-dimensional space is also point O within the hyperboloid, as on Fig. 4.1. Now, "the null cone opening into the future, which issues from the world point P of such a geodesic g with time-like direction, fills the range of action of g" (pp. 230–1). Consider a point P following its geodesic g; the four-dimensional null cone issued from P which covers a part of the hyperboloid as on Fig. 5.1 defines the range of action of the geodesic g. The problem is to find out the relationship between any points following their own geodetic paths. Weyl says:

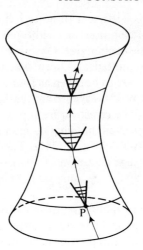

FIG. 5.1. The range of action of a geodesic.

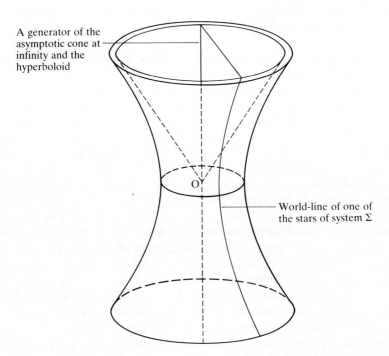

FIG. 5.2. A time-like geodesic in accordance with Weyl's definition of system Σ.

The world-lines of all stars of the system Σ are cut out by planes which pass through a common axis, i.e., a generator of the asymptotic cone (p. 231).

A time-like geodesic is represented in Fig. 5.2 in accordance with the definition of Σ. The light cone issued from a point following this geodesic is formed by two of the generators picked out from each of the two families of rectilinear generators of the hyperboloid at this point. The opening of the cone is at its maximum when it passes through the 'neck', as shown in Figs. 5.3 and 5.4. When P tends towards infinity (in the past or in the future), the cone becomes quite narrow in the neighbourhood of P. In fact, at $t = \pm \infty$ the light cone is *only light*. So, the problem of the relationship between any two points boils down to understanding the general connection between observability and causality. What is that part of the universe which will become visible to an observer from $t = -\infty$? Or, in other words: what lies in the future cone of an observer at $t = -\infty$? At the infinitely distant past, the asymptotic cone connects with the hyperboloid at all points of the circle at infinity, so that any generator of the asymptotic cone is then *also* a generator of the hyperboloid. Consider one point of the circle and a generator issued from this point. Clearly, this generator is the axis common to all planes cutting out the world-lines of the system Σ, i.e. the time-like geodesics. Also, the two straight lines of the future cone at P when P is at $-\infty$ are now one and the same line. Therefore, all the stars of the system Σ had the same future cone at $t = -\infty$, and so they must be causally linked together because of

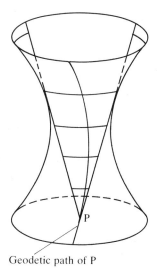

Geodetic path of P

FIG. 5.3. The light cone from P extended towards the infinite future.

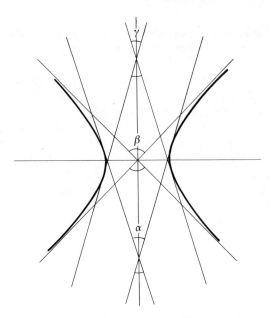

FIG. 5.4. The angle β is the maximal opening of the cone. Any α or γ is smaller.

this; the part of the hyperboloid covered by all cones at $t = -\infty$ is represented in Fig. 5.5.

With this construction we are now in possession of that which was required by the definition of Σ, namely, a family of geodesics which have a common origin. But the fact that this origin is the infinitely distant past has some important consequences as to the possibility of cosmic time for the whole universe. Indeed, from all the planes which cut out the geodesics of a given range of action, there is one which is tangent to the asymptotic cone and which encloses what lies in the future of all the points of the given range (Fig. 5.6). The geodesic cut out of the hyperboloid by this plane comes from the infinitely distant past, but this is true of *any* such plane for all families of planes defining various ranges of action. The remarkable thing is that this 'super' range of action covers only half of the hyperboloid; what was obtained earlier by means of a transformation of coordinates is now obtained in structural terms by means of this 'principle' of causality, i.e. the very definition of the range of action.

Any observer following a geodesic g in the range Σ is 'at rest'. The two lines $L_{-\infty}$ represent, for all the observers of the range, the 'limits of the universe', since these lines recede at the velocity of light from any point within these limits. This is exactly Weyl's hypothesis: all the geodesics of some given range diverge from $t = -\infty$, and the interior of this range is also the *totality* of the

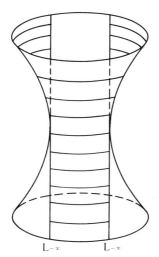

FIG. 5.5. The part of the hyperboloid covered by all light cones existing in the system Σ from infinity in the past.

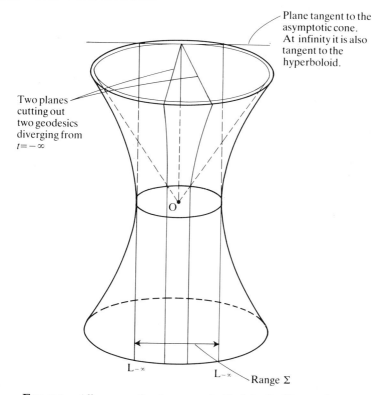

FIG. 5.6. All ranges of action cover half of the De Sitter universe.

spatio-temporal range of an observer. Even though not all the geodesics of the universe belong to some given range, the synchronization of all clocks prevailing in this range will be the same as the synchronization in any other range *provided $t = -\infty$ is common to all observers*.

Eddington reacted quite sharply to this idea, as can be surmised from the important remarks he added to his analysis of the De Sitter cosmology in the second edition (1924) of *The Mathematical Theory of Relativity*. Digging more deeply into the problem of horizon as illusion, he claims that what lies outside the field of observability of a given observer is causally linked to what lies inside:

I have used the rather inaccurate phrase 'experience of an observer' to indicate the lune between his partition $t = \pm\infty$. But this part is not *causally* detached from the rest, and events outside it are just as real as those inside it (p. 166).

But what constitutes the limits of observability cannot exceed the infinite time of an observer at the origin. Therefore what is beyond observability is also something that does not fall within this infinite time. This result is disconcerting:

Events before $t = -\infty$ may produce consequences in the neighbourhood of the observer and he might even *see* them happening through a powerful telescope. Only if he calculated from his observations *when* they must have happened he would find it impossible to assign to them any real value of t (p. 166).

Eddington calls these events "extra-temporal". We must remember his reconstruction of the geometry of the De Sitter model and its properties of motion, where objects may freely enter and leave the observable part of the universe for a given observer. The limits are formed by null-length geodesics passing through H and H', and space intervals (ellipses) approach zero as they get closer to these limits. Now, in the case of very elongated ellipses, the longest parts are almost straight lines so that only the extremities (which are also the remotest points from the horizon) contribute in a significant way to the space to be traversed in order to reach the horizon. In the static frame of reference, the proper time of the hyperbola branch $x = 0$ is selected as comic time (from $t = -\infty$ to $t = +\infty$), so that time becomes indeed infinite at the horizon and any object coming from the origin (at rest), unlike all others, cannot travel through the horizon in finite time. This, in turn, condemns the observer at the origin to eternally wonder at the source and destiny of objects that spend only a finite portion of their proper time within the limits enclosed by the horizon. The 'now' of the observer at the origin is extended in the static frame to the observable part of the universe from this origin, but just as different observable parts overlap only partially, there will be as many different 'nows' as there are observers at their own origin regarding themselves at rest. Similarly, there will be as many $t = -\infty$ as observers at rest, i.e. $t = -\infty$ cannot be in any consistent way something physically common to all observers. There is a paradox with this denying of any physicality to some unique time function intrinsic to the universe, as it would now seem that there could be as many 'universes' as there are 'origins'. In order to extricate himself from this paradox, Eddington finds that differentiating the range of observability from the range of causality (so as to make the latter larger than the former) is the only way of maintaining one and the same universe.

But at this stage of Eddington's argument, it appears that one paradox has been swapped with another. By integrating the singularity in the properties of motion, instead of dissolving it altogether, De Sitter's famous statement is reversed and it must be argued that what happens 'before' eternity (which is now 'outside' time in Eddington's way of presenting the problem) may have an influence on what happens 'during' it, i.e. within time. For some events of the range of causality may act from without, as it were, upon an observer, yet they cannot be referred to in terms of this observer's actual measures. Eddington perceives the source of the difficulty, as he now says that

a static space–time frame is a make-shift contrivance, and leads us to the admission of extra-temporal events as affecting even our experience.

Eddington does not bother with any change of coordinates because the artificiality of the static frame just reflects the fundamental point of his epistemology, the non-being of some hypothetical super-cosmic time. What matters for him is that the revelation of this contrivance points to the need of distinguishing observability from causality in *both* the Einstein and the De Sitter models. For Eddington is still not far from believing that the mass-horizon in the De Sitter model plays a role analogous to the world matter in the Einstein model: in the latter, the enormous but uniform quantity of matter makes it undetectable, just as the nineteenth-century ether escaped observation (1924, p. 168).

Be this as it may, Eddington's conception of the static frame as make-shift contrivance shows how no less paradoxical it would be from now on to reject Weyl's hypothesis than to accept it. By virtue of the hypothesis all events finds themselves synchronized at infinity in the past; there can be no extra-temporal event within any observer's range of observability any more than it would be permitted to have as many $t = -\infty$ as observers at rest. But on this account the status of those objects which lie outside the range of observability is still quite a problem. For if the hypothesis is to be taken at face value, the instant of 'genesis' (at $t = -\infty$) for a given observer must coincide with the 'genesis' of the universe itself: at that instant, there was no elsewhere/elsewhen, so that objects lying outside observability could never come into existence. However, the light cone of a point P within the range of observability but close to either one of the lines $L_{-\infty}$ may still cross the line; conversely, this very same curious interaction is true for a hypothetical point lying just outside the range. The only way of actually 'saving appearances' here is to argue, as Howard P. Robertson did (1933, pp. 71–2), that all bodies which seem to come from elsewhere "must at some previous point have suffered an interaction with other matter which threw them off their natural courses". However, "as we follow them still further back they must, in accordance with Weyl's coherency postulate, approach asymptotically the world line of O". (O being here the observer at rest.) By choosing the word "postulate" to characterize Weyl's hypothesis, Robertson has touched upon the most delicate of all problems, which is the actual sense in which it must be taken as reflecting some intrinsic property of the world.

In order to appreciate the import of the hypothesis as postulate, let us focus on how the 'limits of the universe' behave with regard to the 'limits of observability'. Imagine the original hyperboloid reduced to two dimensions as in Fig. 5.7, in which all points at infinity have been brought back to finite distance. (This representation is suggested by the techniques used in Hawking and Ellis 1973.) Our two lines $L_{-\infty}$ are the actual limits of the universe by virtue of the definition of range of causal action. As for the line $L_{+\infty}$, they are the limits beyond which no observation from O may reach: these parallel generators are the paths of light rays approaching asymptotically (in the infinitely distant future) the world-line of O; in other words, the totality of observable events from O is comprised by the past cone (represented by the two lines $L_{+\infty}$) at $t = +\infty$. The point is that these two limits, of the universe and of observability, do not quite coincide. A point like P' is certainly observable, since it belongs to the past cone at $t = +\infty$, yet it obviously has no causal action on O since it lies outside the 'limits of the universe'. Strangely enough, P' is elsewhere with regard to the 'origin', but not so with regard to the 'history' of the universe.

Among the reactions to Weyl's hypothesis of 1923, there is one which is worth mentioning because it testifies to Weyl's deep and obstinate conviction that no physical cosmology is possible without a speculative assumption of

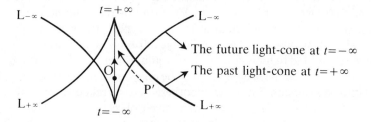

FIG. 5.7. Weyl's principle in a static frame: the limits of observability versus the limits of the universe.

that kind. This is the reaction by astronomer and physicist, Ludwik Silberstein, who polemized about the divergence of geodesics. As he put it, "with this arbitrary hypothesis, and through a number of rather obscure technicalities", Weyl seeks and finds for the redshift "a unique positive sign, yielding always. . .a *positive* 'radial velocity'. Dr Weyl himself insists on this feature as a result of his sublime guess about the remote past of all the stars" (1924a, p. 909). Silberstein argued that the tendency for all particles to scatter would not imply just receding velocities: he sought to re-establish the older idea of some sort of equilibrium between receding and approaching motions. He was wrong, however, not so much because his criticism of Weyl's hypothesis was misplaced, but because he used the incorrect identification by Shapley of globular clusters and extra-galactic nebulae (see Chapter 1). This observational basis for the evaluation of distances, which was soon to be overthrown by Hubble, later turned Silberstein's following statement against himself: "The pencil. . .may as well diverge into the past (. . . $+\infty$ and $-\infty$ can be interchanged), i.e. converge into the future. Whether one or the other or perhaps neither is the case prevailing in Nature, can be decided only by observations on distant celestial objects, not by drawing space-time figures" (1924b, p. 515 n.). (See Weyl's reply in his 1924a, as well as an overall view on the polemic in Silberstein 1930, pp. 179-81.)

The need for making the common origin of a system Σ lie in an infinitely distant past has been explained by Weyl in the following terms:

Stars that do not belong to it [the system Σ] lie beyond the range of influence of [the observer star] A during their early history. On the other hand, it is true that if A' is a star of the system, A ceases to act upon A' from a certain moment of its history on, even though conversely A' remains in the range of influence of A during its entire history (1930, p. 939).

It is for this very reason that, as he says,

Our assumption is that in the undisturbed state the stars form such a system of common origin.

If the difficulty with a point such as P' is that there seems to be more within the range of observability than allowed by the postulated range of causality, the case of a point like P" in Fig. 5.8 reveals how truly necessary Weyl's postulate may be. P" leaves the range of observability of O and crosses $L_{+\infty}$ in a finite fraction of its proper time. Now, if the existence of this point is warranted by its coinciding with O at $t = -\infty$, and if furthermore there is no reason to deny its existence after crossing $L_{+\infty}$, then the redshift of the point must be such that it may cease to be observable *without* losing its causal connection. It is the purpose of the second part of Weyl's article to show that an appropriate re-formulation of the redshift law meets this requirement.

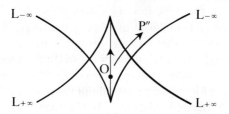

FIG. 5.8. P″ leaves the range of observability but is still causally linked to the universe.

3. The redshift revisited

This reformulation yields throughout space–time those *positive* radial velocities that Silberstein had found so objectionable. In retrospect, from Silberstein's objection we see that the fundamental question of how to connect theory and observation at the cosmological level presented itself during these years in a peculiarly abrupt form. With hindsight this can be articulated in the following way. The Doppler effect, which is observed in the light from distant heavenly bodies, is theoretically 'classical' in the sense that the corresponding velocities conform to the laws of special relativity, i.e. no faster-than-light phenomenon is allowed. Supposing that the velocities increase with distance—as the De Sitter model suggested, and as observation tended to confirm—leads to some abrupt 'edge' at which the maximum receding velocity is that of light itself. From this perspective, the edge cannot be really different from the singularity of the De Sitter model, as Eddington had argued by suggesting a total equivalence of metric and Doppler redshifts. Clearly, however, the causally possible existence of events beyond the edge necessitates another interpretation, which was developed by Weyl in the following way.

He first examines the De Sitter universe in a sort of reduced, three-dimensional model (1923b, p. 231). Of the five dimensions of the homogeneous pseudo-Euclidean metric, we now have x_1, x_2 and x_3. Weyl makes a change of coordinates. If $x_1 = (x_1' + x_2')/2$ and $x_2 = (x_1' - x_2')/2$, then the three-dimensional hyperboloid $x_1^2 - x_2^2 + x_3^2 = 1$ becomes

$$x_1' x_2' + x_3^2 = 1. \tag{5.1}$$

Clearly, $ds^2 = -dx_1^2 + dx_2^2 - dx_3^2 = -dx_1' dx_2' - dx_3^2$. (The primes may be omitted in the following.) Our discussion is summarized in Fig. 5.9.

On the figure, x_2 is the symmetry axis of the system Σ. Figure 5.10 represents what happens in the plane $x_1 x_2$ (or $\xi_1 \xi_2$). The coordinates x_1 and x_2 are

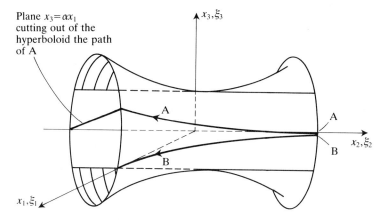

FIG. 5.9. The calculation of the redshift for two stars of the system Σ.

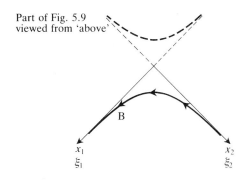

FIG. 5.10. The path of star B in the plane $x_1 x_2$ (or $\xi_1 \xi_2$).

always positive and can thus be re-written under the form e^q, where q is always real. Note that

$$\xi_1 \xi_2 + \xi_3^2 = 1. \tag{5.2}$$

As ξ_3 is zero, we have $\xi_2 = 1/\xi_1$. For the observer star B, $\xi_3 = 0$, $\xi_1 = e^\sigma$, $\xi_2 = e^{-\sigma}$, and $-d\xi_1 d\xi_2 = -(d\sigma e^\sigma)(-d\sigma e^{-\sigma}) = d\sigma^2 = ds^2$, which defines the proper time of B. For A, as $x_3 = \alpha x_1$ and $dx_3 = \alpha dx_1$, we get $x_1 x_2 + x_3^2 = x_1 x_2 + x_3 \alpha dx_1 = x_1(x_2 + \alpha x_3) = 1$ and

$$-ds^2 = dx_1(dx_2 + \alpha dx_3). \tag{5.3}$$

As a result,

$$x_2 + \alpha x_3 = \frac{1}{x_1} \tag{5.4}$$

and if $x_1 = e^s$, then

$$\text{a) } x_2 + \alpha x_3 = e^{-s} \text{ and b) } x_3 = \alpha x_1 = \alpha e^s. \quad (5.5)$$

But is $x_1 = e^s$ compatible with the definition of ds^2, i.e. is it the case that s is the proper time? From (5.3) and (5.4) we have

$$dx_2 + \alpha dx_3 = d\left(\frac{1}{x_1}\right) = \frac{dx_1}{x_1^2},$$

which indeed enables us to verify that

$$dx_1 \left(-\frac{dx_1}{x_1^2}\right) = -\frac{dx_1^2}{x_1^2} = -\frac{[d(e^s)]^2}{e^{2s}} = -\frac{ds^2 e^{2s}}{e^{2s}} = -ds^2.$$

Weyl now has to find the relation holding between s and σ. This relation is physically determined by a ray of light joining A and B at the time of observation. This implies that the interval between x and ξ must be zero. Weyl's way of calculating the relation between s and σ is quite sophisticated. Here is perhaps a more simple route which does not change anything in his intention. The square of the interval between A and B is given by $(\xi_1 - x_1)(\xi_2 - x_2) + x_3^2 = 0$. Now Eqn (5.5a) implies that $x_2 = e^{-s} - \alpha x_3 = e^{-s} - \alpha^2 e^s$. So the expression for the interval becomes $(e^\sigma - e^s)(e^{-\sigma} - e^{-s} + \alpha^2 e^s) + \alpha^2 e^{2s} = 0$ or $e^{2(s-\sigma)} - 2e^{s-\sigma} + 1 - \alpha^2 e^{2s} = 0$. More generally this equation can be written in the form of a quadratic equation in K: $K^2 - 2K + B = 0$ in which $K = e^{s-\sigma}$ and $B = 1 - \alpha^2 e^{2s}$. So that the solution reads

$$K = e^{s-\sigma} = 1 \pm \sqrt{1 - 1 + \alpha^2 e^{2s}} = 1 \pm \alpha e^s.$$

Weyl's hypothesis amounts to supposing that the clocks at A and B were synchronized at $t = -\infty$. Therefore, the fact that observation σ takes place after emission s, leads to $s < \sigma$, that is, $e^{s-\sigma} < 1$. The minus sign is thus prevailing in the solution:

$$e^{s-\sigma} = 1 - \alpha e^s \quad (5.6)$$

Weyl's coordinate-free expression for the redshift comes in at this stage (1923a, pp. 322–3). This expression is a ratio of proper times; that is, $d\sigma/ds$. By differentiating the two sides of (5.6), we have $e^{s-\sigma}(ds - d\sigma) = -\alpha e^s ds$, so that we end up with

$$\frac{d\sigma}{ds} = 1 + \alpha e^\sigma. \quad (5.7)$$

So far, all calculations could be carried out in coordinate-free manner, but the most delicate problem now crops up, which is how to determine the *distance* AB in such a way as to allow the redshift to be measured as a function of this distance. What will be the frame in which the distance is measured?

THE REDSHIFT REVISITED 329

Weyl is quite explicit in assuming that his deduction is not in the least limited to the static field (1923a, p. 232); by which he certainly has in mind the fact that the discussion about redshifts is coordinate-free rather than anything like the possibility of non-static coordinates. But it is precisely at this point that some distinction must come in between the geometrical and the material:

In our hyperboloidic world, an observer of infinite lifetime who lives in a star of the system receives information of only a cuneiform sector of the *material cosmos*, i.e. that portion of the world which covers the range of perturbation (*Störungsbereich*) of the material system (p. 323).

Matter seems to be a sort of perturbation over and above the 'natural' geometry of the hyperboloidic world. The cuneiform sector, however,

lets itself be treated, according to Einstein, in static coordinates; therefore the world is static for the observer, and this makes clear what should be understood by space and time from his viewpoint.

As far as the obtaining of the redshift is concerned, the static frame is nothing more than a convenient way of picking out the set of points which are contemporary of some given set. This frame is always perpendicular to the observer's world-line. Now, the direction of the world-line of B is defined by $(d\xi_1, d\xi_2, 0)$, so that the vectors with scalar product

$$(d\xi_1, d\xi_2, 0) \cdot (x_1, x_2, x_3) = 0 \tag{5.8}$$

define the plane perpendicular to this direction. In fact, the ds^2 is equal to $\vec{ds} \cdot \vec{ds}$, that is, $ds^2 = -dx_1 dx_2 - dx_3^2 = -\frac{1}{2}(dx_1 dx_2 + dx_2 dx_1) - dx_3^2$. Thus (5.8) yields

$$x_1 d\xi_2 + x_2 d\xi_1 = 0. \tag{5.9}$$

On the other hand, following (5.2), we have $\xi_1 d\xi_2 + \xi_2 d\xi_1 = 0$. Inserting the value of $d\xi_2$ in (5.9), $x_1 = x_2 (\xi_1/\xi_2)$ and this again in (5.1): $x_2^2 (\xi_1/\xi_2) + x_3^2 = 1$ or, taking (5.2) into account, $(x_2/\xi_2)^2 + x_3^2 = 1$. As $\sin^2 x + \cos^2 x = 1$, we can finally write

$$x_1 = \xi_1 \cos r, \, x_2 = \xi_2 \cos r, \, x_3 = \sin r. \tag{5.10}$$

The characterization of 'space' by r will soon be justified. It is important to notice the role of Weyl's hypothesis in establishing the parametrization (5.10). Indeed, when $x_3 \to 0$ (and then ξ_3 also tends to zero), that is for $t = -\infty$, r tends towards zero: A and B do coincide at $t = -\infty$. Had Weyl chosen something like $x_3 = \cos r$ and $x_2 = \xi_2 \sin r$, his hypothesis would not have been satisfied. This will have quite important consequences on the comparison between the redshift and the radial velocity of the nebulae.

We will now justify the assertion that the quantity r stands for the distance AB in the static frame, i.e., that it is the 'natural' space variable in this frame. If it does, we should have $-\,ds^2 = dr^2$. (Note that the 'natural' time variable is σ, and $d\sigma = 0$; all points are contemporaneous.) Indeed, by differentiating the three expressions in (5.10), we get $-\,ds^2 = dx_1 dx_2 + dx_3^2 = \xi_1 \xi_2 \sin^2 r\, dr^2 + \cos^2 r\, dr^2$. As $\xi_1 \xi_2 = 1$, this becomes what we wanted, $-\,dr^2 = ds^2$. The spatial position of A, $x_3 + \alpha x_1$, becomes $\sin r = \alpha \xi_1 \cos r$ when expressed in parametric form, that is,

$$tg\,r = \alpha\,e^\sigma. \tag{5.11}$$

Recalling (5.7) we have the result

$$\frac{d\sigma}{ds} - 1 = tg\,r \tag{5.12}$$

in which the left-hand side is nothing other than the coordinate-free expression for the redshift. It is now possible to directly compare this distance function for the redshift with the radial velocity ($dr/d\sigma$) of the nebulae. From (5.11) Weyl derives

$$\frac{1}{\cos^2 r}\frac{dr}{d\sigma} = \alpha e^\sigma, \text{ or } \frac{dr}{d\sigma} = \sin r \cos r \tag{5.13}$$

the fundamental implication of which is:

The displacement (5.12) is thus ($1/\cos^2 r$) times larger than the corresponding radial velocity (1923b, p. 232).

This conclusion directly clashes with Eddington's: the displacement of the spectral lines due to the distance, and predicted by the De Sitter metric, is not equal but *greater* than the displacement due to the Doppler effect. Translating this conclusion into Eddington's own words: the anticipation of the motion of recession can no longer be total. There is a residual effect which does not allow simple elimination of the motion, and this is a *distance effect*.

The effect implicitly provides the desired answer to the problem of how to look at an event such as P'' in terms of 'well-behaved' motion. For when P'' crosses the 'limits of the universe', it does not necessarily overtake the velocity of light; rather, the distance has become so great that its redshift, as observed from O, tends towards infinity.

4. Weyl's principle physicalized: a) Weyl's own approach

Weyl had introduced his new cosmological chapter in the 1923 edition of *Space, Time, Matter* by discussing a theoretical critique of Einstein's cylindrical model. In the Euclidean plane, the relativistic equations of gravitation have an infinite number of solutions as long as boundary conditions at infinity have not been specified. Poisson's equation is not different in this regard, but while the zero potential at infinity is just a convenient way of avoiding trouble in solving this equation, the finite universe of relativity makes the potentials constant throughout space–time, and this constancy is the only solution (1923a, p. 288). Now, if the quantity of matter is made proportional to the radius of curvature, as in Einstein's 1917 model, then it obviously fulfils exactly the same role as infinity in the pre-relativistic universe; in Weyl's language, this world matter creates the homogeneity of the "guiding field" (*Führungsfeld*). Matter creates the metrical field; its distribution defines the world geometry. It is the merit of the De Sitter model, however, to shed new light on how this is possible. For this phenomenon of matter, conditioning the field "does not let itself implement in the sense that 'far from all matter, or when all matter is abolished, there is no guiding field', i.e. the field becomes undetermined" (1923a, p. 296). In fact, the argument that the field becomes undetermined at infinity reminds Weyl of the mistake that Einstein had found so difficult to correct in the years 1913–1915, according to which allowing generally covariant transformations would imply an infinite number of solutions to the field equations (Weyl 1924b, pp. 202–3). Weyl forcefully argues:

To the question as to why the compass of inertia and the compass of the stars go very exactly together, I do not know of any other answer than this: the matter has disturbed only very slightly the state of rest of the 'Father Ether'; the concordance will be all the more exact as there is little matter present (1923a, p. 297).

The overlap between metrical and material homogeneity is a consequence of there being very little matter in the universe. Such a conception derives from Weyl's attempt to unify field physics; his new perspective on the nature of 'boundary conditions' in the De Sitter model boosts his hopes.

Indeed an analogy suggests itself between the postulated undisturbed state of matter at $t = -\infty$ and the correlation of the gravitational with the electromagnetic field that Weyl had so far worked out. In the absence of matter, the electromagnetic field vanishes. But, as Weyl now says, the equations which represent vanishing electromagnetic potentials do not lead to the view that no field at all is present. Rather, they determine this field as being at rest, and the presence of matter is concomitant with the motion of the charges. Correlatively,

The situation of rest of the metrical field is the homogeneous metric as it prevails in the De Sitter hyperboloid (p. 296).

It is therefore in the De Sitter universe that the connection between the electromagnetic and the gravitational potentials shows itself with natural clarity, provided Weyl's hypothesis has been adopted as physical. Or rather, the realization of this connection at the infinitely distant past tends to *prove* the physicality of the hypothesis. The instant at which the light cone is only light is also the only instant which yields the situation of rest in this universe. In other words, the ether at rest, which makes all metrical homogeneity possible, is the physical image of $t = -\infty$; this solves the old problem of boundary conditions because the new standard of rest is nothing 'substantial'. But the actual appearances result from this universe being inhabited, as it were, by what Weyl calls "the spirit of non-rest" (*Geist der Unruh*), which is matter. Thus

the past remains closed behind us; no one enters its petrifying force (*bindenden Kraft*). The world remains open and unpredictable before us in the future $+\infty$. What foolhardiness would it be to further impose boundary conditions (p. 297).

This argument may well represent the culmination of Weyl's painstaking search for a philosophically satisfactory view of physical science, as the evidence accumulated over the years ever since the rise of relativity. It brings together the original relativistic way of thinking about the metric of the world, the topological anathema on closed time-like curves, and the development of unified field physics. For the behaviour of the De Sitter universe testifies in every way to the self-sufficiency of the field equations, that is, their independence from purely mathematical boundary conditions. The sought-for compromise between Einstein's and De Sitter's models has moved from finding the mean between the two (in terms of the quantity of matter that happens to be present in the universe) to establishing what each of them contributes to our understanding of unified field physics. Contingency has been entirely removed: the new solution is independent of the contingent fact that very little matter (or disturbance) is present. What the solution says is that the fundamental separation of homogeneous form (or metric) and chaotic content (which Weyl had identified from the outset with the Riemann–Einstein world view) can be preserved as ultimate foundation of field physics if allowed to be re-thought in terms of rest and motion (or infinitely distant past and physical time). For by dovetailing with general relativity and beyond, the necessary homogeneity of the form would lose the merely ideal character it still had as requirement of special relativity.

It must be noted that the use of static coordinates is also implicitly justified by this argument. Weyl had begun by stating the non-static character of the De Sitter metric as totality. But the static character remains primordial in the sense that the state of rest, as a correlate of metric homogeneity, occurs only at $t = -\infty$. Weyl's principle of causality, formulated as it was here, does not say anything about the infinite future, which is just where the non-static character manifests itself. The principle always sends us back in time to an ideally remote past, when the geometry of the world was still untouched by

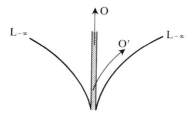

FIG. 5.11. If Weyl's principle is adopted, then the static frame covers only the narrow dotted area around observer O.

the properties of motion inherent in matter. However, as a last comparison with Eddington shows, Weyl's principle certainly reopens the question of the universality of static coordinates.

The existence of the lines $L_{-\infty}$ and $L_{+\infty}$ does not depend on any particular frame. But just how do they exist in the static frame? Strictly speaking the static frame simply merges with the object carrying it. The world continues to appear at rest only within a very narrow neighbourhood around the object, so that finally the static frame covers quite a small portion of the observable universe (the dotted section on Fig. 5.11). The price to pay in order to maintain Eddington's assumption that the frame covers the whole of the observable universe now appears in full light. For an observer carrying his static frame but moving sufficiently fast, as O' on the figure, there is always the possibility of 'overtaking' the frame and falling into a region of imaginary time; this could be avoided only by changing frame at each such displacement. (See the caustic remarks by Schrödinger on this particular point in his 1956, p. 21.) Now, by adopting Weyl's hypothesis, the observer is equipped with only one real time which re-unites all displacements as well as all observers. But this universal time issues from a no less disturbing 'genesis'. It was the staggering project of non-static cosmology to bestow unambiguous physicality on this genesis; by doing so, however, the new cosmology would exist independently and gradually divorce itself from Weyl's original attempt to incorporate unified field physics.

5. Weyl's principle physicalized: b) The adventure of non-static cosmology

As he was developing his new analysis of the De Sitter cosmology, Weyl announced to Einstein in a letter of 22 May 1923 that he was working on the possibility of predicting the sign of the cosmological constant. This was to be understood on the basis of his proposed generalization of gravitation and electromagnetism, and he suggested that it could be done by deducing the field equations from a variational principle. But Einstein was not completely happy with this reasoning about Λ, because he thought the most general theory, of whatever nature, should allow freedom of its sign. And so he replied that if the inclusion of electromagnetic invariants into the field equations would lead to a requirement over the sign of Λ, this would run against the spirit of the relativity theory. What matters, according to Einstein, is the existence rather than the sign of Λ. Accordingly he defended a position conflicting with Eddington's when he said that, by virtue of mutual attraction between material points in the De Sitter universe, "if there is no quasistatic world, then away with the cosmological term" (Einstein to Weyl, 23 May 1923).

But whether we look at cosmology or at the attempt to unify gravitation and electromagnetism, doing away with Λ proved to be a very troubling task. Thus, trying out a unification by throwing Λ away, Einstein was soon to acknowledge that "the term comes in again by a backdoor" (26 May). He was now in a state of despair as he thought that "perhaps the field theory has already yielded up all it could", before concluding that "the mathematics is well and beautiful, but nature still leads us around by the nose". Einstein was followed by Weyl himself shortly after the advent of the formalism of quantum mechanics in 1927. On 3 February of that year he wrote to Einstein that "all the properties I had so far attributed to matter by means of Λ are now to be taken over by quantum mechanics". And he suggested that it was necessary to wait and see if quantum mechanics could be made compatible with any field theory. To which Einstein replied in a way which, as we know, virtually determined the rearguard fighting which marked the rest of his life: "I cannot be happy with the half-causal and the half-geometrical, which is burying one's head in the sand. I still believe in a synthesis between the quantum and the wave conception" (26 April 1927). Some twenty years later, he wrote to Lemaître, one of the chief advocates of Λ for *cosmological* reasons, that the problem is not so much that of a new constant as that of a new unified theory: "one should not forget that the gravitational field with the energy tensor on the right side is only a provisional form which has to be substituted finally by an equation for the total field" (26 September 1947). Conversely, Einstein was not ready to believe "that this step [would] influence the aspect of the cosmological problem". As for Weyl, in the years which immediately followed the establishment of the quantum theory, he tried out various possibilities which would yield the quantization of the field equations, suggesting a view according to which the electromagnetic field should not be combined with gravitation at all; rather it would become "a necessary accompaniment of the matter-wave field" (1929, p. 332). Clearly enough, with regard to the general idea of universal causality, the impact of quantum mechanics was so enormous that the very way of talking about large-scale unifying perspectives had to be re-examined critically. In the context of a discussion over determinacy in quantum mechanics, Silberstein has depicted the uneasy feeling which began to emerge "that the whole history of a single system, either a fragment or the totality of Nature, forms a *unique*, non-bifurcated string of events or states, or that it is uniquely determined, is scarcely anything more than a tautology, amounting to a mere restatement that a system cannot at the same time be in two or more different states" (1933, p. 79). At this level of criticality, it seems that quantum mechanics did provide Silberstein with what was necessary to undermine such a grand hypothesis as Weyl's.

With regard to cosmology, when Einstein prepared in September 1923 his second account of Friedmann's epoch-making paper on the dynamic solutions to the field equations, correcting his first note of 1922 in which he had (wrongly) detected a miscalculation, he was so troubled that in preparing the draft retraction he concluded by saying that the soundness of the dynamic spherically symmetric solution is something mathematical only. He added that "a physical significance can hardly be ascribed to it", and then crossed this remark out before sending the note to print. True, referring much later to Friedmann's work, Einstein emphasized that "it was not influenced by experimental facts", and that "the general theory of relativity can account for [Hubble's experimental discoveries] in an unforced way, namely, without a Λ-term" (1931, p. 235). Interestingly, there is little doubt that Einstein never lost sight of Friedmann's paper, even throughout these obscure years until the establishment of the expanding universe theory, as it is Einstein himself who (some time early in 1929) made Lemaître aware of the Friedmann paper; this is what Lemaître acknowledged in his article that year (1929, p. 216 n.).

The story of non-statical metrics for the universe is one of sudden rediscovery. In drawing up the article 'Space-Time' for the 1929 edition of the *Encyclopedia Britannica*, Einstein could still write, rather imperturbably, that "nothing certain is known of what the properties of the space-time continuum may be as a whole. Through the general theory of relativity, however, the view that the continuum is infinite in its time-like extent but finite in its space-like extent has gained in probability" (1929, p. 108). It was precisely this view which came to be completely challenged early in 1930.

On 19 March 1930, Eddington sent to De Sitter a copy of Lemaître's paper of 1927, with the following note across the top of the front page: "This seems a complete answer to the problem we were discussing." The discussion had started after De Sitter had given an informal communication at a meeting of the Royal Astronomical Society in January 1930, where he expressed his doubts about both solutions A and B. Basing his argument on new estimates of the masses of the nebulae (the data were disputed by Eddington), he proffered the dilemma that the actual universe apparently contained enough matter to make it an Einstein world and enough motion to make it a De Sitter world. Eddington had then hit on the puzzle of "why there should be only two solutions. I suppose the trouble is that people look for static solutions"

(RAS 1930, p. 39). But then he added this comment summing up his position, with an (unintended?) pun: "Solution B is. . .non-static and expanding. But as there isn't any matter in it that does not matter." When Lemaître saw Eddington's comment, he wrote to remind Eddington of his paper of three years earlier. In the meantime De Sitter himself had gone further as he had suggested in a new paper that *"both the solutions A and B must be rejected, and as these are the only statical solutions of the equations. . .the true solution represented in nature must be a dynamical solution"* (1930a, p. 482). And when Eddington sent De Sitter a copy of Lemaître's paper, he had added, "A research student McVittie and I had been worrying at the problem and made considerable progress; so it was a blow to us to find it done much more completely by him (a blow softened, as far as I am concerned, by the fact that Lemaître was a student of mine)." Perhaps Eddington should have said that, as Lemaître had been a student of his, the blow was all the more disconcerting, as McVittie himself reported (see above in the Introduction). What Eddington was working on with McVittie was the question of whether Einstein's cylindrical world is stable, using two papers by Robertson as a basis. In fact, there had been an earlier paper by Lemaître on the dynamical aspect of the De Sitter universe, which was written during his year as Eddington's research assistant at the Massachusetts Institute of Technology in 1925. By the end of 1925, when he returned to Louvain in Belgium in order to take up a lectureship there, Lemaître wrote Eddington (29 November) to thank him for the kind recommendation he had given with regard to his application for the position, and apologized that he could not yet send him a printed copy of the 'Note on De Sitter's Universe' that he had shown him in July.

Quite significantly, just after he had received from Eddington a copy of Lemaître's 1927 paper, De Sitter wrote to Lemaître on 25 March 1930, praising him for his "simple and elegant solution". But he was quick to mention that he too, just as Einstein, had come across (some time in the 1920s) a non-static solution in almost complete form: not from Friedmann's, but one that had been sketched by Tullio Levi-Civita in his lectures on differential calculus. This solution, De Sitter said, "had seemed impossible to me from a physical standpoint". True, Levi-Civita had written down the general equations for all spatially uniform (and static) metrics as early as 1917, and in his lectures he went on to expand these equations to the case of a non-constant curvature of space. Interestingly enough, he found he could derive the Einstein and the De Sitter models as particular cases, but he did not bother with the physical implications of his non-static metrics. On the contrary, he tried to show that the De Sitter original solution could be made just as 'physical' as the Einstein one if the mass-horizon is "taken to refer solely to accidental masses (sufficiently small not to modify the field perceptibly), and not to those uniformly diffused masses which constitute [De Sitter's space-time], the equilibrium of which is automatically assured by the gravitational equations" (1927, p. 435).

It is a conceptually more natural sequence to follow Lemaître's and Robertson's papers first as they dealt directly with the consequences of Weyl's principle. Only later shall we return to the historically pioneering work of Friedmann, which leaves us with a bewildering impression of premature but complete accomplishment.

Eddington may have had one good reason for being oblivious of Lemaître's work: it was Lemaître's purpose, in his first note on De Sitter's universe, to show that the solution "has to be abandoned, not because it is non-static, but because it does not give a finite space without introducing an impossible boundary" (1925, p. 41). The point is that more natural coordinates for this universe would cover the totality of the range of action defined by Weyl. The coordinates are the non-static ones, but the new separation of space and time implied by these coordinates seems to repeat in its own terms a difficulty with space that already existed with the original, static coordinates. For if all points are equivalent in the De Sitter static space–time metric, they are not so with regard to space alone. As Lemaître explains, even though the choice of origin is in itself immaterial, once the choice has been made the origin behaves like a priviledged, absolute centre:

The lines of constant [spatial] coordinates, which are considered as the direction of time, are not geodesics with the exception of the one $x = 0$, which passes through the origin (p. 37).

Lemaître aptly opposes the universe as space–*time* and space alone, it being understood that, as he said a little later in a popular lecture, "the presence of matter determines a natural partition of the universe into space and time" (1929, p. 215). He realizes that this is the fundamental opposition which creates the strange properties of inhomogeneous gravitational field that were *not* initially present in the construction of the metric (as Lanczos had already noted). As he put it, the metric "that De Sitter had adopted does not satisfy the condition of homogeneity with regard to space, in which a non-homogeneous gravitational field is introduced" (p. 215). Just as Eddington had recognized that the properties of motion in the De Sitter universe made its horizon illusory, the task was now to locate and remove the illusion with regard to the centre:

It is clear that such an introduction of an apparent centre in a universe which, by definition, has none, is objectionable for a study of the properties of this universe (1925, p. 37).

Lemaître's new separation of space and time goes like this (see Fig. 5.12). From all ellipses that pass through the 'neck' of the hyperboloid, Weyl

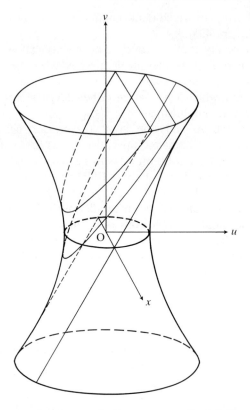

FIG. 5.12. The Lemaître frame of reference covers the totality of the range Σ.

had originally picked out those that contain the x-axis. In such contemporary spaces, time is a function of the ratio v/u. Lemaître's new time function is defined by the family of parabolas

$$v + u = \text{constant}. \tag{5.14}$$

This defines contemporary spaces as space-like geodesics which are all perpendicular to the time-like geodesics. Contemporary spaces are now the parabolas obtained from cutting out of the hyperboloid the planes which are parallel to the x-axis and which form a 45° angle with the u-x plane. In general, therefore, the new time function can be written

$$\tau = f(v + u). \tag{5.15}$$

There is one special problem when the constant in (5.14) is zero, that is, when $\tau = 0$. This yields the pair of parallel generators $x = \pm R$, which leads Lemaître to say:

The origin of time becomes a time absolutely distinct from every other. At the instant $\tau = 0$, all the geodesics of the space are geodesics of the universe; at any other instant τ', there is no geodesic of the space which is a geodesic of the universe. The central point in space is removed, but now a central time has been introduced (p. 38).

So the elimination of the difficulty with space creates one with time. Is the latter difficulty avoidable by specifying the time function in some way? This is what Lemaître tries to do. The spaces ρ are some function g of (x, v, u), and the world lines ρ = constant must be orthogonal to τ = constant. The planes cutting out the world lines ρ = constant pass through the origin O as these lines are geodesics; they are defined by $x/(v+u)$ = constant, and thus $\rho = g(x/(v+u))$. When $v+u = 0$, $\rho = \infty$ and $x = \pm R$, which allows to write $g = R$:

$$\rho = \frac{Rx}{v+u}. \tag{5.16}$$

As for the time function τ, f can be taken so that

$$\tau = \log \frac{v+u}{R}, \tag{5.17}$$

for then the parabola which degenerates in the pair of straight lines is relegated to either the infinitely distant past or the infinitely distant future. By the same token the central time is now removed to either infinity.

Further calculation shows that (5.16) and (5.17) are compatible with this form of elementary space–time interval:

$$ds^2 = -dx^2 - du^2 + dv^2 = -e^{\pm 2\tau} d\rho^2 + R^2 d\tau^2. \tag{5.18}$$

In this metric, space is a function of time, i.e. the coordinates are clearly non-static. This, according to Lemaître, does full justice to Eddington's view on the non-static character of De Sitter's universe. But what he finds objectionable is the price to pay for the elimination of the central time; for space is now Euclidean again, and so, Lemaître argues, "we are led back. . .to the impossibility of filling up an infinite space with matter which cannot but be finite" (1925, p. 41). The only advantage of the model is that it offers quite a straightforward explanation for the redshift in terms of the very expansion of the coordinate distance separating a light source from an observer. And it was this advantage that made Lemaître reconsider his analysis some two years later. For if the radius of curvature of the *Einstein* model is made variable with time, the full homogeneity of the finite space is never endangered and the redshift can be interpreted as a Doppler effect "equal to the excess over unity of the ratio between the radius of the universe when the light is received and the radius when it was emitted" (1927, p. 55). Nor is there any problem with regard to the topology of Einstein's spherical space, for "even if there were no absorption these [ghost-]images would be displaced by several octaves into the infra-red and would not be observed" (p. 59).

Such is the "fair compromise" between the two models that Lemaître wanted. It was by following a very different route that Robertson, quite independently of Lemaître, then hit on a ds^2 for the De Sitter model similar to (5.18). In his view however, the Euclidean character of space was not an impediment. Robertson seems to have worked in total ignorance of Lemaître's papers, yet strikingly enough his argument looks like a response to Lemaître's early doubts about the viability of the De Sitter solution. It proceeds in two stages. Robertson begins by assuming that all such considerations of large-scale space–time are geometrical in essence, as "the aggregation of matter into stars and stellar systems may be ignored". Thus, as the total amount of matter in the universe "has but little effect on its macroscopic properties, we may consider them as being determined by the solution for an empty world" (1928, p. 835). Properties are to be distinguished from the coefficients of the space–time manifold: while the latter are coordinate-dependent, the former are structural. With regard to the ds^2 (5.18) in particular, Robertson finds that the properties of space–time are coordinate, that is, time-independent, which makes not only the geometry of space Euclidean but also the velocity of light independent of direction. A natural consequence of this is the second point of Robertson's argument, which is a transformed viewpoint with regard to the horizon. He finds that going back to an infinite spatial extent in those terms is not an objection of much weight, as "the closed character is maintained in the sense that the only events of which we can be aware must occur within a sphere of finite radius" (p. 837). This sphere defines the range of observability, which Robertson now clearly identifies with the two world lines $L_{-\infty}$ of Fig. 5.7 since, as he says, "all matter and light within our observable universe must have started from the boundary and will be eventually lost to it" (p. 839).

Let us try to make sense of Robertson's idea by looking at a test body in expanding space which would 'resist' expansion and would, therefore, get closer to an observer at the origin. Certainly the 'resistance' could not go for very long, as any time-like geodesic tends to approach asymptotically either one of the straight lines of the asymptotic cone. This means that, at any time, the ratios x/u and x/v, which define the coordinate ρ, tend to become constant, and the logarithmic scale of time sends the test body very quickly (exponentially) to this position. The expansion of space thus takes over any temporary resistance, and Robertson computes a value of *measured* distance beyond which no event may reach the origin. The most delicate point of how to interpret the status of those observationally inaccessible events is simply swept away:

Considerations involving the whence and whither of celestial objects lead to results of a rather paradoxical nature. . .but such long-time predictions cannot be taken too literally, as we have neglected their influence on the line element, an influence which undoubtedly plays an important role in such questions (p. 839).

For Robertson, certainly, the motivation of the whole enterprise of changing to non-static coordinates is the possibility of a "simpler interpretation" and a "convenient method" for the investigation of the De Sitter solution (p. 836). It is his "mathematically equivalent solution" which, he suggests, may overcome the supposedly physical paradoxes inherent in the static version. There is something more, however, beyond this avowed intention. For the restriction of cosmological considerations to the observable, and the concomitant idealization of matter distribution in terms of pure geometry, do imply quite an important advance towards re-unification of the two available models. So much so that, in Robertson's mind, the expression "empty world" does not seem to specifically refer to the De Sitter solution. He calculated a remarkable concordance between the radius of his observable universe, computed from the observed redshifts by means of Weyl's formula, and the radius of curvature of the Einstein universe that Hubble had obtained two years earlier from the computed value of the mean density of matter (p. 845). This, in fact, he took as the major result of his paper.

Just after Hubble had announced his famous correlation between distance and apparent radial velocity for the extra-galactic nebulae early in 1929, Richard C. Tolman published a comprehensive theoretical account of the relations between distance and Doppler effect in accordance with the De Sitter solution. In his paper, Tolman uses static coordinates throughout, apparently ignoring the recent work by Robertson and drawing extensively on the earlier work by Eddington, Silberstein, and Weyl. The contribution is nevertheless highly interesting and suggestive for our purposes. Tolman's attempt really pushes Weyl's hypothesis to its limits, instead of bypassing the issue of causality as Robertson's analysis in terms of non-static coordinates had done. In fact, Tolman's concern is no less than an attempt to remove the last ditch of mystery surrounding the whereabouts of an event such as P' on Fig. 5.7. Obviously, this event does not completely free the universe from the 'many-universes' objection to Eddington's interpretation that Weyl had tried to overcome. According to Tolman, the only way of avoiding the difficulty is by adopting Silberstein's "recognition of the fact that the equations of motion in the De Sitter universe are reversible" (Tolman 1929a, p. 268 n.). This allows him to look closely at what he calls the phenomenon of "continuous entry" or even "continuous formation" of nebulae.

The then recent observations by Hubble yielded an approximately uniform distribution of nebulae as a matter of *fact*, and this, Tolman argues, clashes with the *theoretical* tendency for the nebulae to scatter according to De Sitter's law. This leads him to talk about two alternative scenarios to account for the observed uniform distribution:

Either. . .the nebulae now under examination have entered the range of observation from the outside in recent enough time so that they have not yet been dispersed, or...these nebulae have actually been formed or created relatively recently within that range (pp. 266-7).

And because of continuous dispersion, either one of these processes supplying the nebulae in order to maintain uniform concentration must be continuous as well. Tolman found the "continuous entry" hypothesis quite a natural one—this tallied with his adoption of Silberstein's standpoint—but of course it was not easy to accommodate the observed great preponderance of positive Doppler effects (redshifts) unless severe requirements be assigned to the values of the parameters involved in De Sitter's equations of motion. As for the hypothesis of "continuous formation" of nebulae that would take place within the range of observation, Tolman gave it little credit, describing it as "thin speculative ice" (p. 271).

Resorting to "special acts of creation", as he called them, was thus not to Tolman's taste. Then in July 1929, Weyl wrote his new paper on cosmology in which he discussed both the non-static formalism introduced by Robertson and Tolman's hypotheses which he strongly criticized. On the other hand he convinced himself that Robertson's cosmology was identical to his. Interestingly, even though Weyl acknowledged identity of results with Robertson, he was now quite skeptical that the power of geometrical considerations alone would settle the question of how to understand physically the nature of the systematic and distance-proportional redshift of extra-galactic nebulae. The geometrical account of the redshift, he said, was to be examined only on the basis of its serious virtues as possibility (1930, pp. 936-7)—a word he had used many years earlier while exploring the imaginary worlds of masslessness. He speculated that a physically more promising explanation would develop along the line of the argument Zwicky had just proposed (1929), namely, the so-called hypothesis of tired light: there would exist a mechanism by which light surrenders a small fraction of its energy to nebulae and other matter (intergalactic dust) which it passes on its journey to us; the redshift in the observed spectrum would result from this gravitational interaction. Nothing has ever come to support Zwicky's theory, but Weyl's sudden change of attitude is worth noting. Next, taking the geometrical alternative seriously, Weyl is led to argue that "Tolman's careful investigation shows anew that nothing like a systematic redshift can be derived solely on the basis of the constitution of the metric field in its undisturbed state" (1930, p. 937). Weyl repeats his hypothesis that in the undisturbed state the stars form a system of common origin in an infinitely distant past. The Robertson's coordinates are, to use Weyl's word, the "convenient" way to reflect the properties of this undisturbed state because they cover *the totality of the system* Σ, which is one half of the De Sitter hyperboloid. This avoids the future disturbances being interpreted as either continuous entry or continuous formation, for they all emanate from the same origin. Weyl's principle is now fully physical because the divergence of the world lines $L_{-\infty}$ corresponds to the expansion of space–time itself; at the infinitely distant past there never was any elsewhere nor any elsewhen.

The moral of Weyl's new perspective on his grand hypothesis of 1923 is ambiguous. He was on Robertson's side with regard to the mere convenience of the non-static coordinates, yet his reasons for agreeing with this view were anything but a philosophy of convenience (an emphasis on the privileges of pure geometry, which certainly would have affinities with conventionalism). And, finally, he shared with Tolman, the 'empiricist', a distaste for special acts of creation.

6. Towards the big bang

When Robertson became familiar with Weyl's new paper, just before it was actually published, he also discovered Lemaître's 1925 paper. The impact of these readings was significant, for he went on to write on the foundations of relativistic cosmology from an entirely different perspective. This announced a profound and revolutionary way of thinking, which virtually determined the future course of cosmological science until our day.

Robertson's purpose was now to derive the Einstein and the De Sitter models as particular cases of a class of non-static solutions. In other words, he was trying to establish the most general line element suitable for relativistic cosmology. He approached the question through a method which, however heuristically powerful it proved to be, was to highlight the problematical status of the future theories of the expanding universe, because this method assumes from the outset that geometry dictates physics. That is, while Lemaître was considering that only the matter distribution could impose some natural partition of space–time into space and time, Robertson reversed the strategy and found he could describe "the actual world in terms of coordinates which effect a natural separation of it into space and time and [determine] its ideal background by a single assumption which is but the concrete expression of a uniformity implied by the very concept of a system of relativistic cosmology" (Robertson 1929, p. 827). One of the most interesting aspects of the paper was that it signalled a reaction to a new attempt made by Tolman to account for the distances and Doppler shifts for the extra-galactic nebulae in a more natural way. Tolman made the bold suggestion that perhaps there were metrics other than those of Einstein and De Sitter which could be compatible with cosmological behaviour. By new line elements for the universe, he had in mind "the possibility. . .of modifying the physical assumptions so as to permit a different type of line element" (Tolman 1929b, p. 298). Basically this amounted to criticizing the necessity of Λ as a *positive* quantity in the De Sitter model, for "if Λ were negative. . .R would be imaginary and the universe would not be closed" (p. 302). As for Einstein's universe, Tolman argued that the pressure could be inferred as positive instead of zero if there were a lot of radiation filling the universe. Still pondering on the behaviour of matter, Tolman concluded his paper by noting that his "assumption of a static line element takes no explicit recognition of any universal evolutionary process which may be going on. The investigation of non-static line elements would be very interesting" (p. 304). In fact, this investigation was not to be approached first from the physics of the universe, but rather from the geometry; this task was taken up by Robertson.

Robertson's single and basic assumption relates the homogeneity and isotropy of space to a unique cosmic time function:

Space-time shall be spatially homogeneous and isotropic in the sense that it shall admit a transformation which sends an arbitrary configuration in any of the 3-spaces t = const.,. . .into any other such configuration in the same 3-space in such a way that all intrinsic properties of space-time are left unaltered by the transformation. That is, any such configuration shall be fully equivalent to any other in the same 3-space in the sense that it shall be impossible to distinguish between them by any intrinsic property of space-time (1929, p. 823).

From the outset Robertson realizes that cosmology does not and cannot deal with arbitrary coordinate transformations. Rather, the problem is to find the transformation which leaves unaltered any arbitrary configuration. Only after this has been established may the physical considerations be tackled of the actual matter distribution that defines a configuration. Thus Robertson explicitly moves away from all attempts which first introduce assumptions with regard to the matter-energy tensor and then consider how the field equations may be satisfied. He wants to make full use of "the intrinsic properties of homogeneity and isotropy attributable a priori" to all cosmological spaces, instead of starting (as Einstein and De Sitter did) from "defining manifolds which do possess the desired uniformity" (p. 822).

Robertson formulated his assumption so as to make it compatible with the

"demand that [for] any stationary observer ('test body') in this idealized universe, all (spatial) directions about him shall be fully equivalent in the sense that he shall be unable to distinguish between them by any intrinsic property of space-time, and he shall similarly be unable to detect any difference between his observations and those of any contemporary observer" (p. 823).

This implied an important specification with regard to the global time function, namely, that all space-like hypersurfaces are orthogonal to the time-like hypersurfaces. Armed with the "ideal background", as he called it, Robertson claimed to "have expressed in another way the assumption made by other writers on the subject, above all by H. Weyl, that the world lines of all matter in the universe form a coherent pencil of geodesics" (p. 827). And indeed, Robertson now turned in this paper to the redshift problem, only to find that his assumption yielded a residual Doppler effect in the case of De Sitter's cosmology, an effect which was identical to that calculated by Weyl. Formally, Robertson's most general line element was similar to Lemaître's expression for the ds^2 as given by (5.18).

By proceeding a priori as Robertson did, it can hardly be denied that Weyl's earlier hypothesis lost much of its originally grand metaphysical impact. Robertson claimed to have changed something only in the expression, the formulation of the hypothesis. But this seemingly innocuous change implied no less than a revision of the meaning of 'intrinsic', as the cause of what induces motion was now identified with the properties of geometry rather than with those of matter. Therefore, an analysis of the fundamental concepts implied in relativistic cosmology in Robertson's terms proved to be worlds away from Weyl's, in which the depths of the relativistic world view were part and parcel of the concrete model of the universe. We have here the beginning of a movement which could be called 'despiritualization' of the problem of the universe, in reference to Weyl's desperate eagerness to conceptualize what he called "the spirit of unrest". Perhaps the best illustration of this disenchantment is exemplified by the status now reserved for the 'super-observer': Robertson is almost at the point of using 'observer' and 'test body' as synonyms. As it turned out, in Robertson's later work the preponderence of geometry over physics completed displaced the sense previously ascribed to assumption in cosmology. Thus, in a critical comment on Zwicky's tired light hypothesis, Robertson argued that such *ad hoc* explanations could never alter what was already gained from general relativity, for "so long as the homogeneity assumptions with which we started are not at variance with the observations we may consider relativistic cosmology as a simple corollary of the relativity theory" (1932, p. 226). With 'a priori' now acknowledged as 'hypothetical', our ignorance of the empirical universe was no longer an impediment to the establishment of a definite cosmological formalism.

It is all the more interesting that the transformation began to take place precisely when what was thought to be a completely satisfactory solution of the Einstein–De Sitter controversies emerged. For the 1929 paper by Robertson not only mentions Lemaître, but it also contains the first reference to Friedmann. There is a great deal of irony in Einstein's later judgement (in the appendix added to the second edition of *The Meaning of Relativity*, 1945, p. 127) on Friedmann that "the demand for *spatial* isotropy of the universe alone leads to Friedmann's form". For it was precisely Robertson's claim that Friedmann (like Tolman) had attacked the problem of the most general line element by means of "untenable assumptions on the matter–energy tensor"; these rendered the purely geometrical assumptions "unsatisfactory" (1929, p. 828n). True, Friedmann had carefully based his pioneering work of 1922 on a split between two kinds of assumption, physical versus geometrical. The physical assumptions were:

1) the gravitational potentials satisfy Einstein's equations with the cosmological constant; and
2) matter is pressureless and at rest.

As for the geometrical assumptions, Friedmann stated:

1) spaces are always of constant curvature; and
2) space can be made orthogonal to time.

What Friedmann did assert was that Einstein's and De Sitter's worlds emerge as special cases of the *geometrical* assumptions, and it was for this reason that they included the dependence of space curvature on time. But he found that "no physical or philosophical reasons can apparently be given for the [orthogonality] assumption. It serves only to simplify the calculations" (1922, p. 379). So, apart from the interpretation of this geometrical assumption about orthogonality, Friedmann's calculations were of a generality broadly similar to Robertson's; the main point being that Robertson reinterpreted Friedmann's results by using symmetry arguments.

In general terms, Friedmann's equations define the 'equations of motion' for the universe:

$$\dot{R}^2 = \frac{C}{R} + \frac{\Lambda c^2 R^2}{3} - Kc^2, \qquad (5.19)$$

in which C is a constant ($\frac{8}{3}\pi G\rho R^3$), K specifies the space geometry, and R is the radius of curvature (which varies as a function of the time t). (The differential quotients d/cdt are defined by dots.) Only one model is rigorously static: indeed when $\dot{R} = 0$, we have $K/R^2 = \Lambda = 4\pi G\rho$ and this implies the density ρ to be constant. A positive density yields $K = +1$, which corresponds to Einstein's cylindrical universe. (Note that the possibility $K = \Lambda = \rho = 0$, with R taken to be any constant, yields the Minkowski metric of special relativity.) In the non-static case ($\dot{R} \neq 0$), a series of empty models can be described in accordance with the various values attributable to the relevant variables. The De Sitter model is that for which the cosmological constant is positive and $K = 0$. This, however, is not exactly how Friedmann characterized the De Sitter world; what has just been said about (5.19) is a retrospective presentation from today's standpoint. Friedmann called the De Sitter world "stationary" (p. 382), as a useful transformation of time could make the coefficients of the original *spherical* space independent of time. In fact, spaces with zero curvature were never examined by Friedmann, as if they could not be 'relativistic'. What he found most interesting was that some of the solutions with positive constant curvature correspond to a space which undergoes cyclic expansions and contractions. He stressed that all these calculations were independent of any pre-determined value for Λ, and the purpose of his later paper (1924) was to show that not only the Λ-term was independent of Einstein's theory of the universe but also the finiteness of the universe. This was revealed by the possibility of having a world with constant negative spatial curvature as a solution of Einstein's equations. By contrast with the worlds of constant positive curvature, those of negative curvature were all found to be expanding indefinitely. This fact, Friedmann argued, "demonstrates that the world equations alone are not enough to allow a decision as to whether or not the world is finite". Additional conditions would include the specification of the topology, i.e. "we must know which points in the space are to be regarded as distinct" (1924, pp. 331-2) for otherwise the spatial curvature could not tell by itself whether it is finite or infinite.

Robertson was not happy with the word "stationary" used by Friedmann to describe the coefficients of a line element which is independent of t. Rather, in his terms, "stationary" referred to the most general class of space–times in which any two 3-spaces $t =$ constant are equivalent in the sense of the proposed assumption. That is, a "static" coordinate system has no coefficients that depend on time, while a stationary universe is one in which its "intrinsic properties" are independent of time. This yielded the desired generalization by means of symmetry arguments, for the only possible connection between "static" and "non-static" was now that "a static manifold is stationary but the converse is not necessarily true" (1929, p. 828, n. 6). When it is included among the non-static cases, the De Sitter universe loses the inhomogeneity of its horizon. Then in 1935, Robertson obtained the most general expression for the ds^2 of all universes which are compatible with the Friedmann equation and the assumptions of homogeneity and isotropy. This is:

$$ds^2 = c^2dt^2 - R^2(t)\left[\frac{dr^2}{1-Kr^2} + r^2(d\theta^2 + \sin^2\theta d\phi^2)\right]. \qquad (5.20)$$

When the value of R is appropriately adjusted it can be shown that the values of $K = +1, -1$, and 0 define a closed, open or flat universe. Here Robertson wanted to derive the global time function from a set of assumptions which were mant to supplement his earlier demand of equivalence of all standpoints. Clearly, in this equation, the coordinate distance between two events situated on a given time-like slice of the universe may increase or decrease in the course of time as a consequence of *the geometry alone* (see Torretti 1983, p. 207); this is the virtue of the so-called co-mobile coordinates.

But even before this higher degree of generality was reached, the 1929 paper by Robertson had already set the stage for an entirely different approach, in which the exclusive prerogatives of geometry in dealing with the question of the universe were disputed. With the pre-eminence of geometry, the old cosmological controversies were given a very new twist indeed. Thus, the first hint of geometrical generalization came from Heckmann (1931) who showed that, as far as the compatibility with the observation of the redshift is concerned, open spaces are just as valid as closed spaces. But in his words, and by contrast with Friedmann who had identified physical and philosophical issues, preference for any one of the Euclidean, spherical, or hyperbolic space rests on matters of sheer philosophical taste (p. 126). Heckmann's paper prompted Einstein's first attempt to drop Λ. In a critical look at the hypotheses he had set down in 1917, he now saw the constant density of matter in the universe as a *particular case* of the view that all places in the universe are equivalent (1931, p. 235). On the other hand, an astronomer such as De Sitter was not ready to take part in arguments dealing with geometrical properties of an idealized universe (1934, pp. 598–9). He viewed any speculation on the empty universe as a mere mathematical game and persisted in claiming that it is the physical status of matter and energy in the universe which determines such notions as space and time. In particular, De Sitter saw the cosmic time postulated by Weyl's principle as a *statistical* one because it depends on the actual state of motion of matter. As it turned out, the physical problem of primary interest was then what *causes* the expansion in the first place; it is this problem which was to gather up what remained of Weyl's primordial concern with causality.

This was the very question with which Lemaître had concluded his 1927 paper. He completely reversed the doubts about the questionable physicality of ghost-images in Einstein's cylindrical universe, as he suspected that light emitted by matter would accumulate indefinitely as it traveled round space and the resulting pressure of radiation could initiate the expansion. The inclusion of pressure provided an implicit development of Friedmann's equation. Lemaître showed that Einstein's equations give the following values for the density ρ and pressure p of the matter in space:

$$2\frac{\ddot{R}}{R} + \frac{\dot{R}^2}{R^2} + \frac{1}{R} = \Lambda - \kappa p$$

$$\frac{\dot{R}^2}{R^2} + \frac{1}{R^2} = \frac{1}{3}(\Lambda + \kappa \rho).$$

Combining these equations, we have the equation of energy for the universe:

$$\dot{\rho} + 3\frac{\dot{R}}{R}(\rho + p) = 0.$$

Opinions differed widely as to how to interpret the significance of this equation. One thing was clear: the Lemaître model started from an Einstein static phase and ended as a De Sitter universe as ultimate limit when the expansion has reduced the density so much that it may be treated as empty so far as gravitation is concerned; the pressure exerted by radiation would have initiated the whole evolutionary process. This yielded the long desired compromise between the two extremes. However, Eddington immediately pointed out that from Lemaître's formulas it was deducible that Einstein's world is unstable, *by taking p = 0*; that is, an expanding or a contracting universe is an inevitable result of the cosmological field equations in their original form (1930, p. 673). The question remained whether it could be predicted that expansion rather than contraction would actually occur. To be sure, the virtue of the new theory is that the phenomenon of the recession of nebulae, as Eddington argued, "is not merely consistent with theory, but is foretold by theory". The snag, however, is that we now have a theory which is too broad: "The investigation is incomplete in that we have only been able to study a system of galaxies strewn all over the world. It would be desirable to supplement this by considering cases in which the material system is confined to a part of space" (pp. 677-8). De Sitter, for his part, was very disturbed by the fact that, as he said, it remained to explain "why all velocities are positive and none negative" (1930b, p. 481); contra Lemaître, he realized that the universe could be expanding without radiation at all, so that the cause of the expansion would be Λ only, which De Sitter was now willing to interpret as a real force of repulsion.

Tolman was the first to investigate the possibility that the conversion of matter into radiation could be the cause of expansion. In fact, the relation of matter to radiation in the context of thermodynamics had been long investigated at the astrophysical level, but this had been done in various contradictory ways. A conjecture, now totally abandoned, had been put forward by O. Nernst at the beginning of the twentieth century, according to which radiation could recombine in matter, and William MacMillan had used this in 1925 in order to argue that a static but cyclic universe was possible. His theory was that the radiative energy emitted by the stars could be reconstituted so as to form new stars out of their own ashes. These sorts of 'punctual' cyclic creations had nothing to do with a global variation of the radius of curvature of the universe, for MacMillan understood these phenomena to take place in an otherwise amorphous, infinite, and Euclidean space. (On MacMillan's place in twentieth-century cosmology, see R. Schlegel 1958.) The prevailing opinion, however, was that the universe would simply run down irrevocably towards a state of highest entropy in accordance with the second law of thermodynamics. Eddington, in particular, had argued for quasi aesthetic reasons against the idea of continual re-formation of matter by whatever process—recollection from radiation for instance (1928, pp. 85-6). Several attempts had been made to deduce some observationally interesting consequences from the idea of a thermodynamic equilibrium in the universe, but they met with little success due to the insufficient number of data—in particular the small number of physico-chemical interactions known with some certitude (Zwicky 1928). This question naturally took on a new dimension with the establishment of a non-static metric for the universe. Against

Tolman's premature conclusion, Eddington was quick to show that conversion of matter into radiation tends to retard the expansion and therefore cannot be its cause. The effect of the gradual condensation of the matter into galaxies was then examined by G.C. McVittie (Eddington's assistant) and W.H. McCrea (1931). Their first conclusion was that this also would tend to cause contraction, but shortly after McVittie (1932) obtained the opposite result, so that when the concept of the expanding and evolving universe became widely accepted in the early 1930s, it seemed likely that this was what originally started the expansion. An interesting point was that this cause would be active at the origin only; maintaining the expansion was the job of the Friedmann equation. The random motions of the galaxies would also decrease as the expansion increased, so that if the expansion of the universe is already large (as observation suggested), the motions should correspond to the cosmic repulsion induced by Λ without much masking by other effects. In particular, Eddington, De Sitter, and Lemaître discussed the problem of whether the expansion of space could affect in some way the separation of the galaxies; they discovered that this would be negligible.

That virtually all major physical phenomena would join in the dynamics of space–time itself testifies to the force of the quasi-mathematical method followed by Friedmann and Robertson. Apart from changing the status of assumptions in cosmology, the method has also some interesting consequences with regard to the earlier controversies about static space-time models. Surprisingly enough, from the new viewpoint, we somehow return to the 'many-universes' alternative that Eddington had been compelled to consider before Weyl came up with his hypothesis. This is not simply due to the existence of various geometries and the fact that each geometry seems to define a universe in its own right. It is the very concept of expanding space which poses the problem. The expansion of space clearly abolishes the existence of any possible event outside the two $L_{-\infty}$ on Fig. 5.7. But inflation of space is not a dynamic motion in the classical sense, so that it is not incompatible a priori with faster-than-light phenomena. As expansion goes on and on, we are gradually forced to confront the disturbing consequence that objects separating faster than the velocity of light are cut off from any observability. This fact prompted Eddington to argue that these objects would be

cut off from any *causal* influence on one another, so that in time the universe will become virtually a number of disconnected universes no longer bearing any physical relation to one another (1931a, p. 415).

It is interesting to reflect on the manner in which the arguments have progressed since *The Mathematical Theory*. Eddington seems to retrieve here the argument he had put forward in the second edition of his book in the wake of Weyl's hypothesis: in the non-static interpretation of the universe as

Eddington understands it, events outside the range of observability *are* causally detached from those within this range.

De Sitter, for his part, was quick to go back to his old ideas of "philosophical predilection" or "matter of taste" in cosmological theories, for he chose to emphasize that, as acceptable as the new theoretical solution may be, "there is nothing in our observational data to determine the choice for the representation of the history of the universe" (1930a, p. 218). Attempts to pin down the criterion of selection among the family of world models grew as the expanding universe theory consolidated. Chief among bids seeking to identify this criterion as the means to re-establish universal causality was the elaboration of a concept of singular origin at a *finite* distant past. This is known as the primeval big bang theory, and it was first proposed by Lemaître (1931a and 1931b) as a radical way out.

By 1927 Lemaître knew all the possible solutions for the case of constant positive curvature of space. But, striving for a study of the real universe and not a mathematical structure, he omitted all alternatives to his model describing an expanding universe which, as time goes to minus infinity, approaches asymptotically the Einstein static universe. In his opinion the alternatives provided too short a time scale as compared to that of stellar evolution (1927, p. 58). Referring to James Jean's theory of gravitational instability (Jeans 1918 and 1928), Lemaître undertook to explore and refine the suggestion that the expansion of a universe in equilibrium may be started by the formation of condensations. In doing so, it is certainly not an over-statement to argue that Lemaître did to cosmogony in the early 1930s what Einstein had done to cosmology in 1917. The revolutionary advance was dramatic, as Jeans's *Astronomy and Cosmogony* of 1928 proved to be a culmination point in the long series of cosmogonical speculations that had been going on ever since the early nineteenth century (in the context of thermodynamics especially) outside any definite theoretical *model* of the universe. No more than two years later, Tolman (1930) had already found that the very emission of radiation from the stars (presumably at the expense of their mass) certainly implied time changes in the global gravitational field and hence necessarily a non-static metric for the universe. Jeans set himself the more humble task of verifying Newton's argument to Bentley that a chaotic mass of gas of a more or less uniform density and of very great extent would be dynamically unstable. If it was supposed that in the remote past all matter was uniformly spread through the universe, nuclei of higher condensation would tend to form; Jeans conjectured that the spiral nebulae could be formed by the resulting gravitational instability, but added that "it is improbable that we shall ever be able to prove [it]" (1928, p. 415). All Jeans claimed to do was to develop the mathematical analysis of the process of gravitational instability (p. 351), and in so doing the only part of Einstein's theories that he used was related to the special, not the general theory of relativity. In particular, Jeans posed uncritically that the density of the universe was independent of time. The fundamental result of Lemaître's early investigations about the behaviour of local density fluctuations in a uniform expanding universe was that any general process of condensation (like the formation of stars out of a primeval gas or the formation of galaxies out of a uniform mass of gas or stars) must induce expansion. But again a problem of time scale occurred, as Lemaître calculated that the duration of the formation of condensations was much greater than the duration of the expansion from Einstein's static phase. More importantly, Lemaître realized that it was untenable, physically speaking, that the universe was formerly near equilibrium according to Einstein's model and that it has begun to disperse at an infinitely slow rate, because the dispersion of matter would have had to take place in a perfectly uniform manner. Moreover, the dynamic picture of the universe created a lot of discrepancies with already existing cosmogonic theories derived from the Kant–Laplace paradigm, and many scientists tried hard to reconcile the relatively short time scale allowed by expansion with this kind of model. A discussion between all leading cosmogonists and cosmologists of the time (British Astronomical Association 1931) brought little more than acknowledgement of conflicts. Other difficulties arose from the theory of the evolution of stars. By 1927, Eddington had developed a theory of stellar structure which accounted for the difference in mass of various types of stars (from the red giants to the white dwarfs) in terms of progressive dissipation through the various stages in which we observe them. Again, the loss of mass due to radiation would be insignificant in the time allowed by the expansion of the universe; and the quantitative determination of the time scale required by stellar evolution was brilliantly confirmed by an independent fact, the rotation of the galaxies.

All this led to serious doubts about the correctness and viability of a model which took its expansion from some primordial equilibrium. That is why Lemaître began to look for other types of expanding models (1931a). He could find only one possibility: the models with initial singularity. He had known about this possibility in his 1927 paper, but only now in 1931 did he begin to give it serious consideration. Discussions of that problem with Einstein in 1933 strengthened his conviction that he was on the right path (see Lemaître's report, 1958). Einstein did not like the idea of a singular beginning. Hoping that the initial singularity could appear as a by-product of the symmetry assumptions, he suggested that Lemaître consider a simple anisotropic metric. Lemaître easily wrote down the corresponding field equations and succeeded in demonstrating that, in this particular case, we cannot escape the complete shrinking of space; that is, a singularity is present again (1933, pp. 82-5). This was not, however, a formal proof that the singularity could not be removed by introducing anisotropy, as the metric considered by Lemaître (as he acknowledged himself) was certainly not the most general case. Nevertheless, what it did suggest was that, even in the more general case, the anisotropy is not an effective mechanism for removing singularities. It is a consequence of the Friedmann equations that the radius of the universe *may* pass through the value zero, and anisotropy will do nothing against that. Lemaître noticed that a simple implication of the singularity was that, as we go back in time, matter should have a higher and higher temperature so that only subatomic forces could be capable of stopping the contraction. At some value of the radius of space which Lemaître calculated to be roughly that of the solar system, all astronomical structures (like stars) would be destroyed. In physical terms this very small value would be "like a zero of space", and Lemaître went on to argue that "the undeniable poetic charm" of the periodic solutions (the universe going cyclically through various singular states) would also have to be abandoned. A straightforward development of the theory led to the idea that cosmic rays could be regarded as relics of the hot big bang; in Lemaître's words these were "glimpses of the primeval fireworks".

Eddington did not like Lemaître's idea that the zero of space must be regarded as a beginning. He had already envisaged the possibility before Lemaître systematized it, and he could not see how this could have anything to do with physics. "The difficulty of applying this case is that it seems to require a sudden and peculiar beginning of things" (1930, p. 672). Lemaître countered that there was nothing repugnant in the theory if viewed from the perspective of quantum theory. Indeed, the singularity is compatible with a primeval atom, and this unique atom can be regarded as a unique quantum deprived of the qualities known as space and time. In other words, the big bang is not at odds with a deterministic conception; "the beginning of the world happened a little before the beginning of space and time". And Lemaître thought he could escape the fateful incomprehensibility of this 'before' outside any time reference by calling the principle of indeterminacy to the rescue: "the whole story of the world need not have been written down in the first quantum like a song on a disc of a phonograph" (1931b, p. 706). (See further information on Lemaître in Dirac 1968. For an updated evaluation of the impact of Lemaître's ideas on today's cosmology, see P.J.E. Peebles 1984.)

This is how Lemaître thought of a reconciliation between the methods of geometry and those of physics under the aegis of precisely that which had threatened the substance of Weyl's hypothesis, the quantum theory. But, returning much later to the cosmological problem, Weyl mentioned as postulate the sole topological distinction between past and future of the whole universe (1949, p. 110). Thus a world cannot be closed both spatially and temporally. He conceived of the expansion as a natural consequence of this basic openness, i.e. the non-physicality of closed time-like curves. Talking about the origin as a physical fact, he explicitly concorded with a passage of a 1935 book by Eddington: "When some of us are so misguided as to try to get back milliards of years into the past we find the sweepings piled up like a high wall, forming a boundary—a beginning of time—which we cannot climb over" (1935, p. 60). It is perhaps De Sitter himself who had best summed up the unanimous rejection of the big bang by the creators of contemporary cosmology, when he wrote a confession of ignorance in the face of the dilemma between the two time scales of astrophysics (stellar evolution) and cosmology (nebular expansion): "It appears to me that there is no way out...Our conception of the structure of the universe bears all the marks of a transitory structure" (1933b, pp. 133-4). At most, he was only willing to speculate (1932) that the beginning of the expansion might not coincide in any sense with the beginning of the universe. That is, he tried hard to distinguish a "time of minimum" which would be singularity-free and a "beginning of the world" (1933a, p. 631), basing himself on a concept of beginning which would tally with what present knowledge and theories use as a starting point. Because he laid stress (just as he had already done in 1917) on the fact that expressions such as universe and radius of curvature are metaphorical, he said we should be ready to incorporate outright contradictions (the rate of expansion versus evolutionary changes of stars) within the expanding universe theory (1931, pp. 708-9). So, if De Sitter ultimately remained faithful to his sense of antinomies in cosmology, Eddington was not willing to make any concession to his early intuition of grand synthesis: he took cosmic data as preponderant and argued that "we must accept this alarmingly rapid dispersal of the nebulae with its important consequences in limiting the time available for evolution" (1931b, p. 709).

7. A paradigm and a paradox: the alleged equivalence between Newtonian and relativistic cosmology

The dilemma *has* since been resolved to the advantage of relativistic cosmology: revised estimates of extra-galactic distances by Walter Baade in 1952 have increased the known size and thus also the age of the universe, so that it is now possible to accommodate data from the theory of stellar structure and evolution. In the meantime, however, a number of alternative theories of the universe had been proposed, most notably the so-called kinematic relativity of Milne and the steady-state cosmology of Bondi, Gold, and Hoyle. Even though the primary concern of these theories lay elsewhere, they did contribute to resolution of the dilemma in question. It was the momentous discovery, in 1965, of the isotropic cosmic background radiation which overthrew these alternatives and gave Lemaître's primeval atom theory a decisive advantage over its competitors. But the impact of what is regarded today as totally unorthodox cosmology still makes itself felt at the level of concepts. Therefore, we shall leave here the various problems related to quantitative determinations after 1930 in order to take a close look at the nature of concepts; the intricacies of these quantitative determinations lie beyond the scope of the present discussion. The point is that De Sitter's confession of momentary ignorance does echo in its own way a deep and persistent theme of our science of the whole universe.

Friedmann had concluded his 1922 paper on the prophetic remark that "our knowledge is insufficient for a numerical comparison to decide which world is ours". What could throw light on these questions, he argued, was not so much quantitative knowledge as what he called "the causality problem" (1922, p. 386). Clearly enough, talking about *causality* in a consistent way is a natural implication of a corresponding need for consistency with regard to *observability*, precisely because what we mean by talking about various possible universes suggests that the role of observation in the determination of which universe is ours can no longer be abstracted from an independent meaning attached to observation. The revision undergone by the idea of homogeneity is particularly conspicuous; postulating homogeneity so long as no contrary observation invalidates it is quite different from recognizing homogeneity as a fact of observation. With the rediscovery of the non-static solutions by the end of the 1920s, the two very different approaches to the question of the universe which had dictated the controversial interpretations of Einstein's and De Sitter's models suddenly emerged as the background of an ever more profound conflict, over and above any particular model of the universe. That is, while one school of thought would favour geometry over physics because this tallies with an emphasis on observability, the other school would favour physics over geometry because this tallies with causality and the metaphysical problems associated with it. It is a striking fact of the early history of relativistic cosmology that the authors who had clung for so long to static coordinates were precisely those who were most concerned with the metaphysical issues; the exclusive privilege of geometry was defended by people like Klein or Lanczos who also foreshadowed the relevance of non-static coordinates for the universe. Quite exemplary and perhaps typical in regard to this tension in Weyl's dissatisfaction with the very method (the geometry underlying non-static coordinates) which revealed the essence of his early, grand hypothesis of causality. Now, it is at the level of unification of field physics that Weyl had initiated emphasis on the privilege of geometry, for his gauge-invariance theory made it necessary to interpret Einstein's field equations in such a way that the geometrical side says more about the physical side than was initially thought.

A complete reconsideration of these questions of interaction between geometry and physics at the cosmological level came with the work of Edward A. Milne from 1932. In our context, it is worth the effort to look at the status of non-static cosmology via Milne's early work, for it illustrates the problems implied here by forcing us to return to Einstein's very original motivation for getting into cosmology. And this return has more than mere historical significance for us. For in Milne's early work, we find a suggestion which in fact, among the many controversial contributions he made to cosmology, is taken so seriously today that it has been absorbed as a premise in most scientific approaches to the universe: the idea that a formal equivalence may be made between relativistic cosmology and another cosmology which finds its validation in no less a source than Newton's theory.

Certainly Einstein did not revert to Newtonian cosmology after he had disposed of boundary conditions in general relativity. Nor did he make any serious attempt to revise his first static model, after Friedmann had shown in 1922 that one could dispose of Λ provided a dynamic picture was acceptable. For all this and while Einstein, De Sitter, Eddington, and Weyl were discussing the respective merits of the two known models, attempts were being made to save the integrity of Newtonian cosmology by suppressing the boundary conditions altogether, as Einstein had done in the framework of relativity. In 1922, Charlier reverted to the Newtonian model he had first proposed in 1908 by articulating what became known as a hierarchical model, which drew on the speculations of a number of later eighteenth century thinkers such as Kant and Lambert (see Chapter 1). According to Charlier's model both the optical and the mechanical paradoxes of Newton's infinite universe are overcome when a specific arrangement between the different celestial systems is introduced; an appropriate choice of dimensions between the various clusters (planetary systems, galaxies, clusters of galaxies, and so on, without limit) can make the average density appear to equal to zero. This is primarily a challenge to Einstein's argument in his 'Cosmological Considerations' of 1917 that the mean density must decrease towards zero more rapidly than $1/r^2$ as the distance r from the centre increases. Otherwise, as Einstein contended, the potential could never tend to a limit at infinity and the universe would collapse. Charlier's arrangement provided just what is necessary to avoid both collapse and dispersion. In the view of most cosmologists today, "these models are little more than a curiosity" (W. Rindler 1977, p. 196). The sad fact is that Charlier's universe is not really homogeneous (no volume is large enough to be typical), even though it may be claimed that the indefinite repetition of similar structures makes it tend towards homogeneity. This unwieldy universe is merely an indefinite series of self-contained systems without the slightest evidence of a natural connection between them. Of course, the appeal for a cosmological constant in the relativistic case might seem just as arbitrary as this process of clustering in the Newtonian case. Following this line of argument, one of Charlier's most important supporters, Franz Selety (1922, in particular, pp. 291–2), brought into question once more the validity of Einstein's arguments on boundary conditions. In principle, Selety argues, the possibility of infinite differences in potential has not been ruled out conclusively by Einstein before he turned to his investigation of purely relativistic cosmology. A physics is construable without any need to define the potential at infinity, and in fact if it is construed along these lines then cosmological physics results from the simple extension of local considerations. Indeed, in order to calculate potential differences, one could begin with some equipotential surface (which can be determined empirically) whose value is assigned quite arbitrarily, and then reckon all other potentials either negatively or positively, a procedure quite compatible with the occurrence of infinite differences of potential. In this case, Selety goes on to argue (1923, p. 58), it would still be possible to have star velocities which are small on the average, for very large potential differences would yield large velocities "sufficiently rarely". Selety's dominant idea seems to be that superimposition of a physical meaning onto mathematically specified boundary conditions is scarcely a proper way to convey a just or feasible representation of the entire universe; some element of arbitrariness being unavoidable, this can be located at will, whether in the infinite or elsewhere. And if it is located within finite range, then the calculation of the *density* of matter in the universe would also follow from a limiting process which avoids any postulated difference between an ideal and a local value. Thus, a consistent Newtonian universe (that is, with Euclidean space and with infinite total mass) would result from the compatibility of finite density in all *local* regions and mean zero density for the *whole* universe. Selety further emphasizes that a hierarchical model of that sort rules out statistical arguments altogether, for such distribution would never approximate the uniform distribution and it would never change into anything else.

The hierarchical hypothesis was quite fashionable throughout the 1920s. MacMillan, for instance, had based his above-mentioned suggestion of recombined radiation (the cyclic universe) on this theory. The hypothesis was also entertained quite seriously by those who saw finite space as a possibility rather than a reality: Emile Borel (1923, pp. 242–7) and Bertrand Russell (1925, Ch. 11), among others. Einstein was not moved. His reply to Selety indicates that the hierarchical hypothesis is certainly possible from the standpoint of general relativity, yet it is unsatisfactory if the interaction between matter and space curvature is taken seriously over large portions of space–time (1922, pp. 436–7). Interestingly enough, Selety championed the idea that his hierarchical model did offer an original combination of Newtonian *and* Machian requirements: in the model, all inertial systems are determined by the matter distribution. He took it that Einstein had unnecessarily pushed Mach's critique to mean that the *laws* of nature themselves are so determined (1923, pp. 61–2). Therefore, Einstein's use of statistical arguments in his description of Newtonian cosmology was not to be attacked on the grounds of its (implicit) acceptance of Mach's theory of inertia—as Weyl had done. Rather, the use of such arguments meant the uncritical application of *absolute time* to any possible conception of the universe. Selety developed this point (1922, pp. 326-7) by following the suggestion already made by Felix Klein that the cylindrical world distinguishes the time-axis as absolute. By contrast, there would be no reference system in the hierarchical model with respect to which the whole matter of the universe is at rest. In Selety's ideation, this tallied better than Einstein's own model with the relativistic requirement regarding immateriality of the reference system. Replying to this particular point, Einstein felt the need to justify that his static model in no way promoted an absolute time *à la* Newton (1922, p. 438): the specification of such a universal time function in his model is not a hindrance to the principle of relativity and the freedom of coordinates, for the principle speaks about the general laws of Nature, not an actually existing system. Einstein was definitely loath to overestimate any link between the two. But again in his own reply to Einstein, Selety (1924) argued that if physics is to be ultimately a description of the concrete cosmos, then only dropping the quasi-static condition would restore the originally postulated meaninglessness of absolute time. 'Non-static' meant for Selety: free of privileged time direction. And he went on to define as the only viable cosmological models those which would incorporate and do full justice to the principle of *special* relativity.

Be this as it may, suppressing both the boundary conditions and Λ in the Newtonian static model, would result in the process of clustering described by Charlier and Selety. What remained was to evaluate the potential consequences of dynamism as something capable of restoring unity—that is, unity in the sense of relativistic cosmology. A consideration of this question leads us to the work of Milne, with its far-reaching result, that this very unity *is* compatible with a 'Newtonian' dynamism. Ultimately, this work also raises the issue of greatest significance that Einstein diligently skirted: do material content and general behaviour of the universe determine the laws of nature, or is it that the laws themselves determine this content and this behaviour?

It was in 1934 that Milne first articulated this most striking paradox of twentieth century cosmology: the relativistic laws of the universe could be derived in the simplest possible way from Newton's theory. Milne was keen to demonstrate that "an analyst of Newton's period. . .would have secured all the results yet capable of observational test" (1934, pp. 71-2). Such an analyst would have been led to predict, on theoretical grounds alone, a velocity-distance proportionality very similar to Hubble's law and indeed compatible with the entire picture of a non-static universe, whether expanding or contracting. The equivalence has been developed in the following way. Milne's pivotal idea is that a smoothed out universe can be represented as an expanding cosmic ball. This is a ball of matter expanding in empty Euclidean space (pp. 67ff.). The density of the ball is uniform and consists of particles in a state of free fall within the gravitational field produced by the ball. Let $M(r)$ be the mass contained in the sphere of radius r, and v the escape velocity from that mass. The minimum velocity that will enable a particle to reach infinity is the parabolic velocity of escape:

$$(1/2)v^2 = GM(r)/r \qquad (5.21)$$

where G is the Newtonian constant of gravitation. As Milne proceeded to show in a joint paper with W.H. McCrea (1934), it is clear that a velocity greater than the parabolic velocity of escape defines hyperbolic paths, while a velocity less than that of the parabolic velocity defines elliptical paths. The latter instance is the only one that predicts a collapse of the cosmic ball. Now, Eqn (5.21) also represents the fact that the total energy of the system, kinetic ($v^2/2$) and gravitational ($-GM/r$), remains constant if an appropriate constant is added to the right-hand side. The value of this postulated constant will now determine the behaviour of the cosmic ball: a positive or negative value means that the particles will follow hyperbolic or elliptical paths, while a zero value will result in a parabolic case. By recourse to the Hubble law of velocity–distance proportionality and the scale factor R, Milne was able to show that the most general situation is defined by nothing other than (5.19), the Friedmann equation. Of course, the interpretation of the constant K differs in each theory. While the Newtonian theory interprets it in terms of the total energy of the particles, the relativistic models take it as a specification of the curvature of space. Nevertheless, as far as prediction of the overall history of the universe is concerned, the equivalence seems to be total.

The Hubble law appears to be the key to the equivalence. Milne saw this very clearly, since he managed, in his book which followed soon after the publication of the equivalence paper, to deduce the law from purely kinematic arguments (1935, pp. 79-80). His demonstration is most remarkable and it may be reconstructed in the following way (see also J. North 1965, p. 160). Two viewpoints O and O' necessarily lead to the same picture in a uniform universe. Imagine some event or object vectorially related to O and O' by $\vec{r}$ and $\vec{r}\,'$. Let $\vec{a}$ be the vector for O to O'. We have $\vec{r} = \vec{a} + \vec{r}\,'$. If v' is the velocity seen at O', then $v'(\vec{r}\,') = v'(\vec{r} - \vec{a}) = v(\vec{r}) - v(\vec{a})$. Because of the uniformity, the function v' must depend on its argument $\vec{r}\,'$ precisely as the function v depends upon its argument r. It follows that $v'(\vec{r}\,') = v(\vec{r} - \vec{a}) = v(\vec{r}) - v(\vec{a})$ for all $\vec{r}$ and $\vec{a}$. Thus the function v must be a linear vector function of its argument. Further, isotropy requires the multiplier of $\vec{r}$ to be a scalar, let us say H. This scalar may, of course, be dependent upon the epoch of the universe, the time t. Finally, we have

$$v(\vec{r}) = H(t)\vec{r}, \qquad (5.22)$$

which is the Hubble law.

Isotropy and uniformity are used in this reasoning which leads to derivation of the Hubble law. Such a revival of Newtonian cosmology does indeed bring to the fore a concept which Milne himself sees as purely relativistic, tracing it back to Einstein's cosmological paper of 1917. This is the 'cosmological principle', which is accepted today by virtually all cosmologists, whatever their theoretical persuasion. In fact, soon after Milne's derivation, Robertson established the relativistic metric (5.20) which is compatible with the Friedmann equation and the assumptions of homogeneity and isotropy. Enlarging Milne's original argument, he included under the name 'cosmological principle' *all* assumptions regarding the geometry. And independently of Robertson, A.C. Walker (1935 and 1936) achieved the same result, but this time the work was done in the explicit context of Milne's cosmology. It should be noted that in the integration of (5.20), the theory of gravitation is relevant only in determination of $R(t)$; the geometry, K, is fixed quite independently. The problem is thus to find a dynamic theory of the function $R(t)$. In relativistic cosmology, the equations of Einstein's gravitational field theory are applied here, but predictions of large-scale correlations of the universe are identical with those of the scalar $H(t)$ in Eqn. (5.22) where only the concept of Newtonian 'absolute' time and Euclidean geometry for vector addition are brought into play.

The equivalence creates a serious problem of interpretation. In a recent textbook of cosmology, Edward Harrison expresses his puzzlement at this: "In all its applications, Newtonian theory is only approximately true, and yet in this most unlikely of all instances it yields the correct answer" (1981, p. 283). In short, the real problem is: Why does some form of (mildly) modified classical cosmology yield the right answer? Or, to be more precise: Why did we have to wait for general relativity before a consistent Newtonian cosmology could emerge? The problem had already been given a decisive impetus when Schücking managed "to discuss the question why such a simple and beautiful theory as Newtonian cosmology was not already formulated centuries ago" (1967a, p. 270). To be sure, even though Milne was purely polemical when he said that an analyst of Newton's period would have established all the results yet capable of observational test, the validity of the equivalence is widely accepted today and we can speak of such a reconstruction in terms of a "neo-Newtonian" model (the expression is that of J. North 1965, pp. 180–5). The point is that an interplay between 'concept' and 'world model' is entirely characteristic of present-day attitudes to equivalence, as exemplified in the work of Heckmann and Schücking. (See their detailed exposition of Newtonian cosmology in 1959, pp. 491–9, as well as some pertinent remarks in L. Sklar 1976, pp. 5–6.) In order to demonstrate the consistency of Newtonian cosmology, such writers introduce concepts borrowed quite explicitly from relativity theories, and by so doing they go a good deal beyond Milne's original strategy. So, armed with the notion of local inertial frames, it is easy enough to 'trade-off' non-inertiality and gravitational fields so as to circumvent the traditional problem of boundary conditions. As Sklar says with some emphasis, "a convenient comparison of related Newtonian and general relativistic models is easily made in its terms".

The convenient method is perhaps not so far from something like an expedient. When Newton himself faced the conclusion that his own theory would predict a collapsing universe, he seems to have envisaged the *force* of gravitation as being deliberately limited by the will of God. In order to get round what is in effect the same problem, it is only a short step to propose the limited *descriptions* of the large-scale properties of the phenomenon of gravitation, which now hold sway in contemporary versions of Newtonian cosmology. By contrast, Milne's original interest in the equivalence was primarily of epistemological scope. Einstein had found it unacceptable that the intensity of the Newtonian gravitational field on the surface of a sphere would increase indefinitely with its radius, but Milne countered by arguing that here Einstein was confusing concepts and observable entities: "the notion of the 'intensity' of a gravitational field", according to Milne, "is a pure concept", whereas the mathematical proof appeals to yet another concept, that of lines of force (1935, p. 300). Einstein showed Newtonian cosmology to be impossible by exhibiting the paradoxes of a physical picture of the infinite, but his boundary conditions were primarily hypothetical and were therefore physically meaningless. Milne preferred to test the validity of any model by referring to "the accelerations actually undergone by the particles present and capable of observation by other particle-observers". This sends us straight to the heart of his theory.

The original comparison between the two theories was first made by Milne in relation to the so-called Einstein–De Sitter model of the universe, and this is certainly no accident. In 1932, Einstein and De Sitter were sufficiently reconciled to write a joint paper proposing a new model of the universe. Einstein had shown in his 1931 paper that dropping the static condition was enough to have a positive uniform density of matter with zero pressure and with a zero value for the cosmological constant. And there is no point in denying that in his joint paper with De Sitter he does his best to beat yet another retreat. In this new theory, our two authors claim that "there is no direct observational evidence for the curvature [of space]. . .It is therefore clear that from the direct data of observation we can derive neither the sign nor the value of the curvature". This led them to suggest that perhaps "it is possible to represent the observed facts without introducing a curvature at all" (Einstein and De Sitter 1932, p. 213).* In a September 1932 manuscript,

---

*'Reconciliation' between Einstein and De Sitter may well be here a manner of speaking only. Eddington had heard the paper they read together at California Institute of Technology, and he reports the amusing but interesting story that "Einstein came to stay with me shortly afterwards, and I took him to task about [the dropping of Λ]. He replied: 'I did not think the paper was important myself, but De Sitter was keen on it'. Just after Einstein had gone, De Sitter wrote to me announcing to visit. He added: 'You will have seen the paper by Einstein and myself. I do not myself consider the result of much importance, but Einstein seemed to think that it was' " (Eddington 1940, p.128).

Einstein wrote that a non-zero density of matter does not necessarily imply a spatial curvature (the assumption of the 1917 paper De Sitter had tried to refute). On the contrary, it was a spatial *expansion* (1933, p. 109) and therefore the consequent model is that of a Euclidean expanding space, which will yield the simplest of all Friedmann universes (with $\Lambda = 0$). The option chosen is a telling sign of the status of the theory behind the newly discovered physics of non-static metrics. The Friedmann equation (5.19) relates temporal variation of the scaling factor to the density of matter, the geometry of the universe, and (very possibly) the cosmological constant. The difficulty is that a redshift measurement only tells how much the universe has expanded since the epoch of emission. Distances, recession velocities, and the lookback time all depend on the geometry of space and on how the scaling factor changes with time. As long as no test is available which can single out one of these factors and provide an independent measurement of it, a choice must be made at the outset. The Einstein–De Sitter model opts for a simple geometry (so disarmingly simple in fact that Friedmann himself does not seem to have realized that $K = 0$ is possible as a non-static case); as the geometry is fixed, corresponding distances, recession velocities, lookback times, and the age of the universe can be calculated. In the book he was writing at the time, De Sitter proffers the opinion that "we shall never be able to say anything about the curvature without introducing certain hypotheses" (1933b, p. 117).

Milne seems to conceive no higher aim than attacking the cosmological problem in just that hypothetical and speculative spirit. But an essential solution, as distinct from a momentary one, can be found only if the very concept of space is seen as a mathematical construct devoid of any physical significance (see Milne and McCrea, p. 64). The choice of a particular space is not constrained in any way by physical considerations, but is simply a matter of convention. Following Poincaré, Milne opts for Euclidean space. In contrast to the Einstein–De Sitter selection of this space, Milne's procedure restricts the class of possible universes by overcoming the need for arbitrary reduction in the number of independent variables. In short, the guiding line of reasoning in Milne's treatment of the problem is the theory of kinematic relativity which he had just begun to elaborate a year earlier (1933). This theory is supposed to be an account of the actual systems presented to observers, and the treatment of those systems denies the validity of any assumed theory of gravitation. More fundamentally, Milne believed that the basic principles of cosmology ought to be referred to the actual aspects and motions of the universe rather than to the "laws of nature" (see Bondi 1960, p. 124). From this point of view, kinematic relativity would then emerge as a theory quite separate from either the relativistic or the Newtonian. In other words, Milne's original construction of the equivalence was aimed at neutralizing both rival theories.

Because no content can be attached to the phrase "space of nature", the comparison between Newtonian and relativistic cosmologies focuses on the possibility of a locally Newtonian *time* in relativistic cosmology (Milne and McCrea, p. 71), with the proviso that in dealing with Newtonian time one "assumes the usual definition of simultaneity by means of light-signals" (p. 65). The result is a solution to the miscomprehended problem of motion versus uniform distribution in a Newtonian universe. Milne's cloud of freely moving particles conveys the idea of matter and motion as given *simultaneously*, in contrast to the more classical concepts of motion as disturbing a pre-existing uniform distribution or pre-existing inhomogeneity as precluding the motion which would effect a return to uniformity (this latter view was just Newton's argument to Bentley: see Chapter 1). Both of Einstein's objections fade away: there is no unique centre and the infinite is not a 'place'. The demand of a perpetual interaction between particles and all forms of radiation is declared by Milne to be a metaphysical demand (1935, p. 301). Milne shows "that Newtonian systems can be constructed in which the relative accelerations are small near the observer and everywhere finite in his experience", thereby narrowing the field of causal interaction to the sphere of observability of a given observer. This is a new kind of demand which is derived from the special theory of relativity, a theory which Milne indeed claims to have incorporated, to a large extent, in his kinematic theory. Still, Milne is clearly not at ease in doing so because he reverts to the large-scale point of view in a moment of muddleheadedness: the notion of curvature of space, he says, "is merely a mathematical device for describing on both the large and the small scale what is equally well described on the small scale (locally) in Newtonian terms, and what is formally describable on the large scale in the same terms" (p. 316). The recourse to a merely formal equivalence when the large scale is contemplated is, in fact, the crucial give-away which highlights the central issue.

Indeed, what is today known as the Milne–McCrea theorem, which allows the scale factor $R(t)$ to satisfy the same equation in both the Newtonian and the relativistic theories (see D. Sciama 1973, p. 114), gave rise to interesting disputes before it became endorsed unanimously by the scientific community. Bondi (p. 178) first emphasized the ambiguity in the definition of inertia according to the theorem, since all reference systems are assumed by it to be both inertial and accelerated relative to each other. He tried to bypass the difficulty by arguing that "as long as we assume that each observer only uses his own system no difficulties or contradictions arise". Whatever the merits of this resolution may be, while Milne seems not to have perceived the problem, Bondi can hardly be said to have resolved it. Indeed, Milne's problematic solution is one version of the clash between the Newtonian and the relativistic class of allowable transformations of coordinates, while Bondi's answer simply discards the problem, failing to wrestle with it at all. Milne's original intention was to abandon the relativistic transformation of coordinates and replace it by transformations from observer to "equivalent" observer, where equivalence was to be "defined in terms of observations and tests which the observers can actually carry out" (1935, p. 5). He did not address Bondi's problem, because his own idea of transformation between equivalent observers dominated all of his conclusions, and it is a concept in which any distinction between inertial and accelerated observers becomes irrelevant. Working explicitly within the framework of general relativity, David Layzer (1954) dealt explicitly with Bondi's statement of the problem, and found that all neo-Newtonian derivations of the Friedmann equation are invalid because the Milne-McCrea model is incompatible with the Newtonian conception of gravitation. Layzer demonstrates that the Bondi problem evaporates only in the case of unaccelerated expansion (p. 269). This case takes its bearing from the proposition that "all forms of Newton's law of gravitation depend on the idea that the specific gravitational force at every point is determined by the instantaneous distribution of matter in the universe, and is independent of the state of motion of the matter". Now, because the distribution of matter *taken alone* defines no preferred direction in space, it follows that the potentials are constant and identical everywhere. As a consequence, expansion of Milne's cloud of particles is not accelerated. Layzer could have concluded just as easily that no motion at all is possible, since he quite deliberately treats the *distribution* and the *motion* as each being on a different footing. The distinction is contrary to Milne's approach, and highlights the fact that Milne is interested in much more than the mere *form* of Newton's law. Not surprisingly, then, Layzer reveals the invalidity of the Newtonian derivation of the Friedmann equation by underlining its incompatibility with the following pair of statements:

1. The relative *accelerations* are equivalent to gravitational forces, as required in the Milne-McCrea theorem.
2. The instantaneous *distribution* of matter determines the gravitational force at any given point.

It should be clear that it is the distinction between distribution and motions which dictates the criticism. Of course, it would be futile to deny that Layzer's arguments are mathematically sound. The point is that, by clinging to methods of mathematical physics acknowledged as valid in normal circumstances, they fail to take account of Milne's intention.

In response to these arguments, McCrea rehearsed a more radical view of the subject (1955, p. 273—Milne had died five years earlier). If we take a system with uniform density, McCrea says, and "if the gravitational force is to be defined in the present (Newtonian) manner, then it does not *exist*"; the fact that the force shall be zero in that case cannot be inferred, since the very concept of force becomes nugatory. Responding primarily to Layzer, McCrea manages to perceive the limiting role of traditional mathematical physics in Milne's general approach. By the same token, McCrea solves the

Bondi problem by realizing that there is only *one* ideal Newtonian reference frame, that which is located at the 'true' centre of the cloud. All properties relative to any observer moving with the material of the universe could be deduced by the resulting motion relative to that ideal frame, the resulting motion being obtained from purely Newtonian kinematics (p. 272). McCrea's new achievement, then, was the revelation that the distribution/motion simultaneity is necessarily dependent on the possibility of a particular unique centre.

This conclusion is far-reaching because the idea of a unique centre resurrects the old question of the universe having a determinate *boundary*. Layzer had shown that the Newtonian theory is quite satisfactory for resolution of the dynamics of a uniform system with zero pressure, whether the system is an island-like universe or space itself is expanding, that is, the dynamical properties are rendered without approximation by that theory. Yet, the equivalence collapses when a uniform, *unbounded* distribution of matter is in question. As I show in Chapter 2, it is notoriously the case that a unique solution for the potential cannot be derived from Poisson's equation when boundary conditions remain unspecified. Milne perceived the problem quite clearly, since he first constructed the equivalent of the Einstein–De Sitter model by stating the parabolic velocity of escape from a certain sphere. In writing that equation, Milne warned, "we are not using the notion of gravitational potential, here inapplicable, but are employing simply an integral of the equation of motion with a particular value of the constant of integration" (1934, p. 68). In fact, the particular value was zero, though it was later shown to be positive in the case of hyperbolic paths, and negative with elliptical orbits. Milne's supposition was that the conditions at infinity are simply compatible with the assumption that the matter outside the sphere can have no influence on the motions inside it (as in the well-known theorem of Newtonian mechanics). Layzer, on the other hand, emphasized the Newtonian-relativistic equivalence in a bounded system by recalling two theorems of Bondi. One of these states that by neglecting the influence of exterior matter we are enabled to determine *without approximation* the interior motions via Newton's theory. Without the very existence of some frontier as a necessary precondition, the equivalence will not be applicable. In the light of these criticisms, McCrea went on to modify the whole position. "We can suppose", he wrote, "any observer to have a finite range of observation. So, in particular, we can take the extent of the system to be arbitrarily large compared with this range" (1955, p. 272). As a result of this arbitrariness, only a small number of observers will perceive the edge of the world, and so "the difference between an arbitrarily large system and an unbounded system is scarcely significant". At the same time, McCrea argued that a consistent picture of a truly unbounded system would demand an altogether new meaning for the concept of a gravitational field.

The status of the whole question is undoubtedly best apprehended by referring to the actual *differences* between the Newtonian and the relativistic models of the universe. In fact, the *physics* of the two universes are not exactly the same, as Bondi has emphasized with some vividness: "the difference between relativistic and Newtonian theories is governed, in cosmology, by the ratio of pressure to density, and. . .the ratio of gravitational potential to rest mass is important mainly in local applications" (1960, p. 104). Bondi obtains this result by comparing the respective derivations of the Friedmann equation. A dynamic interpretation of the Robertson-Walker metric (5.20) implies a simplified form of the material tensor, where the material density and the isotropic pressure p are functions of the time alone. A straightforward consequence of the resulting relations is the equation $dE + pdV = 0$, verifying the law of conservation of energy. Now, a consequence of equation (5.19) using the *Newtonian* equations is $dE = 0$, which shows that any full equivalence between the two theories demands that the pressure be very small with respect to the energy due to matter. This, indeed, is quite a good approximation of observed *reality* and has prima facie nothing to do with *theory*: the kind of approximation involved here is not of the usual type between a Newtonian and a relativistic equation, since nothing enables us to say that the pressure is a purely relativistic effect. A suitably generalized discussion would therefore have to include one of a precisely opposite kind, and as it happens, it has been carried out by G.C. McVittie. He is not at all concerned with any deduction of the Friedmann equation from Newtonian postulates. On the contrary, he adopts the opposite strategy of seeking a Newtonian approximation to the relativistic formulae, explicitly given at the outset. His discussion is not even limited to a priori uniform models: the demand is primarily for a theory of spherical symmetry, simply because "we can observe the universe from one point in it only, namely the earth" (1954, p. 173). Uniformity is introduced much later for the sake of comparison in terms of putative equivalence. His approximations are twofold: the constant of gravitation in the field equations (powers of κ higher than the second are neglected), and the velocity of light (c is identified with an infinitely large constant, since this is the only value which remains unchanged in the transformations from one inertial system to another). McVittie proceeds to articulate a Newtonian approximation of the relativistic equations in the case of spherical symmetry, and he finds that density and pressure gradients are included in the corresponding formulae. Another restriction is the limitation to uniform models: at this stage, the impossibility of using a co-moving coordinate system in the Newtonian instance (there is an absolute space in which matter moves) reveals itself as crucial. For that reason we find that the density, pressure, and radial velocity of the cosmic fluid are independent of each other—a significant contrast to the relativistic case. The important point is that, whereas the density and radial velocity are known functions of absolute time, the pressure remains an arbitrary function of it, and therefore "it cannot be shown that a relativistic model in which $p = 0$ corresponds to a Newtonian model in which $p = 0$ also" (p. 180). There is much significance here in McVittie's preference for the term 'analogy' rather than 'equivalence'.

Thus, in the relativistic case, density and pressure are interdependent, in accordance with the already established superiority of general relativity over Newtonian physics. (Relativity links previously unrelated laws of nature, such as the identity between the inertial and the gravitational mass, Poisson's equation and the conservation laws.) Yet, a unique and dynamical interpretation of $R(t)$ cannot be inferred from that interdependence, even in the case of uniformity. This certainly exemplifies the a priori irrelevance of general relativity in Einstein's original form to any cosmological consideration, a fact that Weyl had already pointed out at the very beginning of his investigations in 1918. On the other hand, the independence of pressure and density in the respective Newtonian models does imply that $R(t)$ is *predetermined*, to use McVittie's term (p. 179). This, in turn, leads him to argue that the cosmological constant cannot be regarded as a force of repulsion in the relativistic case; rather, it is a trivial additive constant in the pressure.

Of course, Milne himself had already shown that the cosmological constant is not necessary for prediction of motion in a Newtonian universe, just as Friedmann did in the case of a relativistic universe. So the difficulty raised by the existing parallel, which has been deemed to convey a true equivalence, is of a quite general order: what exactly does dynamic cosmology add to the underlying assumptions of early, pre-dynamic cosmology?

In 1917, Einstein discussed the Newtonian model solely on the grounds of its unsuitability for predicting a definite size (periphery) of the universe, whether finite or infinite. Yet, because of the requirement of static equilibrium, the cosmological constant could appear to be the bridge between the "infinite extension of central space" in the Newtonian model, as he called it, and the "self-contained continuum of finite spatial volume" according to the relativistic concept. Einstein can see clearly enough that the crux of the difference between the two models comes from their object and purpose rather than from the formal nature of Λ: while Newtonianism constitutes a view of the *centre*, relativistic cosmology involves a leaning towards the periphery. While still ignoring all of the non-static forms of the metric, Weyl has contributed most to our understanding of the reversal which this decision implies, albeit by showing that Λ itself has the effect of closing space rather than simply making possible the quasi-static distribution of matter. But the advent of the non-static forms, together with the denial of the cosmological constant as necessary condition, tends to blur Einstein's original emphasis. Thus, the equivalence as worked out by Milne is ratified by the apparent failure of relativistic cosmology to predict a definite size for the universe. The theoretical postulate of equally possible universes seems to be taken as a proof that the ultimate choice cannot be dependent upon some kind of absolutely objective agreement between theory and observation. In Milne's view, this fact alone condemns the relativistic theory as too loose and, if left to its own devices, sorely in need of the greater stringency which he thought he had found.

At this point, Milne's formulation of the concept of an expanding cosmic cloud acts as the cornerstone of the whole question. In replacing extension with expansion, the dynamic picture of the universe erases the salient distinction between centre and periphery; the expansion proceeds from a centre which is not only the origin but also (and for that reason) the condition of the periphery. As Weyl would say, the universe expands just because it is topologically open. In reference to metrical relations, however, the relativistic version cannot be written off as failed cosmology, in that each model or solution taken in isolation does prescribe some determinate size of the universe. For instance, the Euclidean geometry of the Einstein–De Sitter model is compatible with a definite, albeit an infinite, geometry. Milne's criticisms have their pertinence at precisely that level: space so conceived cannot be a part of physics at all. In consequence, his aim of showing the equivalence between the two cosmologies, so far as the *observable* part of the universe is concerned, seems to be based on a notion of observability which is very far from any theory of relativity. In Milne's own theories as well as in his criticism of traditional cosmology, observability is taken as the defining quality of a cosmological *model* or construct. Milne repeatedly asserts that the relativistic phenomena of fresh particles entering an observer's field of view is no different from the creation of matter within experience (1935, p. 9 e.g.). Kinematic models always exhibit within the field of view of any particular observer, at any epoch of observation, all the particles already in existence. Clearly, the actual overlapping of observability and causality neutralizes the kind of concern for limitation that relativists may have in mind, because this overlap is a quality of the model, not of the universe. At the other extreme, it is quite significant that most post-Einsteinian relativists would see observability as the very subject matter of cosmology, identifying the *universe* itself with what may be seen happening in it. Thus, the relativistic response to the tension between theory and observation created by the rise of dynamic cosmology is to say that it is observation, not theory, that needs greater stringency. For instance, Layzer holds the view "that the defining properties of the universe have the status of natural laws", in the sense that this "encourages us to construct theories that are especially vulnerable to observational disproof" (1967, pp. 237–8). Accordingly, Layzer's argument to the effect that Newtonian and relativistic cosmology cannot be equivalent in the case of an *infinite* universe has the effect of reinforcing this specific kind of 'definition' of what the universe 'is'. Similarly, when Schücking discusses the problem of the universe as an unbounded totality, emphasizing that we cannot neglect the gravitational field contributed by all matter, he reshapes the problem so that it becomes a problem of operational definition of the inertial system. Now, clearly, such a definition can be given only in the absence of matter (something that Newton had already done in his thought experiment of two rotating globes in an otherwise empty universe), and that is why he introduces the relativistic notion of local inertial systems before he finds the appropriate law of transformation for the potential, ensuring overall finite solutions for allegedly Newtonian models (1967a, p. 274 and 1967b, p. 223). This fits in with his general definition of the particular subject matter of cosmology, that is, "all circumstances of which we have positive knowledge" (1967b, p. 221).

In contrast, Einstein's original discussion arose from a clear division between two kinds of problem, where definition of the inertial system and a consideration of the universe as a whole were conceived of as two quite distinctive endeavours; their only connection being simply that they are two factors which compel us to abandon the framework of classical physics. True, it is particularly the paradigm of the first *finite and static* universe which reveals the pertinent difference between Newtonian and relativistic cosmologies. A determinate size, like a finite one, is in absolute opposition to the arbitrarily large universe envisaged by McCrea, where the proportion of non-typical observers located near the periphery is declared to be minimal in any case. It was primarily his refusal to fall into such a confusion which prompted Einstein's drive towards a new cosmology in the first place. If McCrea's small number of atypical observers is to be completely passed over in order to preserve a positive meaning for the equivalence, then the emphasis on what is observational as the only possible plank for a scientific cosmology does no better justice to Einstein's original breakthrough than it does to Milne's wrestling with the ghosts of theory. A constant pressure, as we have seen from Einstein's first tentative steps, whether zero or any other value, demands a determinate size for the universe in accordance with the original construction of a spatially finite totality. When boundary conditions are abandoned in Newtonian theory (for instance by recalling that the matter lying outside a large spherical volume has no gravitational effect on the matter inside it), neither the definite structure nor the definite size of the material universe can be inferred. From that viewpoint, Einstein was certainly quite consistent when, in his 1917 paper, he apparently omitted to consider the equivalence in terms of the elimination of boundary conditions. This elimination is endemic to general relativity, since it articulates the idea of a universe of determinate size. Furthermore, the non-static condition does nothing by itself to abrogate general relativity's privilege to articulate such determinacy. It is only when it brings into play the idea of topological identity between different metrics or initial singularity (metrical breakdown) that the distinctive features of the periphery are absorbed in the centre. Again Einstein seems to have had good reasons to refuse what ran against his original motivation.

8. Weyl's principle and the 'many-universes' problem

Over and above these considerations, there is an originality in Milne's paper on equivalence which remains unassuagable. For Milne seems to have done to relativistic cosmology what De Sitter had done to general relativity prior to the very advent of Einstein's model. Milne's idea is that the same equation (the Friedmann equation) is apt to receive quite different interpretations, but this is reminiscent of De Sitter's early claim that different equations (Newton's and Einstein's equations of gravitation) would yield the same type of solution, were the relativists to maintain the requirement of determinate, universal values for the constants of integration occurring in the solutions of their differential equations. Both De Sitter and Milne strived to restore what they claimed to be the essence of general relativity, De Sitter with the intention of enhancing it and Milne in order to discredit it, but with the common rsult of forcing relativity to exhibit the true nature of its foundations.

Before proceeding to this latter problem, we may note that any evaluation of the degree of comparability between Newton and Einstein is not simply a matter of cosmology, as De Sitter's early critique drops a hint of something more. In 1929, Elie Cartan developed a neo-Newtonian theory of gravitation by redefining the preferred class of motion in that theory as that of *free falls*. Using the locally Euclidean affine connection of relativistic space-times, Cartan was able to show that from this connection, together with Newtonian absolute time and a non-relativistic spatial metric, it is possible to derive an equation of the form of Poisson's equation. Of the corresponding non-relativistic space-time with such local inertial frames, it can be said that general relativity modifies the *metric* only. Cartan's theory was successfully developed by P. Havas and A. Trautmann in the 1960s, so that in general terms both relativistic and Newtonian theories of gravitation use a four-dimensional space-time manifold, the essential distinction being that Einstein's theory is metric whereas Newton's is only affine (see Havas 1967 and Trautmann 1966). The upshot is a serious blow to Einstein's project of constructing a completely dynamical picture of space-time: the space-time continuum can be said to 'exist' independently of the existence of the gravitational field, for the case of no gravitational field is physically meaningful in that it is represented by an integrable affinity (and gravitation itself corresponds to a non-integrable affinity). Extending his views to a still higher level of generality, namely topology, Cartan understood this blow as meaning that cosmology can no longer be a kind of extension of local considerations to the global. Rather, as he put it, "the search for local laws of physics cannot be dissociated from the cosmogonic problem. We cannot say that one precedes the other; they are inextricably linked to one another" (1932, p. 18).

That the status of relativistic cosmology as a distinctive theory of the universe becomes quite precarious has also been emphasized with remarkable lucidity by Milne. By injecting into his interpretation of general relativity his own philosophical inklings about the physical reality of space, Milne has not only paved the way for equivalence, he has also pointed to the major conceptual difficulties involved in all relativistic images of the universe. It is no doubt true that all manner of thoughts about space as an a priori entity tend to erode relativistic cosmology and to deprive it, if not of all validity, at least of any pre-eminence among theories. Even relativistic cosmology does not escape some kind of a priori, as the conspicuous example of the Einstein-De Sitter model shows unambiguously the amount of independence from geometry with respect to other variables, in the sense that the curvature is not directly apprehensible. The resulting equivalence to a sophisticated form of Newtonian cosmology demonstrates the incapacity of cosmological theory to secure the meaning of its own basic concepts. In terms of Poincaré's epistemology as used by Milne, this has the consequence that some element of convention does play a role in both world pictures, as far as the fixation of that tricky entity space is concerned. But this relative autonomy enjoyed by space is undeniably reminiscent of the role of absolute space in Newton's original theory. And so, the equivalence may be pushed to metaphysical implications which exceed the language of conventions.

For there is more to come. Milne's alleged equivalence forces relativistic cosmology to exhibit another of its basic ambiguities, i.e., time. What Milne first shows in the Einstein-De Sitter model, and later (with McCrea) in all other varieties of relativistic world model, is that the equations describing the behaviour of a particle with fixed (co-mobile) coordinates are formally identical with the Newtonian equation describing the distance of a particle from the origin as a function of time. In the Newtonian case, the time indicated by a clock in motion accompanying a given particle is the same as indicated by the clock of any distant observer—it being assumed that the ordinary definition of simultaneity with the help of light signals is applicable. In the relativistic case, it is clear that the cosmic time of any event does not coincide with the epoch allotted to it by a distant observer using the same definition of simultaneity. It is essentially the assumption of *homogeneity* of the universe, as it is developed in relativistic cosmology, which in itself suggests the equivalence. Milne defines homogeneity by referring to two particles of a homogeneous distribution (1935, p. 61). Let us call them P and Q: the density distribution is homogeneous if, for any pair (P, Q), the density in P is the same as the density in Q, that is if $\rho(P) = \rho(Q)$. In the case of non-static homogeneous systems, the situation is far from simple. Suppose an observer O of such a system measures the density distribution of particles, at an arbitrary point P at the epoch t of the event E in P. He finds $\rho = \rho(P,t)$. From the point of view of the experience of O, the system will be homogeneous if the density in P at epoch t is equal to the density in another point,

Q, at the same epoch t of the event in Q. Thus, for O, $\rho(P,t) = \rho(Q,t)$. Another observer of the system, let us say O', in motion with respect to O, will allot to the events E_P and E_Q in P and Q at time t of O two different epochs t'_P and t'_Q. Thus, O and O' will find equal densities in P and Q, but at different times; a homogeneous system for O will not be so for O'. Of course, the reason is that no objective simultaneity exists between two events in two different places for two separate observers.

Relativistic cosmology overcomes the difficulty by postulating another time, say τ, instead of t. This new time denotes the proper time elapsed at each particle from a common origin: a system is homogeneous if, for any pair of points P and Q, $\rho(P,\tau) = \rho(Q,\tau)$ where ρ is computed by the observers in P and Q at the epochs indicated by their own clocks at these points. This, of course, is justified by Weyl's principle. Its definition is given by Bondi in the following terms: "The particles of the substratum (representing the nebulae) lie in space–time on a bundle of geodesics diverging from a point in the (finite or infinite distant) past" (1960, p. 100). All geodesics of the substratum intersect only once, at the very moment when they define the zero of time (see G.J. Whitrow 1980, pp. 290–1). This allows all clocks carried by the fundamental particles to be synchronized. Milne thinks this homogeneity provides quite a *conventional* definition, since it looks as if the previous difficulty has been simply reversed. Instead of having O' measuring equal densities at different times in two points P and Q, we have a density which changes from point to point, different in P and Q at the same epoch of the experience of O'. This situation suggests to Milne, despite the numerous claims that the *facts* of the expanding universe are simply expressed by Weyl's principle, that the homogeneity assumption can have no counterpart in 'reality', i.e. that the cosmological principle cannot be a law of nature. Even if the absence of objective simultaneity is taken as something to be overcome at any cost, it is certainly more natural, Milne says, to adopt a straightforwardly Newtonian concept of time and, along with it, the Newtonian theory as the sound basis of a cosmology. This, in fact, is what Milne himself endeavoured to do after he had established the equivalence, and while he was arduously engaged in the process of formalizing kinematic relativity (1944).

From this viewpoint, the equivalence between Newtonian and relativistic cosmology only reinforces the conviction that cosmic time is indeed a necessary ingredient in the formalization of a relativistic cosmology, however alien to general relativity and congenial to Newton's theory the notion of universal synchronization might seem. In Milne's radical terms, "if a *rational* understanding of the universe is possible, it ought to be possible to set up a consistent system of time-keeping throughout the universe" (1952, p. 50). The common origin postulated by Weyl's principle is clearly not an unambiguous, 'operational' definition of cosmic time, as the very terms used by Bondi reveal. Just as in Weyl's original paper of 1923, one has no hint as to the meaning of finite past *or* of an infinitely distant past, i.e. whether the common origin is a datable event (in the astrophysical sense) or some sort of chronogeometrical source. The fact that it does not, after all, matter where *observations* are concerned, reflects the conventional part of relativistic cosmology: it is as though Weyl's attempt to conceive of his principle in quasi metaphysical terms could work only in the static frame. Quite significantly, however, an attempt to make Weyl's principle 'real' instead of merely 'conventional' *has* been carried out. This is the steady-state cosmology first developed by Bondi, Gold, and Hoyle in 1948, which looked to Milne as "the ultimate irrationality" (1952, p. 23). In our context, this theory can be understood as a way of complementing the latter view on observability with the issue of causality. That is, there is nothing like a unique origin underlying all datable events. Again, the theory pushes the problems inherent in the static frame to their limit. Tolman had already remarked in 1929 that an alternative view to seeing an event like P' on Fig. 5.7 as entering the range of observability was the phenomenon of *creation*. Milne did interpret all phenomena of this sort in exactly the same way as Tolman, because his basic idea was that the range of observability and the range of causality should overlap completely in any satisfactory cosmology. And already in 1930, when Eddington was working on the instability of Einstein's static model, he conjectured that "unless a theory is invented which provides some force opposing the recession, there is no evading the rapid departure of nebulae from our neighbourhood" (1930, p. 677). This theory was invented: steady-state cosmology challenges the sacrosanct principle of conservation of energy, and postulates continuous creation out of the 'elsewhere/elsewhen' area so as to compensate for the recession of galaxies.

Still in the context of the equivalence problem, John North has remarked that the absence of boundary conditions in the classical equations allows for as indefinite a number of density distributions as are compatible with a given gravitational field (1965, p. 183). He says that "it might be reasonably objected that a theory which can explain a whole range of possibilities can have little predictive power in the relevant respects". Yet, the contrast with any other type of multiple solution is difficult to assess here. In particular, the range of possibilities predicted by relativity does not entirely escape some form of logical problem which affects all procedures spoken of so far. The problem has been given a very clear form by Bertrand Russell in one of his most striking pieces of argument. Referring to the Machian conception of dynamics, Russell equates its logical basis with the idea "that all propositions are essentially concerned with actual existents, not with entities which may or may not exist" (1937, p. 493). But there is no logical contradiction in the possibility of laws being applied to universes which do *not* exist. In the calculations of the distribution of matter, "it can be no necessary part of their *meaning* to assert the existence of the matter to which they are applied". The consequence of this is that the universe is brought into being as many times as there are possible distributions of matter.

The fabric of universes is not even limited to matter distributions. By explicitly denying the *principle* dictating the anathema on closed time-like curves, Kurt Gödel proposed in 1949 the first exact solution of Einstein's field equations depicting a completely homogeneous universe filled with pressureless matter of constant density, in which the matter rotates rigidly, relative to the local compass of inertia. Gödel conceived of rotation as a consequence of the denial that cosmic time would be everywhere orthogonal to the three-spaces. He began with a space–time model Euclidean in its topology and with a Minkowski metric, and he went on to assert the existence of a vast class of rotating solutions with pressureless matter of variable density and with homogeneous and finite three-space (1949a). Gödel's strategy is not just opposed to Weyl's; it is also just the inverse of Milne's. Gödel derives a possible structure of the physical universe from a space–time metric compatible with general relativity, instead of starting from a requirement about this structure. The rotating solution has created much confusion, but it is now generally admitted, after H. Stein (1970), that the arguments used in Gödel's construction are entirely correct. It is interesting to reflect that both Weyl and Milne pushed the relativistic way of thinking to its limits, and each had a profoundly disturbing effect on the original concepts, yet Gödel succeeded in disturbing 'orthodox' relativistic cosmology *from within*, without claiming any new principle. Both Weyl and Milne had used the special relativity background in order to restore unity (metrical for Weyl, kinematical for Milne) that was deemed to be threatened by the interaction of metric and matter density in general relativity. Now Gödel used precisely this background in order to demonstrate the possibility of anomalies of temporal order in general relativity. Quite significantly, a central claim made by Gödel was that the static or non-static character of his solutions is immaterial to the failure of the absolute cosmic time function to exist (1949b, p. 562).

Proceeding from the possible to the real, as Gödel does, is an uneasy method yet "the mere compatibility with the laws of nature of worlds in which there is no distinguished absolute time...throws some light on the meaning of time also in those worlds in which an absolute time *can* be defined" (p. 562). Certainly there is no way of denying our freedom to start from a possible universe compatible with a metric which was originally designed to fit our unique and really existing universe. But it seems equally certain nature itself does impose some restrictions over and above this freedom. As Einstein remarked in his reply to Gödel, in the case of a signal being sent from one point to another along a time-like path, "there exists no free choice for the direction of the arrow" (1949, p. 687). However, we can hardly know of what nature has to say when "the points, which are connectable by a time-like line, are arbitrarily far separated from each other"

(p. 688). If the 'earlier–later' relation breaks down in this case then we would have to face the apparent absurdities that Weyl had already described in *Space, Time, Matter*, that is, one would be able "to travel into the near past of those places where he has himself lived" (Gödel 1949b, pp. 560–1). Gödel discards the feasibility of a voyage that could send an observer back to his own past, but this only reinforces his conviction, by contrast as it were, that "it cannot be excluded a priori. . .that the space–time structure of the real world is of the type described" (p. 561). Of course, Weyl did acknowledge this, but it is also precisely why he felt the need to distinguish what he called "phenomenal" from "cosmic" time.

The issue hinges on the nature of becoming. Gödel's point is that becoming cannot be forced into any space–time background, contrary to Weyl's claim. For two very different things are involved in the space–time topology which most naturally lends itself to the homogeneity required in Weylian cosmology—that of special relativity. Firstly the relativity of simultaneity, which introduces "the non-objectivity of change" as Gödel calls it (p. 560), and secondly the equivalence of all observers in accordance with the principle of relativity. So, while the relativity of simultaneity denies any validity to objective becoming, it would be wrong, so Gödel argues, to believe that the postulated equivalence of inertial observers also enables us to recover equivalence with regard to a unique, absolute time function. The latter claim may be true as far as the laws of nature are concerned, "but this does not exclude that the structure of the world (i.e., the actual arrangement of matter, motion, and field) may offer quite different aspects to different observers" (p. 559, n. 6). And in fact, "the observer. . .plays no essential role in these [latter] considerations". The consequence of Gödel's discovery of possible anomalies in the global temporal order is that (as Einstein acknowledged in his reply) a return to the original motivation for using cosmology is unavoidable: it would seem that safe physical grounds are salvaged if, again, we start from the local and gradually extend to the global. In Einstein's language the very idea of points arbitrarily far separated from each other indicated rupture with his ongoing faith in general relativity as providing the only way of domesticating the self-enclosed universe.

Physicists in their vast majority, however, are not likely to be very fond of such isolated anomalies. There appeared in 1965 a paper which can be seen as the standard formulation of the present-day ratifying of the idea of equivalence between Newtonian and relativistic cosmology—and this, of course, also means ratification of the intuitive idea of absolute time lapsing objectively. Newtonian mechanics, as a general and loose term, is supposed by the authors of this paper to provide "neither a crude approximation to the correct relativistic calculation, nor a cooked-up montage cleverly contrived to look like the real thing" (Callan, Dicke, and Peebles 1965, p. 105). That Newtonian mechanics is entirely adequate is a conclusion the proponents reach in a rather revealing manner. Of course, no one wants to deny that the classical concepts fail to describe the dynamics of vast regions separated by velocities comparable to the velocity of light, and this is the authors' first admission. The calculation of $R(t)$ is possible in Newtonian terms only for two neighbouring galaxies, but the argument then moves without the slightest logical transition to the assertion that the calculation is "also applicable to the whole universe" (p. 108). Implicitly recalling one of the Bondi theorems already worked out by Layzer, the conclusion is that isotropically distributed matter has no physical effect on the interior of a large spherical volume, irrespective of the Newtonian or relativistic nature of these effects. But more importantly, the authors made a frontal attack on the puzzling question of a possibly devastating effect of the expansion of space. They substitute the expression "expansion of the universe" for "expansion of space", claiming that only the former concept can release physics from the spell of a mysterious power that initiates the pulling apart of galaxies. Motion *in* space rather than *of* space is the key to the proposed substitution. This has been completely overthrown by the developments of cosmology that took place shortly after 1965, in conjunction with the promised unification of all aspects of physics under the guidance of a comprehensive model of the very early universe. In this model, pressure appears to be extremely significant. This model does not simply tend to undermine any full equivalence between Newtonian and relativistic cosmology, it is also supposed to answer once and for all the daunting question of what caused the expansion of the universe. The examination of this claim forms the background of our epilogue.

References

Borel, E. (1923). *L'espace et le temps*, F. Alcan, Paris.
Bondi, H. (1960). *Cosmology*. 2nd ed., Cambridge University Press.
British Astronomical Association (1931). The evolution of the universe: contributions to a discussion, Suppl. to *Nature*. **128**, 699–722.
Callan, C., Dicke, R. and Peebles, P.J.E. (1965). Cosmology and Newtonian mechanics, *American J. of Physics*. **33**, 105–8.
Cartan, E. (1932). *Le parallélisme absolu et la théorie unitaire du champ*, Hermann, Paris.
Charlier, C.V.L. (1922). How an infinite universe may be built up. *Archiv för Math. Astr. och Fys.* **16**, No. 22.
De Sitter, W. (1930a). The expanding universe: discussion of Lemaître's solution. *Bull. Astr. Inst. Netherl.* **193**, 211–18.
De Sitter, W. (1930b). On the distances and radial velocities of extra-galactic nebulae, and the explanation of the latter by the relativity theory of inertia. *Proc. Nat. Acad. Sci.* **16**, 474–88.
De Sitter, W. (1931). Contribution to the British Astronomical Association.

De Sitter, W. (1932). On the expanding universe. *Proc. Kon. Akad. Wet. Amst.* **35**, 596-607.
De Sitter, W. (1933a). On the expanding universe and the time-scale. *Monthly Not. Roy. Astr. Soc.* **93**, 628-34.
De Sitter, W. (1933b). *Kosmos*. Harvard University Press, Cambridge, Mass.
De Sitter, W. (1934). On the foundations of the theory of relativity, with special reference to the theory of the expanding universe. *Proc. Kon. Akad. Wet. Amst.* **37**, 597-601.
Dirac, P.A.M. (1968). The scientific work of Georges Lemaître. *Pontificae Academiae Scientiarum Commentarii*. Vol. II, No. 11, 1-20.
Eddington, A.S. (1920). *Space, time, and gravitation*. Cambridge University Press.
Eddington, A.S. (1923). *The mathematical theory of relativity*. Cambridge University Press.
Eddington, A.S. (1924). *The mathematical theory of relativity*. Second edn. Cambridge University Press.
Eddington, A.S. (1927). *Stars and atoms*. Clarendon Press, Oxford.
Eddington, A.S. (1928). *The nature of the physical world*, Cambridge University Press.
Eddington, A.S. (1930). On the instability of Einstein's spherical world. *Monthly Not. Roy. Astr. Soc.* **90**, 668-78.
Eddington, A.S. (1931a). Council note on the expansion of the universe. *Monthly Not. Roy. Astr. Soc.*, **91**, 412-16.
Eddington, A.S. (1931b). Contribution to the British Astronomical Association.
Eddington, A.S. (1935). *New pathways in science*, Cambridge University Press.
Eddington, A.S. (1940). Forty years of astronomy, In *Background to Modern Science*, (eds., J. Needham and W. Pagel) pp.115-42. Cambridge University Press.
Einstein, A. (1922). Bemerküng zur F. Seletyschen Arbeit: Beiträge zum Kosmologischen Problem. *Annalen Phys.* **69**, 436-8.
Einstein, A. (1929). Space-Time. In *Encyclopedia Britannica*, 14th ed., Vol. 21, pp. 105-8.
Einstein, A. (1931). Zum kosmologischen Problem der allgemeinen Relativitätstheorie, *Sitz. Ber. Preuss. Akad. Wiss. Berlin*, 235-7.
Einstein, A. (1933). Sur la structure cosmologique de l'espace. (French trans., M. Solovine) Hermann, Paris.
Einstein, A. (1945). *The meaning of relativity*. 2nd ed. Princeton University Press.
Einstein, A. (1949). Reply to criticisms. In *Einstein, philosopher-scientist* (ed., P.A. Schilpp) Open Court, LaSalle, Ill.
Einstein, A. and De Sitter, W. (1932). On the relation between the expansion and the mean density of the universe, *Proc. Nat. Acad. Sci.* **18**, 213-14.
Ellis, G.F.R. (1975). Cosmology and verifiability. *Qu. J. Roy. Astr. Soc.* **16**, 245-64.
Friedmann, A. (1922). Uber die Krümmung des Raumes. *Zeitschr. Phys.*, **10**, 377-86.
Friedmann. A. (1924). Uber die Möglichkeit einer Welt mit konstanter negativer Krümmung des Raumes. *Zeitschr. Phys.*, **21**, 326-32.
Gödel, K. (1949a). An example of a new type of cosmological solutions of Einstein's field equations of gravitation. *Rev. Mod. Phys.* **21**, 447-50.
Gödel, K. (1949b). A remark about the relationship between relativity theory and

idealistic philosophy. In *Albert Einstein, philosopher-scientist* (ed., P.A. Schilpp) pp. 555-62. Open Court, LaSalle, Ill.

Harrison, E.R. (1981). *Cosmology*. Cambridge University Press.

Havas, P. (1967). Foundation problems in general relativity. *Delaware seminar in the foundation of physics*, **8**, 124-48.

Hawking, S.W. and Ellis, G.F.R. (1973). *The large scale structure of space-time*. Cambridge University Press.

Heckmann, O. (1931). Über die Metrik des sich ausdehnden Universums, *Nachr. Ges. Wiss. Göttingen*. pp. 126-130.

Heckmann, O. and Schücking, E. (1959). Newtonsche und Einsteinsche Kosmologie. In *Encyclopedia of physics* (ed., S. Flügge), Vol. 53, pp. 489-519. Springer-Verlag, Berlin.

Hubble, E. (1926). Extra-galactic nebulae. *Astroph. J.* **64**, 321-69.

Hubble, E. (1929). A relation between distances and radial velocity among extra-galactic nebulae. *Proc. Nat. Acad. Sci.* **15**, 169-73.

Jeans, J. (1918). *Problems of cosmogony and stellar dynamics*. Cambridge University Press.

Jeans, J. (1928). *Astronomy and cosmogony*. Cambridge University Press.

Lanczos, C. (1923). Über die Rotverschiebung in der De Sitterschen Welt. *Zeitschr. Phys.* **17**, 168-89.

von Laue, M. (1923). Book review of Eddington's Mathematical Theory of Relativity. *Die Naturwissenschaften*, **20**, 382-4.

Layzer, D. (1954). On the significance of Newtonian cosmology. *Astron. J.*, **59**, 268-70.

Layzer, D. (1967). A unified approach to cosmology. In *Lectures in applied mathematics* (ed., J. Ehlers) Vol. 8, pp. 237-258. American Math. Soc., Providence.

Lemaître, G. (1925). Note on De Sitter's universe. *J. Math. Phys.* (MIT), **4**, 37-41.

Lemaître, G. (1927). Un univers homogène de masse constante et de rayon croissant. *Ann. Soc. Sci. Brux.* **47(A)**, 49-59.

Lemaître, G. (1929). La grandeur de l'espace. *Rev. Quest. Sci.* **16**, 189-216.

Lemaître, G. (1931a). The expanding universe. *Monthly Not. Roy. Astr. Soc.* **91**, 490-501.

Lemaître, G. (1931b). The beginning of the world from the point of view of quantum theory. *Nature*. **127**, 706.

Lemaître, G. (1933). L'univers en expansion. *Ann. Soc. Sci. Brux.* **53(A)**, 51-85.

Lemaître, G. (1958). Rencontres avec Einstein. *Rev. Quest. Sci.* 129-32.

Levi-Civita, T. (1917). Realtà fisica di alcuni spazi normali del Bianchi. *Rendiconti della R. Acc. dei Lincei*. **26**, 519-31.

Levi-Civita, T. (1927). *The absolute and differential calculus*, (trans. M. Long) Blackie and Sons, London.

MacMillan, W. (1925). Some mathematical aspects of cosmology. *Science*, **62**, 63-72, 96-9, 121-7.

McCrea, W.H. (1955). On the significance of Newtonian cosmology. *Astron. J.* **60**, 271-4.

McVittie, G.C. (1932). Condensations in an expanding universe. *Monthly Not. Roy. Astr. Soc.* **92**, 500-18.

McVitte, G.C. (1954). Relativistic and Newtonian cosmology. *Astron. J.* **59**, 173-80.
McVittie, G.C. and McCrea, W.H. (1931). On the contraction of the universe. *Monthly Not. Roy. Astr. Soc.* **91**, 128-33.
Milne, E.A. (1933). World-structure and the expansion of the universe. *Zeitschr. für Astroph.* **6**, 1-90.
Milne, E.A. (1934). A Newtonian expanding universe. *Qu. Journal of Math.* **5**, 64-72.
Milne, E.A. (1935). *Relativity, gravitation, and world-structure*. Clarendon Press, Oxford.
Milne, E.A. (1944). On the nature of universal gravitation. *Monthly Not. Roy. Astr. Soc.* **104**, 120-35.
Milne, E.A. (1952). *Modern cosmology and the christian idea of God*. Clarendon Press, Oxford.
Milne, E.A. and McCrea, W.H. (1934). Newtonian universes and the curvature of space. *Qu. J. of Math.* **5**, 73-80.
North, J. (1965). *The measure of the universe*. Clarendon Press, Oxford.
Peebles, P.J.E. (1984). Impact of Lemaître's idea on modern cosmology. In *The big bang and Georges Lemaître* (ed., A. Berger) pp. 23-30. Reidel, Dordrecht.
Rindler, W. (1977). *Essential relativity*. 2nd ed., Springer-Verlag. Berlin.
Robertson, H.P. (1928). On relativistic cosmology. *Phil. Mag.* **5**, 835-48.
Robertson, H.P. (1929). On the foundations of relativistic cosmology. *Proc. Nat. Acad. Sci.* **15**, 822-9.
Robertson, H.P. (1932). The expanding universe. *Science*, **76**, 221-6.
Robertson, H.P. (1933). Relativistic cosmology. *Rev. Mod. Phys.* **5**, 62-90.
Robertson, H.P. (1935). Kinematics and world-structure. *Astroph. J.* **82**, 284-301; **83**, 187-201; **83**, 257-71.
Royal Astronomical Society (1930). Report of the meeting in January 1930. *The Observatory*. **53**, 33-44.
Russell, B. (1925). *ABC of relativity*. Allen & Unwin, London.
Russell, B. (1937). *The principles of mathematics* 2nd ed., Allen & Unwin, London.
Schlegel, R. (1958). Steady-state theory at Chicago. *Amer. J. of Physics*, **26**, 601-7.
Schrödinger, E. (1956). *Expanding universes*. Cambridge University Press.
Schücking, E. (1967a). Newtonian cosmology. *Texas Quarterly*, **10**, 270-4.
Schücking, E. (1967b). Cosmology. In *Lectures in Applied Mathematics* (ed., J. Ehlers) Vol. 8, pp. 218-36. American Math. Soc., Providence.
Sciama, D. (1973). *Modern cosmology*, Cambridge University Press.
Selety, F. (1922). Beiträge zum kosmologischen Problem. *Ann. Phys.* **68**, 281-334.
Selety, F. (1923). Erwiderung auf die Bemerkungen Einsteins über meine Arbeit: Beiträge zum kosmologischen Problem. *Ann. Phys.* **72**, 58-66.
Selety, F. (1924). Unendlichkeit des Raumes und allgemeine Relativitätstheorie. *Ann. Phys.* **73**, 291-325.
Silberstein, L. (1924a). Determination of the curvature-invariant of space-time. *Phil. Mag.* **47**, 907-18; **48**, 619-28.
Silberstein, L. (1924b). *The theory of relativity*. 2nd ed., MacMillan, London.
Silberstein, L. (1930). *The size of the universe*. Oxford University Press.
Silberstein, L. (1933). *Causality*. MacMillan, London.

Sklar, L. (1976). Inertia, gravitation and metaphysics. *Phil. of Science.* **43**, 1–23.
Smith, R.W. (1982). *The expanding universe.* Cambridge University Press.
Stein, H. (1970). On the paradoxical time-structures of Gödel. *Phil. of Science.* **37**, 589–601.
Tolman, R.C. (1929a). On the astronomical implications of the De Sitter line element for the universe. *Astroph. J.* **69**, 245–74.
Tolman, R.C. (1929b). On the possible line elements for the universe. *Proc. Nat. Acad. Sci.* **15**, 297–304.
Tolman, R. (1930). The effect of the annihilation of matter on the wave-length of light from the nebulae. *Proc. Nat. Acad. Sci.* **16**, 320–7.
Torretti, R. (1983). *Relativity and Geometry.* Pergamon, Oxford.
Trautmann, A. (1966). Comparison of Newtonian and relativistic theories of space-time. In *Perspectives in geometry and relativity*, ed. B. Hoffmann, Indiana University Press, pp. 413–25.
Walker, A.G. (1935). On Riemannian spaces with spherical symmetry about a line and the conditions for isotropy in general relativity. *Qu. J. Math.* Oxford Ser., **6**, 81–93.
Walker, A.G. (1936). On Milne's theory of world-structure. *Proc. Math. Soc. London* (2) **42**, 90–127.
Weyl, H. (1922). *Space, Time, Matter.* English trans. by H.L. Brose on the Fourth German edition. Methuen, London.
Weyl, H. (1923a). *Raum, Zeit, Materie*, Fifth edn, Springer, Berlin.
Weyl, H. (1923b). Zur allgemeinen Relativitätstheorie, *Phys. Zeitschr.* **24**, 230–2.
Weyl, H. (1924a). Observation on the note of Dr Silberstein: Determination of the curvature-invariant of space-time. *Phil. Mag.* **48**, 348–9.
Weyl, H. (1924b). Massenträgheit und Kosmos. *Die Naturwissenschaften.* **12**, 197–204.
Weyl, H. (1927). Philosophie der Mathematik und Naturwissenschaften. In *Handbuch der Philosophie* (eds. A. Bauemler and M. Schröter). Oldenburg, Munich.
Weyl, H. (1929). Gravitation and the electron. *Proc. Nat. Acad. Sci.* **15**, 323–34.
Weyl, H. (1930). Redshift and relativistic cosmology. *Phil. Mag.* **9**, 936–43.
Weyl, H. (1949). *Philosophy of mathematics and natural science*, Engl. trans. of his (1927), Princeton University Press.
Whitrow, G.J. (1980). *The natural philosophy of time.* Clarendon Press, Oxford.
Wirtz, C. (1918). Über die Bewegungen der Nebelfäcke. *Astr. Nachr.* **206**, 109–16.
Wirtz, C. (1922). Notiz zur Radialbewegungen der Spiralnebel. *Astr. Nachr.* **216**, 451.
Wirtz, C. (1924). De Sitter's Kosmologie und die Bewegungen der Spiralnebel. *Astr. Nachr.* **222**, 21–6.
Zwicky, F. (1928). On the thermodynamic equilibrium in the universe. *Proc. Nat. Acad. Sci.* **14**, 592–8.
Zwicky, F. (1929). On the redshift of spectral lines through interstellar space. *Proc. Nat. Acad. Sci.* **15**, 773; and *Phys. Rev.* **33**, 1077.

Epilogue

Proliferating inventions

Ever since Einstein constructed a relativistic model of the universe in 1917, our understanding of what can meaningfully be said about the universe has completely changed. We can see the difference between cosmology before and after Einstein, and so establish a gauge for Einstein's originality by contrasting two positions on Olbers' paradox, one of the universe's most resilient theoretical aporias. We touched on this paradox in Chapter 1. Early this century astronomer Simon Newcomb complained that Olbers' paradox could only characterize what the universe is not; when Hermann Bondi gave the Halley Lecture at Oxford University in 1962, however, he set out to argue that Olbers' background assumptions constitute a theory in themselves, no matter how far out of step this theory is with possible observations. According to Bondi, "we know, as a result of Olbers' work, that whatever may be going on in the depths of the universe, [the theory] cannot be constructed in accord with his assumptions. By this method of empirical disproof, we have discovered something about the universe, and so have made cosmology a science". Hence, we can deduce facts as significant as the expansion of the universe on the grounds that the universe is not infinitely bright.

Before Einstein's insights it would have been inconceivable that pure theory could be used to forecast the observational data of the universe as a totality. In fact, the expanding universe theory was in no way a response to Olbers' paradox; the bearing it might have on its solution is only incidental. Twentieth-century cosmology did not originate in the confrontation between hypothetico-deductive reasoning and observation, as Bondi's after-thought would imply; rather, it emerged in a *tour de force* of conceptual invention attendant on a series of brash epistemological questionings. Einstein's starting-point was Mach's critique of the Newtonian theory of mass. Towards the end of the nineteenth century, Mach attempted to define what for Newton was an undefined primitive, the concept of inertial mass. Defining inertial mass is bound up with identifying its source, and Mach's explanation was that inertia rests in the gravitational effect of distant masses. Some physicists pointed out, however, that such positing of a universal causality was at variance with the very idea of field and its exclusion of action-at-a-distance. And so there was a serious problem of interpretation for general relativity as a new field theory of gravitation. In so far as general relativity aimed to revamp Mach's critique of the old theory as a basis for a new one, Einstein had to work out (he saw the necessity after several discussions with De Sitter) a matter distribution in the whole of space–time that could be compatible with a field–law, that is, a law in $1/r$. The finite model offered just the right distribution of sources of the field. It was the *pièce de résistance* of Einstein's solution to a problem which had haunted him from the beginning. For immediately after establishing the theory of special relativity in 1905, he faced two alternatives in conceiving gravitation: either he could include it in some way within a special-relativistic background, or it would constitute a new background altogether, namely, non-Euclidean geometry.

Einstein's fundamental idea was that any new theory of gravitation should be an extension of special relativity (rather than a radical revision). The problem with the non-Euclidean approach, however, was that allowing generally covariant equations would open the gate to a virtually infinite number of solutions, and Einstein had been intimidated by this for a while. His solution was to reconsider what is meant by a gravitation-free case. Nature does not allow just any value for the potentials of gravitation: the elevator in free fall already imposes a Euclidean connection between the various bits of space it sews together. Moreover, at an infinite distance from all masses, there would be a pseudo-Euclidean continuum. By stipulating these boundary conditions, Einstein was finally able to show that general covariance does not affect the structural features of space–time—it is compatible, for instance, with the differential effects of rotation. And so cosmology was the realization that any form of boundary condition remains purely mathematical and lacks any significance as an entity of field physics: we cannot assert with certainty the *physical* existence of the pseudo-Euclidean continuum at infinity.

This discovery was radically unsettling, for (throughout the nineteenth century) large-scale mechanics presented itself as an account, growing in sophistication, of the inadequate fit between Newtonian physics and any view of the universe as homogeneous, infinite, and Euclidean. We should not downplay Einstein's inventiveness even considering the whole of classical physics, for it offers a powerful report of the peculiar situation of the mechanical world view in the seventeenth and eighteenth centuries. Pascal's formulation is perhaps the most insightful on this question: "since all things are cause and effect, supported and supporting, at first or second-hand, and since all are united by a natural and imperceptible chain which links all together, the most distant and most different . . . these extremes meet and combine by the very reason of their distance apart". This looks very close to something like a Mach's principle in the most general sense, but Pascal adds the further thesis that these extremities "find each other in God and in Him alone" (in 1950, pp. 27 and 23). Einstein's cosmology is the ultimate attack on the resolution of these questions of extremities by means of theology, an attack inaugurated early in the nineteenth century, though with less scope and impact.

The invention of cosmology as a science was not the end of epistemological debate, however, and the number of conceptual critiques it continues to spawn is proof of its vast heuristic power. And so, just after the publication of Einstein's first cosmological memoir, Ehrenfest saw troubles with what Einstein had let stand of relativity's original insight as a physical theory, that is, the postulated equivalence of all observers. Since the theory of special relativity had done away with the concept of ether, motion with respect to the ether could not exist. Moreover, absolute acceleration would also have to be conceived as taking place with respect to nothing, and it was general relativity which went on to extend the class of equivalent motions to acceleration. Cosmology obscured the issue once more, however. As Ehrenfest wrote to Einstein: "Now one can no longer say that they [the equivalent observers] move with respect to nothing, for now they move with respect to an enormous something!". Ehrenfest's typically hot-blooded letter continues: "Einstein, my upset stomach hates your theory—it almost hates you yourself! How am I to provide for my students? What am I to answer to the philosophers?!!" (Ehrenfest to Einstein 9 December 1919. In M.J. Klein 1970, p. 315). Einstein's reply focused on the need to distinguish the motivations of epistemology from those of physics when it comes to forming a theory. Special relativity had its impetus in the empirical fact that the velocity of light is constant in all inertial frames, while the principle of general relativity was epistemological in that it evolved from the observation that inertial and gravitational masses are numerically identical (pp. 315-16). We are now reasonably confident that Mach's principle is not essential to general relativity; more especially, there seems to be no necessary connection between Mach's ideas and the finiteness of the universe, as Einstein himself was to acknowledge and as De Sitter had argued from the first. Of course, this does not stand in the way of the principle once and for all: it could still play an insightful role in the elaboration of other theories of gravitation. (H. Goenner reviewed this question in 1980.) It does mean, however, that we must try to understand why and how the invention of the finite universe survived what seems on the face of it a failure to meet its primary goal.

In fact, classicism rode its last stand not in Einstein's ambiguous attempt to refine Mach's ideas but in his tacit assent to the only pre-Copernican concept that Newton's revolution had not ruptured, the universe's globally static nature. Here, too, Einstein resisted any pressure (from De Sitter and after him repeatedly from a few others) to identify the cosmic time function of his cylindrical model with a return to the absolute time of classical physics. For he thought there was one acquisition of relativity theory which had priority over all others: the denial of time as an independent dimension, which implied formidable resistance to the idea that physics (precisely because it incorporates a notion of field expressed in differential equations) should assume an influence of the distant past over the present. Weyl was the first to give this a shake up, but perhaps the greatest lesson of his intellectual reflections on cosmology is that the word "non-static" has always referred to something basically alien to general relativity as a *physical* theory. He made this quite clear when he introduced the term in the fourth and fifth editions of *Raum, Zeit, Materie* (1921/23a), well before anything was known about non-static *coordinates*. He called De Sitter's universe non-static "as a whole", but continued to treat it as static coordinates because no coordinate system could ever cover the whole of such a universe—'whole' implying here a perfect overlap between the range of observability and the range of causality. However, when he came to know in 1930, of the non-static coordinates used by Robertson, he conceded that these cover the whole of the observable universe, but with the lament that the new scheme turned the universe into something geometrical. What was at stake was the way Robertson arrived at the cosmic time function he needed to account in a 'Copernican' way (that is, in accordance with the assumption that all viewpoints are equivalent) for the most salient observation, the redshift phenomenon of the extra-galactic nebulae. In Robertson's work, which was a development of Friedmann's and also (implicitly) of Lemaître's, time is made orthogonal to space. It is only in recent times that Hawking and Ellis have clarified the necessary properties of a global time function if it is to be physical: we need so-called 'stable causality' to make sure that 'true' time resides in the cosmos itself and not in human consciousness.

Before tracing this development in some detail, we should outline Weyl's reasons for doubting the physicality of the non-static universe. Geometry was reconceived as non-static because this worked well with a principle of *observability*, but this shunted off the problem of *causality* Weyl had envisaged. True, his first impression from the observation of large redshifts among extra-galatic nebulae was that the universe revealed itself as chaotic, very far from the highly organized structure that the geometrical interpretation was to propound later. In Weyl's terms, postulation of equivalent viewpoints throughout space–time would presuppose an over-arching notion of causality, and his grand hypothesis (1923b) on universal causality (that the particles representing the nebulae lie in space–time on a bundle of geodesics diverging from a common point in the past) was a fundamental attempt at finding the only possible connection between observations and causes. More ambitiously, the hypothesis also aimed to identify the ultimate sense of boundary condition for the universe together with the conditions of intelligibility that would surround any possible unification of field physics: this was the gauge-invariance theory unifying gravitation and electromagnetism. The advent of quantum mechanics, however, toned this programme down. At the cosmological level, geometry and physics then parted ways to the point that Weyl's principle was now absorbed as a premise of the geometrical universe.

Lemaître's model of the primeval atom, sketched in 1931, was a first attempt to physicalize these conditions of intelligibility by incorporating quantum theory. By bringing the point of common origin Weyl had postulated back from the infinitely distant past, it provided a concrete image of both the universe itself and our science of it. The detection of isotropic microwave background radiation in 1965 was interpreted as the decisive observational evidence in favour of Lemaître's scenario, since this radiation was identified with the appropriate relics of the big bang. One result of this was that theoreticians could discard most of the other models entertained in the meantime. In fact, by the late 1960s Lemaître's early conclusion that there must be an intial singularity became irresistible. Lemaître once claimed that even asymmetries (a suggestion of Einstein) would not remove the singularity; investigations of Penrose (1965), Hawking (1967), and Geroch (1968), to name a few, have proved that if we speak of self-gravitation of our universe in the past, the gravitation would be of such magnitude that singularities would come to be a constitutive property of space–time. Their investigations even suggest a physical explanation for the observed homogeneity of the universe by resolving the 'many-universes' problem which had loomed to Eddington and Weyl as early as 1923. In what we call Lemaître's 'standard' big bang model, an original state of infinite density was followed by spatially uniform expansion. The problem is that the microwave background radiation was emitted at a time when some parts of the expanding universe were already separated by distances prohibiting causal interaction and communication by light.

The so-called 'inflationary' model of the expansion (A. Guth 1981) fleshes out the standard model by looking at the very short period of time just before this degree of expansion took place. The new model offers an origin for the expansion which hinges on the role of negative pressure (see Chapter 3). The point is that in its pre-inflationary phase the universe was so minuscule that the many-universes problem did not operate: all parts were causally interacting. It was also at this time that all the fundamental forces of nature operated as a single super-force. Granted this, the residual possibility of many universes is *entirely* dependent on observability. The existence of various disconnected parts to the universe does not jeopardize a concept of causality fixed in the pre-inflationary phase. Moreover, the inflationary model helps explain the basic observation we recognized from the outset: that space is almost perfectly Euclidean over very large portions of space–time. The model gives an account of just why the amount of energy density released entailed this type of space: that is, the metric homogeneity, which had been thought in separate geometrical terms ever since the failure of Einstein's first model, is now tied up again with a physical justification. When all is said and done, quantum cosmology investigates the very instant in which all things have their origin, that is, what happened during a length of time so short that things were not yet determined—in accordance with indeterminacy allowed by quantum theory. At this level cosmologists are toying with probabilities, the probabilities that the universe did this rather than that in the interval of uncontrollable quantum fluctuations. (The topic was initiated by E.P. Tryon in 1973.)

The development of cosmology in the last two decades offers a physical interpretation of Weyl's principle which had never before been pushed so deeply. By contrast, then, very little remains of the other postulates that had been put forward by alternative, post-relativistic cosmology (kinematic relativity and steady-state); those postulates are invalidated, and such an unorthodox relativistic cosmology as Gödel's is simply left aside. Instead, basic ingredients like homogeneity, isotropy, quasi-Euclideanity, cosmological constant, and singularity, all of them primitive elements in the various earlier attempts at making general relativity cosmological, now appear to be elements of physics itself. Viewed from such a historical perspective, this movement shows up more surely than ever the connection between cosmology and the fundamental problem of delimiting geometry's place in the unification of physics. The Eddington–Weyl controversy of 1923 is certainly the first explicit instance of this fundamental problem: it is illustrated in the way cosmology makes use of physical theory to make intelligible that which is beyond observability (see M. Munitz 1986, pp. 54–6). In fact, the same sort of problematical distinction (though expressed in very different terms) had dictated the early controversy between Einstein and De Sitter. More generally, while De Sitter and Eddington denied that the idea of 'super-observer' had any meaning in the physical realm, Einstein and Weyl did show, each in his own terms, the impossibility of dispensing with it. The super-observer can travel at will through all the space and time of the universe; he is the one who defines the operations of a causality which, by rights, extends over any given range of observability.

Now, when we consider the nature of the physical and mathematical concepts with which these models of the universe work, we find that the crucial idea lies in the way the distinction between the local and the global is articulated.

Local versus global

From the mid-nineteenth century until this day, geometry has been engaged in the pursuit of two very different kinds of ideas, the local viewpoint and the global viewpoint. The former is Riemann's: it specifies an (infinitesimal) element and proceeds with a step-by-step progression towards the link between various such elements, until an idea of the whole emerges. The latter seeks to characterize the whole independently of its parts and tends to define the elements in terms of functional properties. The global concept is that of

Felix Klein in accordance with his Erlangen Programme. It defines a geometry as the study of those properties that are preserved under certain transformations; these transformations form a group of transformations of the space in its totality. Whatever the properties of invariance (metric, affine, projective) that can be established with respect to a certain group of transformations, the important point is that all spaces studied in this manner are characterized by their homogeneity, i.e. the group operates in the same way at all points of the space. On the other hand, the local concept is deprived of such homogeneity: the Riemannian spaces are characterized by a quadratic differential form which defines the squared distance between two neighbouring points; this form is a generalization of the Euclidean expression of the distance in finite geometry. Each neighbourhood is thus like a piece of Euclidean space, and these pieces may be 'sewn' together in an infinite number of ways. To the extent that the physics of relativity was based on these two ideas, local and global, it offered a unique coalescence of very distinct strategies. Thus, the distinction between the two types of geometry becomes quite conspicuous in that Kleinian geometry is that of special relativity: it is based on consideration of the invariants of the Lorentz group in four-dimensional Minkowski space–time; but general relativity is a Riemannian geometry in which the $g_{\mu\nu}$ are specified by the distribution of matter at each point of space–time. Restrictions and limitations were brought in by Einstein in order to view general relativity as a smooth extension of special relativity. Locally, the space–times of general relativity were declared to have a Euclidean affine connection. Globally, if we write the metric for the universe on the basis of assumptions such as homogeneity, we still need boundary conditions in order to find solutions to the system of equations with partial derivatives. These conditions are not easily absorbed in the postulation of globally finite space, for they may still sneak in the form of Mach's principle as De Sitter had quickly shown.

It appears that the topological viewpoint is most decisive in tackling the problem of the nature of boundary conditions. Weyl was the first to investigate it, and again it comes to the fore in recent grapplings with the origin of the universe. In all other types of geometry, one important invariant under the permitted transformations is the preservation of straight lines as straight lines. In taking the step to topology, this invariant is abandoned: in determining which properties of figures are topological, any one-to-one bicontinuous transformation is permitted. Intuitively, this means that such transformations and their inverses map each point to a unique image point, and points which are 'near' remain 'near', that is, *neighbourhoods* are preserved. There is an obvious relevance of these geometrical considerations to our physical problem, since they offer the most consistent way of thinking of the interdependence between the local and the global. The very idea of invariance of neighbourhoods entails both of them at one stroke, invariance referring to global, and neighbourhoods to local.

From the mathematical point of view, the search for a connection between the structure of totality and the properties of elements, that is, the reunion of topology and differential geometry, culminated in the formulation of the Hopf–Rinow theorem (see H. Hopf 1932). This theorem tells us the extent to which the topological structure of totality can somehow be 'reflected' in the properties of its parts. It provides the bridge between two inverse problems: metricizability (what are the metrical properties prescribed by the topology of some given surface?) and extendibility (what can be inferred about the whole surface from knowledge of a small part?). There is one restriction of fundamental importance in the theorem: it tells us how the choice of ds^2 is constrained by the topology of global space and vice versa *only if the surface under consideration is already complete*, that is, when it cannot be extended metrically by adding a certain number of points. An incomplete surface would be one that cannot be extended beyond some termination point, i.e. it presents a singularity.

Study of the analogy between these mathematical cases and physical reality is quite difficult; the subject has certainly not been exhausted and is still an active topic of research. The reason is that the Hopf–Rinow theorem was formulated for all Riemannian spaces, while relativity physics deals with pseudo-Riemannian space–times (that is, the sign of one of the dimensions must differ from the others). This introduces the idea of causality, which first appeared in special relativity as the light cone and as the postulation of the velocity of light as a maximum. General relativity was built on the basis that its 'causal' structure of space–time would manifest locally the same qualities as the flat space–time of special relativity, so that the significant differences would occur globally because of non-trivial topology, space–time singularities, or tipping over of the direction of light cones as one moves from point to point in the vicinity of a strong gravitational field so that continuous designation of past and future would no longer be possible. Clearly, Weyl was the first to point out these cases of independence of cosmic-scale considerations from general relativity in its original form. He attempted to overcome indetermination by subjecting the global application of the theory to a Kleinian analysis of invariants—an identical separation of space and time would be necessary at each point of space–time. He went on to discover that a higher conception of cosmic time, over and above the individual proper times of particles, is required in order to achieve this. In fact, the impact of this result has been appreciated to its full extent only in the last two decades or so. And a unique coalescence between the ideas of invariance and causality is now at work in the development of unified field theories. Invariance (or symmetry) principles are the key to unification of the three fundamental forces of the microphysical world. Causality principles dominate the large-scale approach in terms of a history of the universe, and include the weakest of all forces, gravitation.

Thus, in relativistic cosmology, the principle of the velocity of light as a limit is violated (expanding redshifts may exceed this velocity), yet causality is maintained in the form of Weyl's postulate. This postulate defines a sort of 'super' special-relativistic light cone, as the two $L_{-\infty}$ represent an event horizon. Hawking and Ellis (1973) generalize this by first defining a causal curve as any curve lying within the light cone (including its boundary, the null geodesics of light). Then they define strong causality: for every neighbourhood O of a point p, there exists a neighbourhood V of p contained in O such that no causal curve intersects V more than once; if a space-time violates strong causality at p, then near p there exist causal curves which come arbitrarily close to intersecting themselves. But an even stronger version of causality is needed, because examples can be constructed in which strong causality is satisfied and yet a modification of $g_{\mu\nu}$ in an arbitrarily small neighbourhood of two or more points produces closed causal curves. This stronger version is referred to as *stable causality*. Further, the future (past) domain of dependence of a set of points S is such that every past (future) inextendible causal curve through p intersects S. In other words, the full domain of dependence represents the complete set of events for which all conditions should be determined by a knowledge of conditions on S. Now, a space-time equipped with a set of points for which the full domain of dependence is a complete surface (in the sense defined above), is said to be globally hyperbolic. It is essentially this hypothesis of global hyperbolicity that translates Weyl's early hypothesis of causality, for in such a space-time the whole past and the whole future of the universe should, in principle, be determinable from the instant at which observations are made. Moreover, even if one still wishes to consider non-globally hyperbolic space-times, one still can apply all theorems proven in the globally hyperbolic case to any suitably defined 'interior' region of these space-times.

This seems to be how global considerations have acquired a sort of priority over local ones. Without global hyperbolicity, it would no longer be possible to think of any intrinsic relation of metric to topology in the framework of general relativity. Hawking and Ellis justify their anathema on closed timelike curves in a way which virtually expresses the essence of Weyl's principle:

the existence of such curves would seem to lead to the possibility of logical paradoxes: for, one could imagine that with a suitable rocketship one could travel round such a curve and, arriving back before one's departure, one could prevent oneself from setting out in the first place. Of course there is a contradiction only if one assumes a simple notion of free will; but this is not something which can be dropped lightly since the whole of our philosophy of science is based on the assumption that one is free to perform any experiment (1973, p. 189).

The freedom to perform any experiment anywhere at any time reflects the need for super-observers, yet the authors assume but "a simple notion of free will". In fact, the preference for space-times free of such pathologies is closely related to the occurrence of incomplete curves (singularities of space-time) in the case of gravitational collapse—the Hawking–Penrose theorem. And our philosophy of science is all in order, as Hawking and Ellis say, since when we drop the condition which guarantees our ability to talk meaningfully about the relation between metric and topology, that is, in the case of incompleteness, physics still provides firm ground, namely, the physics of initial singularity.

The philosopher's universe and the physicist's

Obviously not all philosophers would rejoice at the model proposed. The argument that alternative topologies are still compatible with the same set of observations has been taken by some people as representative of the 'conventional' part of physical theory. That is to say, other scenarios are certainly acceptable and any final discrimination between them is beyond our epistemic range. Arguments of this sort tend to posit a globally 'anomalous' space-time and proceed to show that this is harmless at the local level of our observations. But physics does not operate like that for a very profound reason. Just because we are still free to choose alternative topologies does not mean that general relativity is always prevented from ascribing the world's true nature, if only for the reason that global considerations were not included in topological terms when general relativity was first erected. If we say that the diversity in speculation about the global temporal order indicates something conventional about physical theory, we neglect a more important facet of conventionality, one which bears not so much on its effect on already constituted theory as on the way ideas are developed in the first place. At most, the residual diversity only shows up the relative independence of the global from the local in the *original* construction of the theory. And so, when a philosopher like Sklar writes that "since space-times with almost closed causal curves are themselves compatible with general relativity, we can attack the causal theory of topology within general relativity by using its own space-time models" (1977, p. 267), the force of such an attack against the causal theory can be re-evaluated if we realize the crucial fact is that general relativity was not built on explicitly global considerations. It is still the case that global considerations, speculative as they are, are vouchsafed by the hypothesis of a unique time function. For this time function, our historical study demonstrates, is central to and reflects the metaphysics of relativity. The metaphysical dimension inherent in the early foundations of relativistic cosmology can appear in the form of conventions only when the ultimate concern with what constitutes a 'world' is separated from the business of physics. So, what this metaphysics says is not fixed *just as* one non-pathological model of the universe cannot be selected once and for all.

Einstein's first model was the object of an extraordinary series of rebuffs. This very resurrection of theoretical and observational cosmology in the early twentieth century can in fact be understood as a reaction to a nineteenth century anathema on non-strictly-physical speculation within the self-imposed limits of physics.

Cosmology has always drawn a peculiar status from the notion that somehow the universe in which we live must be unique. But modern science cannot even imagine a unique phenomenon; it always proceeds with categories of phenomena. Munitz puts it like this: the paradox is that "insofar as the universe, by definition, is regarded as a unique object or question, it cannot be treated as an instance of some recurrent phenomenon. Cosmology does not undertake to establish laws about universes" (1962, p. 37). The emergence of non-static cosmology suggestively complicates the eternal question of how 'the' universe compares with any other object, and in the same move it promotes cosmology to the rank of a science. For it was Lemaître's original insight that the two models of Einstein and De Sitter can be reduced to one if they are inscribed in the order of cosmic history. Furthermore, non-static cosmology renders 'multiple' the 'one' universe by considering various theoretically predictable stages of a unique space–time background. This is how physical laws came to be viewed as applicable to the whole in the same sense that local laws are applied to collections of objects within the universe. So much so that, in recent times, cosmologists have proposed the apt idea of a "cosmic box". This is a self-influencing region of the universe, constructed by allowing for the time effects predicted by general relativity. What happens in such a sample region over a long period of time must be identical with what happens elsewhere; hence, "a partitioned universe behaves in exactly the same way as a universe without partitions" (E. Harrison 1981, p. 267). This means that everyday physics can be used to determine all the consequences of the expansion of the universe, and so reduces the universe to its observable parts, and discharges us from the obligation of choosing a model before we can apply cosmology. It is really not sure, however, that this procedure does away with the 'many-universes' problem. Pushing Lemaître's idea to its final conclusion (as quantum cosmology does) it appears that once again our science toys with the old paradox. In the very early quantum era of the universe, a vacuum fluctuation made real particles out of the virtual ones which existed, for want of a better term, in the time allowed by the uncertainty relations, but this fluctuation is repeatable. That is, there could be as many universes as there are fluctuations. The idea of quantum fluctuation arose from close scrutiny of what underlies any possible process of measurement and its limits. Here we encounter again what lies beyond any limit that physics may set for itself.

And again, any procedure for extending the local to the global finds limits already independent of quantum restrictions in the residual freedom of starting out from alternative global hypotheses. This can be carried out at the pre-topological level, as it has been acknowledged and investigated by Ellis himself in a 1978 paper which epitomizes, as few could, the major issue arising from the way theoretical problems of early relativistic cosmology have been tackled. For, over and above any possible degree of equivalence between neo-Newtonian and relativistic cosmology, Ellis discovers a remarkable correspondence between static and the Friedmann–Robertson–Walker non-static models from within the framework of general relativity itself. He argues that a spherically symmetric static space–time would, in fact, account for all those observations which have come to favour the non-static models over any other. In this static space–time, however, the existence of microwave background radiation is seen as making impossible the rejection of a (hot) singularity. But the peculiar, non-Weylian feature of the model is that *not all* action emerges from the singularity. Only the past null cone from any point lying on a world-line (which does not necessarily coincide with the singularity at some point in the past) is said to refocus at the singularity. This results in an Einstein cylindrical model which is *not* homogeneous, in which the redshift is accounted for in terms of *gravitational* redshifts. And an equivalence between the orthodox non-static models and this static space–time arises in the sense that what is accounted for in terms of a time variation in the non-static case gets interpreted in terms of spatial variation of properties in the static case. What we have here is strikingly similar to Weyl's transitory suggestion of 1921, when he still sought a compromise between the Einstein and the De Sitter models without bringing the concept of common origin to the fore. In Ellis's terms, the proposed model has, among many attractive features, the property of having a singularity "sitting 'over there' where it can influence, and be influenced by, the universe continually" (1978, p. 92)—something like a Machian concept of inertia extended so as to encapsulate the 'origin'. But above all, this is an even more striking return to De Sitter's scepticism, as it is precisely the principle of uniformity set up by the standard models that Ellis wishes to call into question. Ellis uses De Sitter's own terms when he writes that the assumption of spatial homogeneity, which forms the basis of the claim that the universe is expanding, "is made on philosophical rather than observational grounds" (p. 93). Where Ellis diverges from and pushes further than De Sitter is when he is ready to fix as a meaningful research programme some basic epistemic indeterminacy as a feature of our cosmological models themselves (1975, p. 258). All of which would seem to open on to an extraordinarily sophisticated learned ignorance.

Surely enough, the idea of a static universe has never been quite put away ever since the emergence of the non-static solutions. It has persisted, however timidly, in meddling with the growing body of well-established cosmological doctrine, and only by disturbing counter-example. There is no doubt that the steady-state theory was already a disguised return to the static idea. Even in the late 1930s a physicist and philosopher like Sambursky (1937, p. 335), starting with the idea that if the radius of the universe increases with time "no answer could be obtained regarding the question of expansion by measuring physical processes at different times at one and the same point of space", suggested that an alternative explanation of the nebular redshift could be formulated in terms of shrinkage of the universal lengths of atomic physics. At the same time, Hubble continued to express redshifts in terms of velocities "as a matter of convenience only" (1936, p. 123), as it seemed to him that the remarkable uniform distribution of nebulae up to remotest distances would re-open the case of the static universe.

Pondering on the way out of the dilemma cosmologists faced in the early period of static cosmology, James Jeans explains the new form our science of the universe assumes: "The question at issue was no longer whether the actual universe was an Einstein universe or a De Sitter universe, but rather how far it had travelled along the road which begins with an Einstein universe and ends with a De Sitter universe" (1960, p. 84). Thus, the question is no longer what the universe 'is', but what it 'becomes'. While Jeans saw the promise of reconciliation between being and becoming in the fact that the uniqueness of the universe makes its final state inherent in the present state ("just as this present state was inherent in the universe at its creation" (1931, p. 701)) , De Sitter was perhaps the closest to truth when in the views he expressed towards the end of his life, he saw the two as fundamentally antithetical. As early as 1917, he had sifted the "philosophical predilections" out of the cosmological problem in general relativity. What has never been in question is that these obscure but fundamental predilections are mixed up with the more or less explicit presuppositions of the non-static solution. And in response to the allegedly new cosmological consistency of relativity put into play by the solution, he replied, in one of his last papers (1932, p. 604), that the universe poignantly resembles a "hypothesis" in the mathematical sense.

References

Bondi, H. (1962). The Halley Lecture at Oxford University.
De Sitter, W. (1932). On the expanding universe. *Proc. Kon. Akad. Wet. Amst.* **35**, 596-604.
Ellis, G.F.R. (1975). Cosmology and verifiability. *Qu. J. Roy. Astr. Soc.* **16**, 245-64.
Ellis, G.F.R. (1978). Is the universe expanding? *Gen. Rel. and Grav.* **9**, 87-94.
Geroch, R. (1968). What is a singularity in general relativity? *Annals of Physics.* **48**, 526-40.
Goenner, H. (1980). Mach's principle and Einstein's theory of gravitation. *Boston Studies in the Philosophy of Science.* **6**, 200-15.
Guth, A. (1981). Inflationary universe: a possible solution to the horizon and flatness problems. *Phys. Rev. D.* **23**, 347-68.

Harrison, E. (1981). *Cosmology*. Cambridge University Press.
Hawking, S.W. (1967). The occurrence of singularities in cosmology. III. Causality and singularities. *Proc. Roy. Soc.* **A300**, 187-201.
Hawking, S.W. and Ellis, G.F.R. (1973). *The large-scale structure of space-time*, Cambridge University Press.
Hopf, H. (1932). Differentiale Geometrie and topologische Gestalt. *Jahresbericht der Deutschen Math. Vereinigung*, **41**, 209-29.
Hubble, E. (1936). *The realm of the nebulae*. New Haven: Yale University Press.
Jeans, J. (1931). Contribution to a discussion of the British Astronomical Association on the evolution of the universe. *Nature*. **128**, 699-722.
Jeans, J. (1960). *The universe around us*. 4th ed., Cambridge University Press.
Klein, M.J. (1970). *Paul Ehrenfest*. North-Holland, Amsterdam.
Munitz, M. (1962). The logic of cosmology. *Brit. J. Phi. Sci.* **13**, 34-52.
Munitz, M. (1986). *Cosmic understanding*. Princeton University Press.
Pascal, B. (1950). *Pensées*, trans. H. F. Stewart, Routledge and Kegan Paul, London.
Penrose, R. (1965). Gravitational collapse and space-time singularities. *Phys. Rev. Letters*. **14**, 57-9.
Sambursky, S. (1937). Static universe and nebular redshift. *Phys. Rev.* **52**, 335-8.
Sklar, L. (1977). Facts, conventions, and assumptions in the theory of space-time. In *Foundations of space-time theories*. Minnesota studies in the philosophy of science. Vol. 8, pp. 204-73. University of Minnesota Press, Minneapolis.
Tryon, E.P. (1973). Is the universe a vacuum fluctuation? *Nature*. **246**, 356-7.
Weyl, H. (1921). *Raum, Zeit, Materie*. J. Springer, Berlin. Fourth edition.
Weyl, H. (1923a). *Raum, Zeit, Materie*. J. Springer, Berlin. Fifth edition.
Weyl, H. (1923b). Zur allgemeinen Relativitätstheorie. *Phys. Zeitschr.* **24**, 230-2.

Index

absolute
  acceleration 44-5, 225-6
  motion 44, 59-60
  position 88
  rest 47, 57ff
  rotation 59-60, 104
  space 44-5, 47, 60, 84, 105, 108-9, 112, 128, 136, 225, 227, 253
  world 64ff
acceleration 72, 116
action
  in cosmology 392
  in general relativity 246, 250, 252
  range of 316ff, 337
action and reaction
  in general relativity 112, 214, 222, 245-6, 253
  in Newton's theory 84
action-at-a-distance 53, 90, 143, 214, 258
Adams, W. S. 39
Adler, R. 71
aesthetic 148, 164, 349
affine connection/structure 69-71, 76-7, 287, 370, 387
Aiton, E. J. 51
d'Alembert equation 63
Alexander, H. G. 45, 49, 104, 225
Andrillat, H. 201
angular momentum 52
anisotropy 352-3
anthropic principle 17
antinomy 56, 117-19, 133, 145, 179, 194, 200, 230, 233-4, 236, 354
  Kant's first antinomy 226-9
antithesis 2-3, 393
a priori 17, 74-5, 120, 128, 178-9, 247, 249, 254, 265, 275-6, 344-5, 371, 375
Archimedes 2
Aristotle 26
assumption
  in cosmology 4, 144, 159, 177, 250, 324-5, 343-6, 360, 381
  in general relativity 76, 82, 389
  in Newton's theory 46
atom (primeval) 353-4, 384-5

Baade, W. 354
Bachelard, G. 234

Barrow, J. 17
becoming 66, 246, 290-1, 313, 375, 393
Bentley, R. 29-31, 33, 39, 49, 51-2, 90, 147, 351, 363
Berendzen, R. 36
Berkeley, G. 105
Bernouilli, D. 51
Bernstein, J. 22
Bessel, F. W. 39
Besso, M. 10, 12, 27-8, 91, 135, 142, 145, 147, 151-2, 155, 160-1, 173, 190, 213
big bang 190, 213, 353-4, 385
Birkhoff, G. D. 79
black holes 91
Boltzmann's constant 148
Boltzmann's law 151
Bolyai, J. 85
Bondi, H. 75, 234-5, 354, 362-6, 372-3, 376, 381
Bonola, R. 85
Borel, E. 357
Born, M. 110
boundary conditions 54ff, 77, 79-81, 100-2, 128ff, 153ff, 177ff, 190-3, 223, 228-9, 250-1, 256, 331-2, 356-7, 387
Bruno, G. 35
Buchdahl, G. 53
bucket experiment 43-4, 84, 127-8, 210, 225-6

Callan, C. 375
Campbell, W. W. 40, 42
Carnot, S. 34
Cartan, E. 255, 370
Cassirer, E. 119-20
causal structure 250, 388
causality 63, 76, 89, 102-3, 112-13, 118, 127, 142, 245, 247, 250, 286, 296, 315ff, 350, 355, 384, 388
  stable 384, 389
  strong 389
cause 6, 43, 48, 52, 104-8, 112-13, 285-6
centre/periphery relationship 30, 46-7, 50, 90, 147-52, 162-3, 189, 208, 235, 337, 363, 365, 367-9
centrifugal force 28, 84, 104-5, 108, 112, 115, 211, 254
Cepheids 40-1, 311

Chamberlin, T.C. 40
Chandrasekhar, S. 22
Charlier, C.V.L. 50, 146, 356, 358
Christoffel symbols 119, 249-50
Clarke, C.J.S. 190, 302
Clarke, S. 44-5, 49, 56, 103-5, 119, 225-6, 228-9
Clausius, R.J.E. 34
Clerke, A. 36, 38
Clifford, W.K. 85
clocks
  in cosmology 192-3, 203, 206-7, 371-2
  in general relativity 79
  in special relativity 66-7
closed space 155, 255-6
closed time-like curves 250, 252, 263, 266, 315, 353, 374, 389
Cohen, I.B. 29, 45-6, 52
coincidence in space-time 74, 100, 177, 254
collapsing universe 6, 29, 35, 150-1, 159, 361
comets 29, 31-3, 51
compass of inertia 331, 374
condensation 281, 350-2
Conduitt, J. 31-2
conformal structure 314
consciousness 4, 66, 244-6, 291, 293, 384
  intentionality of 244
conservation laws 70, 165-6, 258, 267-9, 366-7
consistency (logical) 166, 175-6, 201, 221-2, 259
constant of integration
  in cosmology 152, 161, 213
  in general relativity 123ff
constant curvature 156-7, 166, 212, 214, 268, 346-7
continuity/discontinuity distinction 68, 75-6, 198-200, 207-8, 223
  of matter 281-2
  in space-time 187, 263-5
contraction (length) 61-2
conventionality (*including* convention, conventionalism) 67, 88, 137, 285, 287, 341-3, 362, 371-3, 390
coordinates
  arbitrary 126-7, 177, 344
  Cartesian-Galileian 58, 118, 253
  co-moving 347, 366, 371
  static/non-static 242-3, 276-8, 300, 313-14, 323, 332, 336ff
Copernican revolution 26, 39
Copernican system 115
Coriolis force 211
cosmic ball 358-9, 368
cosmic box 391
cosmic repulsion 6, 163-4, 213, 219, 297, 349

cosmic time 21, 77, 157, 191, 237ff, 269-70, 273-6, 301, 304, 320-3, 371-2, 383-4
cosmogony 351-2, 370
cosmological constant 6-8, 160-6, 181-2, 191, 194-200, 212-24, 255-6, 259-62, 269, 286, 296-7, 334, 346-7, 367
  its sign 333-4, 343
covariance 64, 73, 129, 132, 145, 258, 267, 331
Coxeter, H.S.M. 270
creation 32, 51, 166, 190, 291, 312, 368
  continuous 342, 373
Curtis, H. 36, 41
curvature (instrinsic) 3, 69, 79, 85, 119-20, 122, 286, 295, 359, 361-2
curvature (negative) 202-3, 347
curvature tensor 69-70, 119, 287
curvature/matter relationship 181-2
cyclic universe 32-3, 279-80, 346, 349, 353

Davies, P.C.W. 10
deduction (and deductive reasoning) 210, 284, 287-8
deductive cosmology 17
density of the universe 35, 164, 181-2, 194, 214-15, 255, 265, 348, 352, 357
Descartes, R. 51, 257
De Sitter, W. 1ff, 42, 120ff, Ch. 3, 233-4, 237, 266, 268-9, 288, 297, 311, 323, 326, 335-6, 348-51, 354-6, 361-2, 370, 382-3, 386-7, 392-3
Dicke, R. 375
dilatation (of time) 61-2, 192
dimensionless numbers 285-6, 293-4, 306
Dirac, P.A.M. 283, 353
distances
  to nebulae/galaxies 40ff, 91, 354
  to stars 38ff, 87-91
distant masses 113ff, 210-11, 381-3
  as supernatural 133, 182
  as world-matter 181, 196, 206-7, 288-9
  world-matter vs. ordinary matter 195ff, 294
distribution vs. constitution 37ff, 144
distribution vs. motion 363-4
distribution of matter
  as a problem 135, 156-7, 166, 254-5, 340-1, 382
  uniform/non-uniform 28-9, 35, 38, 45, 68, 116, 127, 146ff, 194-5, 219-20, 298, 307, 341-2, 357, 392
disturbance/rest distinction 5, 316, 325, 331, 342
Doppler redshift 141, 204-5, 298-300, 330, 339
Dreyer, J.L.E. 42

INDEX

duality
 source/field 166, 222
 space/matter 148ff, 228-30, 253
dynamic cosmology (before relativity) 50-1
dynamics
 Einstein's 62
 Newton's 25, 43-4, 59, 105

Earman, J. 86, 101, 141, 292
Eckhart (Meister) 245
Eddington, A.S. 6ff, 18, 34, 40, 66, 71, 75, 86, 99, 120, 122, 128, 133, 155, 164-6, 212, 215, 221-2, 234-5, 256, 267, 282ff, 311ff, 322-3, 326, 330, 332ff, 356, 361, 373, 385
Ehrenfest, P. 99, 138, 142-3, 146, 155, 178, 189, 224, 383
Einstein, A. 2, 5, 8ff, 27ff, 41, 43, 58ff, Ch. 2, Ch. 3, 233-4, 237, 242-3, 252, 254-63, 269-70, 274-5, 278, 288-9, 295, 306, 333-6, 345, 348, 352, 356-8, 361-2, 367-70, 374, 381, 385-7
Einstein-De Sitter model 97, 361-2, 365, 368, 371
Einstein's law of gravitation 285, 360, 382
Eisenhart, L.P. 156
eleatics 2
electromagnetism 10-11, 57ff, 221, 292
 in relation to gravitation 212-13, 250, 257-61, 277, 287-8, 331-2
electroweak theory 91
Ellis, G.F.R. 190, 212, 302, 316, 324, 384, 389-90, 392
empty space (-time) 55, 60, 62, 70, 227, 261, 290, 296, 340-1, 348
 denial of 214-15, 222-3
empty universe
 in pre-relativity physics 56, 114-15
 in relativity physics 121, 165, 181, 191ff, 242, 274-6, 279
energy 11, 34, 91, 165-6, 206, 268, 349, 359, 366, 385
 equation of 348
energy-momentum tensor 69-70, 74, 81, 255, 366
 of Einstein's universe 131-2, 157
entropy 291, 349
epistemology 9, 88, 98-100, 109, 112-13, 116-17, 133, 173, 215, 252-3, 288, 295, 301-2, 307, 361, 383
equivalence
 of all places/viewpoints 344, 348, 384
 of Newtonian and relativistic cosmology 358ff
 of observers 364, 384
 of reference systems 154

Erlichson, H. 61
eternity 5, 33, 208, 266, 291, 323
ether
 in general relativity 5, 214, 218, 222, 224, 288, 331
 in Newton's theory 31
 in special relativity 57ff, 136, 138
Euclid 2
Euclidean space 35, 65, 85, 339-40, 385
Euler, L. 52
evolution (of the universe) 216-17, 233, 281, 348ff
expansion of space 2-3, 7, 224, 339-40, 342, 362
 cause of 348ff, 376
expansion vs. contraction of the universe 349-50, 359
expansion/extension distinction 368, 376
explanation 6, 91, 112-13, 125-6, 133-4, 136, 143, 165, 176, 200, 285-7, 300, 345
extendibility 388
extrapolation 174, 193, 200, 235

fact/theory distinction 153, 155-6, 159, 161-2, 173-4, 243
Faraday, M. 10, 71
Feinberg, G. 22
Fichte, J.G. 245
fiction (fictive, factitious) 112, 120, 196, 268, 276
field
 electromagnetic 10, 223, 260, 292, 331
 gravitational 28, 53ff, 68, 214, 218, 222-3, 381-3
 guiding 253-4, 331
 large gravitational 190, 252, 256
 limits of the concept of 223, 247, 259, 317, 334
field equations 69ff, 101ff, 109, 131, 163, 212, 250, 255, 268-9, 286, 332, 355
finite universe 155ff, 382-3
finite/infinite universe 28-9, 45ff, 55, 133, 142, 145, 227, 347, 356-7, 369
Fitzgerald, G.F. 61
Fizeau's experiment 60
form/content distinction 126, 152, 176, 227, 229, 237, 253-4, 257-9, 332, 358
Foucault's pendulum 59-60, 211
Fourier, J. 34
free fall/particles 149, 370
Friedmann, A.A. 13-16, 22, 58, 100, 107, 110, 118, 221, 223-4, 277, 279, 335-7, 345-7, 350, 355-6, 362, 367, 384
Friedmann's equation 346-8, 350, 353, 359-60, 362, 364, 366, 370

398   INDEX

Galileo, G. 25, 71, 143
Galilean group 246
Gamow, G. 163
gauge 258-9, 287, 295-6, 355
  calibration of 277
Gauss, C.F. 86
Gauss's theorem 28, 147
geodesics
  in cosmological models 201-5, 238,
    240-1, 304, 317ff, 338-9, 372
  and causality 248-51
Geroch, R. 385
ghost-images (in Einstein's universe) 77,
  216, 218, 315, 339, 348
Giedymin, J. 87
global hyperbolicity 389
global properties of space-time 249-51, 279,
  387
globular clusters 41
Glymour, C. 86, 101, 141, 292
God 25, 29-30, 33, 44-9, 90, 222, 225-6,
  361, 382
Gödel, K. 374-5
Gödel's cosmology 19, 374-5, 386
Goenner, H. 383
Gold, T. 354, 373
Goldberg, S. 19
Gonseth, F. 270
gravitational
  force 5, 53ff, 89
  potential/field 28, 53ff, 69ff, 257-61,
    277, 361, 364-5, 370
Grommer, J. 153, 223
Grossmann, M. 72-3, 101
Grünbaum, A. 79
Guth, A 213, 385

Halley, E. 33, 49
Harrison, E. 360, 391
Havas, P. 370
Hawking, S. 212, 324, 384-5, 389-90
Hawking-Penrose theorem 390
Heckmann, O. 347-8, 360
Heine, H. 229
Helmholtz, H.L.F.von 155
Heraclitus 2
Hermann, A. 120, 220
Herschel, J. 50
Herschel, W. 36, 41, 43
Hertzsprung, E. 40
Hetherington, N.S. 13, 40-2
hierarchical universe 50, 146, 356-7
Hilbert, D. 267-8
Hins, C.H. 99
history 166, 246
Hölderlin, F. 5

Holmes, A. 33-5
Holst, H. 224
Holton, G. 103
homogeneity 5, 7, 50, 56, 149, 316, 355-6,
  371-2, 385
  of metric 237, 242, 253-8, 331-3, 344,
    385, 387
  vs. non-homogeneity 392
homogeneity/isotropy relationship 156-8,
  286, 344, 360
Hopf, H. 388
Hopf-Rinow theorem 388
horizon
  event 189-90, 206, 389
  particle 189-90, 206
  of visibility 203
Hoskin, M. 27, 37, 39
Hoyle, F. 354, 373
Hubble, E. 1-2, 12, 36-7, 41-2, 311-12,
  325, 335, 341, 393
Hubble's law 1, 42, 311, 342, 358-9
Huggins, W. 37
Humboldt, A. von 37
Husserl, E. 244-5
Huyghens, C. 38
hyperbolic geometry 87
hyperboloid
  one or two sheet 183ff
  De Sitter's 238-40, 303-4, 317ff, 337ff
hyperweak force 11, 91
hypothesis (in cosmology) 317, 325, 331-2,
  342, 362, 384, 391-3

idealism 4, 7, 245, 283
ideality (idealization) 55, 109, 116-17, 128,
  201, 220, 332, 341, 344
inertia
  as interaction 106, 219
  law of 5, 104
  relativity of 106, 144-5, 154
  of test body 154, 182, 195
inertial
  field 5, 72, 83, 196-200
  force 43-4, 116
  frame/system 108, 125, 200, 224, 289,
    299, 357, 363, 368
inertial vs. gravitational mass 71-2, 89, 117,
  166, 253, 366-7, 383
Infield, L. 164
infinite universe 28-9, 44-5
infinitely small vs. infinitely large 142-3,
  236
infinity 128, 130, 133, 135-6, 148, 150, 193,
  356-7
  in past time 319ff
inflationary universe 213, 385

INDEX

innate gravity 45, 52
interaction (duality) matter/metric 214, 218, 253–4, 357, 374
interior vs. exterior 54ff, 134, 144, 147–9
interval (infinitesimal) 70, 257–8, 286, 295, 315
intuition 210, 228, 245, 282, 295
invariance 64, 74, 125, 132, 248, 258, 268, 387–8
inventing 9–10, 293
island universe 37–8, 40–2, 146ff, 177
isotropy 56, 131, 144, 159, 345, 359

Jaki, S. 27, 33, 49, 56, 162
Jammer, M. 86
Jeans, J. 120, 351, 393
Joule, J. 34

Kahn, C. 99
Kahn, F. 99
Kant, I. 38, 50–3, 58–60, 62, 83–4, 105, 119–20, 145, 179, 226–9, 283, 356
Kelvin (Lord) 34
Kepler's laws 43
kinematic relativity 354, 362–3, 368
kinematics (classical) 43, 59, 365
kinetic energy 148, 218, 359
Kisiel, T. 244
Klein, F. 157, 161, 216, 221, 237, 261, 263, 266–78, 282, 304, 355, 358, 387
Klein, M.J. 99, 138, 383
Klein, O. 118
Kockelmans, J. 244
Kohlschütter, A. 39
Kopff, A. 110
Kretschmann, E. 145, 224
Kubrin, D. 30, 32

Lambert, J.H. 50, 53, 356
Lanczos, C. 68, 221, 237, 277–82, 304, 307, 312–13, 337, 355
Laplace, P.S. 36–7, 40, 50–2
Laplace's equation 54–5, 70
Larmor, J. 61
Laue, M. von 69, 282, 312–13
laws (equations)
  vs. real structure 358, 362, 375
  vs. solutions 74–5, 123ff, 156, 159, 175–6, 251, 256, 331, 358
Layzer, D. 364–5, 368, 376
Leavitt, H. 39
Leibniz, G.W. 33, 44–5, 49, 103–6, 119, 136, 225–6, 229

Lemaître, G. 14–15, 20, 22, 164, 209, 233, 235, 276–7, 297, 301, 307, 334ff, 384–5, 391
Lense, J. 211–12
Levi-Civita, T. 206, 287, 336
light
  faster than 326, 350
  tired 342, 345
  velocity of (in cosmology) 201–5, 262, 302, 389
  velocity of (in relativity) 57, 60–2, 65, 138
light-cone
  in cosmology 317ff, 389
  in general relativity 77, 250, 314, 388
  in special relativity 65
Lightman, A. 34
Lobatchevski, N. 85
local inertial frames 360, 369–70
local/global distinction 49, 53, 76, 89–90, 106, 113, 133, 139, 145, 149–51, 194ff, 209, 211–12, 250–2, 254, 277, 363, 370, 375, 386–90
Lorentz, H.A. 11, 61–2, 64, 99, 137–8, 143, 196, 222, 247, 267, 314
Lorentz group/transformations 61–2, 126, 241–2, 268, 387
Lundmark, K. 311

Mach, E. 10, 83–4, 88–90, 100, 103, 105–7, 112–13, 115–16, 127, 131, 135–6, 172–3, 195, 207, 210–11, 214, 357, 381
MacMillan, W.H. 349, 357
mass-horizon 242, 262ff, 276–7, 280–2, 299, 304–5, 312, 336
Maxwell, J.C. 10, 57, 71, 86, 149, 247, 258
Maxwell's equations 61, 64, 69, 299
Maxwell's law of distribution 217, 255
Mayer, J.R. 34
McCrea, W.H. 164, 350, 359, 362–5, 369, 371
McGuire, J. 33
McVittie, G.C. 14, 164, 336, 350, 366–7
Mehra, J. 110, 266
Merleau-Ponty, J. 21, 27, 33, 36, 98, 179, 226, 283
metaphysics
  Eddington 283–4
  Kant 58, 60
  Leibniz 45, 103, 226
  Milne 363
  Newton 49, 136
  of general relativity 120, 390
  of relativistic cosmology 193, 355, 371, 390
  Weyl 4, 345, 373

metric coefficients
  of Einstein's universe 131-2, 160ff
  of rotation 127
metric structure 76-7
metrical field 253, 276-7, 331-2
Michelson–Morley experiment 61-2
microwave background radiation 354, 385, 392
Mie, G. 11, 68, 213, 224, 247
Milky Way 38, 40-2, 49, 91, 143, 195, 204, 219-20, 256, 298, 311
Miller, A.I. 60, 137
Milne, E.A. 17, 158, 354-74
Milne–McCrea theorem 363-4
Minkowski, H. 62ff, 77, 90, 137, 143, 167, 236, 286, 290
Minkowski metric 78-82, 97, 127, 133, 290, 346
Minkowski space-time 63ff, 131, 140, 246, 387
Misner, C. 106, 211
Morando, B. 36
motion (equation of) 233, 296-7, 341, 346, 365
Moszkowski, A. 92, 227, 230
Moulton, F.R. 40
Moyer, D. 284
Munitz, M. 386, 391

Narlikar, J. 201
nebulae (extra-galactic) 3, 6-7, 36ff, 50, 204-5, 218-20, 294-5, 298
  velocities of 40-2, 205, 217, 300
neighbourhood 70, 118-19, 248-9, 387, 389
Nernst, O. 349
Neumann, C. 50, 146, 159
Newcomb, S. 35, 38, 50, 381
Newton, I. 10, 25ff, 38-9, 43ff, 55, 57, 59, 82, 84, 90, 103-5, 108-9, 112, 114, 116, 126-7, 134-6, 143, 172, 214, 253, 351, 358, 361, 363, 369, 381, 383
Newton's law of gravitation 43-4, 53ff, 71, 119, 162-3, 358-60, 364, 370
Newtonian
  cosmogony 31-3
  cosmology 27ff, 146-52, 222, 228, 252-6, 356ff
  mechanics 5, 12, 25, 30, 50ff, 103ff, 365, 375-6
  space-time 136
non-Euclidean geometry 73, 85ff, 142, 266
non-linearity 71
non-static universe 242-3, 343ff
  De Sitter's universe as 297, 337-40
Nordstrom, C. 68
North, J.D. 13, 21, 86, 97, 100, 120, 150, 279, 296, 301, 359-60, 373

Norton, J. 68, 72
nothing(ness) 7, 55, 82, 131, 133, 155, 165, 193, 227, 229, 383
novae 32, 41

observability
  in classical physics 45, 105-6, 115
  in cosmology 203, 300, 304, 316ff, 340, 350-1, 355, 368, 384, 386
  in general relativity 113, 116-17, 121, 125, 128, 133-4, 136, 253
  in special relativity 66
observation/theory distinction 138, 174, 254, 265, 326, 341, 345, 351, 355, 366, 368, 381
Olbers' paradox 48-50, 146, 381
Oort, J.H. 99
open space 223
  in topological sense 315-16, 353, 368
operational(ism) 139, 283, 369
origin 4-5, 49, 52, 105, 256, 302, 316ff, 342, 350-4, 373, 392
  of inertia 129, 134, 200, 218, 253-4
  of time 189-91, 338-9, 354

Pais, A. 98, 223, 266
parallax 39, 87, 89, 174, 202, 204
parallel postulate 85-6
Pascal, B. 382
past history (of particles) 260, 292, 302, 319ff
path-dependency 258, 292
Pauli, W. 72, 110, 150
Pease, F. 205
Peebles, P.J.E. 97, 130, 219, 353, 375
Penrose, R. 22, 385
periodic coordinate 279-81
Perrine, C.D. 284
phenomenal/cosmic time 77-8, 243-4, 375
phenomenology (Husserl's) 244-5
Planck, M. 68, 89, 104, 209
Platonism 9
Plummer, H.C. 138
Poincaré, H. 67-8, 86-9, 137-9, 155, 173, 213, 221, 285, 362, 371
Poisson, D. 53
Poisson's equation 53-5, 63, 69, 128-9, 131, 162, 166, 255, 331, 365, 367, 370
positivism 9, 34, 120
possibility/reality distinction 135, 137, 139, 201, 213, 261, 286, 342, 373-4
postulate
  of the absolute world 66
  mathematical vs. material 191-2, 200
  in relativity theory 67ff
  Weyl's principle as 324, 353

INDEX

pressure 81, 97, 159-61, 164, 182, 196-200, 213, 215, 220-1, 264-5, 281, 343-4, 348, 366-7
principle
  of causality 247-8, 316ff
  of classical relativity 58ff, 246, 253
  cosmological 8, 17, 144, 157-8, 360, 372
  of equivalence 69, 72-3, 101, 135
  of general covariance 73-4, 100-2, 118-19
  of general relativity 99ff, 107, 117, 375, 383
  Mach's 100, 103, 106, 117, 129, 132-3, 143-5, 154, 156, 159, 166, 191-3, 214, 230, 252, 256, 288-9, 382, 387
  of special relativity 17
  of sufficient reason 103-4, 106, 226
  Weyl's 4, 17, 19-20, 22, 316-23, 342-3, 348, 372-3, 384, 386, 389
projective geometry 270-4
proper time 82, 238, 279, 302, 304, 328, 372
pseudo-Euclidean metric/continuum 118, 128-9, 249
Ptolemaic system 115-16
Pythagoreanism 9

quantum
  cosmology 385-6, 391
  mechanics/theory 2, 7, 11, 26, 68, 74, 212-13, 221-2, 247, 334, 353, 384-5

radiation 146-50, 344, 348-52, 363
radius of curvature 2, 91-2, 120, 160ff, 204-5, 229, 256, 288, 290, 292, 296, 311
  as a function of time 278-9, 281, 339, 346
Rankine, W. 34
realism 2, 283-4
receding vs. approaching motions 203, 297, 325, 349
redshift
  gravitational 86, 141, 298, 392
  in expanding space 313, 339, 384, 389
  in static metrics 204-5, 217, 294, 297-301, 312-13, 326-30, 393
reference (space of) 109-12
Reichenbach, H. 120
relativity
  general theory 67ff, 99-107, 123ff, 387
  special theory 58ff, 84, 107, 136, 351, 358, 363, 374, 383, 387
rest (absolute) 5, 39, 43, 47, 57ff, 116, 159, 222, 301
Riemann, B. 142, 155-6, 158-9, 254, 258, 386
Riemann tensor 69

Riemannian
  geometry 251, 257-8, 315, 387-8
  manifold 156-7
  metric 70, 88, 101, 121
rigid (measuring) rods 63, 66, 88, 156, 189, 258
Rindler, W. 22, 44, 70, 82, 189, 303, 356
Ritchey, G.W. 40
Ritz, W. 138-9
Roberts, I. 37
Robertson, H.P. 15, 158, 279, 324, 336, 340ff, 360, 384
Robertson-Walker metric 16, 347, 366
Rosse (Earl of) 37
rotating
  axes 122ff, 139
  disc 73, 121-2
  globes (Newton's experiment) 44
  nebula 52
rotation
  in general relativity 100ff, 110ff, 122ff, 210-11, 254
  in Newton's theory 59-60, 83-4, 103ff
  in relativistic cosmology 288, 374
Russell, B. 357, 373

Saccheri, G. 85
Sambursky, S. 392
scalar 56, 281
scale factor 359, 363
scepticism 8, 120-1, 193, 208, 233, 392
Schaffer, S. 33
Scheiner, J. 40
Schlegel, R. 349
Schlick, M. 120
Schrödinger, E. 238, 302, 333
Schücking, E. 360, 368-9
Schur, M. 157
Schur's theorem 157-8
Schwarzschild, K. 78-82, 87
Schwarzschild's solutions 79-82, 84, 140, 155, 159-60, 176, 194-5, 201, 208-9, 261-4, 270, 275, 279, 297
Schweikart, F. 86
Sciama, D. 363
Seeliger, H. von. 50, 146, 163
Selety, F. 222, 356-8
Sen, N. 282
Shapley, H. 41, 294, 325
signature 160, 182, 249
Silberstein, L. 212, 325-6, 334, 341-2
simplicity (logical) 9, 58, 61, 125, 164, 285
simultaneity 66-7, 175, 363, 371-2, 375
singularity
  apparent vs. intrinsic 186-90, 209, 266-7, 270ff, 280, 302, 305

singularity (*cont.*)
  initial 352-3, 369, 385, 390
  matter/particles as 222-3, 247, 277, 280
  of space-time 183ff, 190, 206-7, 222-3, 262, 268ff, 278ff, 289, 385, 388, 392
Sklar, L. 59, 105, 114, 360, 390
Slipher, V. 40-2, 153, 294, 311
Smith, R. W. 21, 41, 294, 311
solar system 25, 29ff, 36, 43ff, 49, 51, 134
Sommerfeld, A. 120, 220, 222
source variables 55, 71
Spencer, H. 33
Spencer-Jones, H. 99
spirit
  Newton 52-3
  Weyl 5, 332, 345
stability 51-2, 56, 195
  vs. instability 51, 351
Stachel, J. 68-9, 73, 121, 284
stars (fixed) 26, 30, 43-4, 48, 57, 104, 106, 108-9, 114, 116, 125, 254
  total mass of 139-41, 174
static space-time 66-7, 78-9, 132, 163, 260, 299
static/non-static universe
  in pre-relativity physics 25ff, 32ff, 49
  in relativity physics 2-3, 15, 21, 34, 174, 194-5, 198-200, 203, 209, 242, 274-7, 306, 329, 335ff, 358, 367-9, 373, 383-4, 392-3
stationary
  field 254-5, 346-7
  system 279
statistical
  mechanics 148, 150-2, 348, 357
  equilibrium 216-17, 254-6
steady-state cosmology 17, 19-20, 373, 392
Stein, H. 374
Steinhardt, J. 213
stereographic projection 177ff
stress-energy tensor (in cosmology) 196-200, 212, 215-16
strong force 11
super-observer 293, 301, 345, 386, 389
supersymmetry theories 64
symmetry
  in cosmological solutions 261-3, 282, 302, 346, 366
  distribution of matter 28, 39, 42, 56, 135, 140, 147
  laws of Nature 64, 286-7, 296
  in solutions of general relativity 78-9
synchronization
  in cosmology 322, 372
  in general relativity 79

Synge, J. L. 69, 82, 97, 149

thermodynamics (second law of) 33-5, 291, 349
Thirring, H. 210-12
Thorne, K. 106, 211
time
  absolute/relative 130, 137, 143, 178ff, 203, 266, 293, 358
  cosmic vs. non-cosmic 275-6, 306
  curvature of 129, 142, 179, 234
  extra-mundane 225, 233
  imaginary vs. real 179, 183ff, 273, 279-80, 290-1, 322-3, 333
  indeterminate 242, 272-5, 304
  Newtonian 358, 363, 372, 375
  orientability 247, 250-1
  subjective (as aspect of human experience) 224, 244-7, 291, 293
time function (as separate from space) 224, 251, 254, 293, 306, 313, 337-9, 344, 347, 374-5, 390
time-like (vectors, trajectories) 248-52, 260, 314, 318ff, 374-5
time scale 351-4
time-travel, paradoxes 375, 389
Tipler, F. 17, 190, 302
Tolman, R. 80-1, 138, 341-3, 345, 349-51, 373
Tonnelat, M. A. 122
topological structure 76, 250, 314-16, 347, 370, 387, 390
  elliptical vs. spherical 216, 263, 269-70, 315
topology
  in relation to causality 315-16, 353, 390
  in relation to metric 314-15, 369, 388-90
Torretti, R. 13, 98, 156, 210, 266, 270, 347
total mass (of the universe) 181, 256, 259, 265, 276, 290
transformation of coordinates 238, 278, 302, 320, 344, 363-4
Trautmann, A. 370
Truesdell, C. 52
Tryon, E. P. 386

uncertainty relations 391
unified field theories, unification
  in pre-relativity physics 53
  in relativity physics 2, 10, 12, 67-8, 91, 103, 212-13, 257-61, 277-8, 287-8, 332-4, 388
uniform
  distribution of matter 35ff, 56, 116, 359, 366
  field 54

INDEX

universe
   cylindrical 157-8, 177ff, 268, 330-1
   as non-static 279, 339
   definite size of 161, 202, 204, 367
   history of 12, 106, 254, 292, 324, 334-5, 351, 359, 388
   hyperbolodoical 183ff, 270-4, 317ff
   motion of 45, 60, 84, 225-8
   one/many distinction 139, 213, 323, 341-2, 350, 368, 373-4, 385, 390
   rotating 210-11, 374

vacuum 213
   fluctuation 391
du Val, P. 270
velocity-distance relationship (as distinct from Hubble's law) 203-4, 294-5, 311-12, 330
violetshift 141, 174, 194, 294, 298-9, 312
virial theorem 218
Vizgin, V. P. 244
Voltaire 29-30
vortex theory 51

Walker, A. C. 360
weak force 11

Weinberg, S. 11, 79
Weyl, H. 3ff, 17-18, 66, 68, 70, 74, 75-7, 112, 143, 194, 212, 217, 221, 234-82, 286-90, 292, 295, 298-9, 301-2, 304, 307-8, 313ff, 355-7, 367-8, 373-5, 384-6, 388, 392
Wheeler, J. A. 71, 106, 211
Whitrow, G. J. 97
Whittaker, E. T. 138, 257
Wirtz, C. 311-12
world
   Einstein 156
   Galileo 253-4
   Minkowski 63ff, 90
   Weyl 390
world-lines 17, 64
   of De Sitter's universe 240
   of Einstein's universe 157-8
Wright, W. 205

Yolton, J. W. 283

Zwicky, F. 342, 345, 349